Soil Analysis

Sampling, Instrumentation and Quality Control

A Word about the Cover Illustration

I have the notion (and I enjoy persisting with this notion) that the shapes liked by living matter are everywhere the same, true for all small objects or large geographical areas. In this spirit, I have desired in these landscapes to confuse the scale in such a manner that it will be uncertain whether the painting represents a vast area of mountains or a tiny parcel of land. I feel that, having found these rhythms of matter and being provided with any object, the painter could endow that object with life.

Many persons have imagined that because of a disparaging bias I like to show unfortunate things. How I have been misunderstood! I had wished to reveal to them that these things they consider ugly or have forgotten to see are also great wonders.

Jean Dubuffet, commentary on his paintings
'Population on the soil, 1952'
and 'Fruits of earth, 1960'.

Soil Analysis
Sampling, Instrumentation and Quality Control

Marc Pansu
Jacques Gautheyrou
Jean-Yves Loyer

Translated by
V.A.K. Sarma

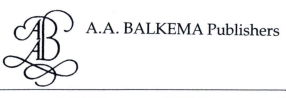
A.A. BALKEMA Publishers

LISSE   ABINGDON   EXTON (PA)   TOKYO

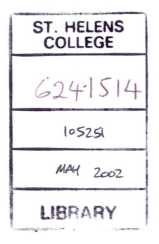
Ouvrage publié avec le concours du Ministère français chargé de la Culture - Centre national du livre
Published with the support of the French Ministry of Culture

Translation of: *L'analyse du sol Echantillonnage, instrumentation et contrôle*, © Masson, Paris, 1998.
ISBN : 2-225-83130-0.

Cover photograph: 'Population on the soil', Jean Dubuffet, 1952,
oil on hardboard, 66 × 81 cm,
© Adagp, Paris 2000, Photo Laurent-Sully Jaulmes,
Musee des Arts Décoratifs Paris, all rights reserved

Translation team: Dr. V.A.K. Sarma
Margaret Majithia

Published by: A.A. Balkema, a member of Swets & Zeitlinger Publishers
`www.balkema.nl` and `www.szp.swets.nl`

ISBN 90 5410 716 2

Printed in India

Foreword

Soil science has made remarkable progress over the years. Once upon a time 'soil' meant only the thin layer of earth playing the role of storehouse of nutrients necessary for the growth of crop plants. Then 'pedology'—soil science—became recognized as a unique scientific discipline with ever-expanding applications. Indeed, today, soils are viewed as natural objects with properties that depend not only on the duration and intensity of the primordial processes that have shaped them, but also on the nature of those set in motion by mankind.

It is readily obvious that some attributes of soil are relatively stable while others change rapidly not only in space, but also in time. It is therefore appropriate, as is done for water and air, to monitor and assess the quality of soils. Soil analysis is a must and this book examines every aspect of such analysis in detail.

The pre-eminent role played by IRD (French Research Institute for Development formerly ORSTOM) in the development of knowledge pertaining to soils of the Mediterranean and intertropical regions is well known. Thanks to a wide range of maps, monographs and scientific papers on soil genesis, the results of investigations of the pedologists of this institute are also well known. Contrarily, what is neither known nor recognized is the fact that these innumerable studies of inventory and functioning would never have seen the light of day had not IRD at its command in France and in partner countries the infrastructure necessary for competent analysts to develop the tools and know-how indispensable for practical soil analysis. The idea of writing this book was conceived in this community and it is thanks to the perseverance of many of its members that the task has been timely completed.

During their careers, Jacques Gautheyrou and Marc Pansu, chemists and research scientists IRD and Jean Susini, recently deceased—initiated this publication, placing their experience at the service of soil chemistry and pedology. Responsible for the research and analytical laboratories, they did not confine themselves to simply producing results. They also contributed greatly to the adaptation and improvement of the performance of analytical techniques, and maximal application of the development of scientific instrumentation in the domain of physical methods of analysis with, among other objectives, the goal of quality assurance—one of the criteria of industrial analysis and scientific research.

Jean-Yves Loyer, also of IRD, specialized in the improvement of saline soils and the development of Mediterranean and tropical regions is the third author of this book. He better than anyone else can vouch for the importance of soil analysis in the earth sciences. Further, all three authors received advice from researchers, scientists and technicians not only of IRD, but of CIRAD, CNRS and INRA. The collaboration between scientists responsible for analysis and agricultural researchers who interpret the analytical data, makes this book unique. This collaboration is further evident through the author's publications in the area of analytical methodology and its application to soil science.

The book consists of two parts. Part I covers methods of sampling, field examination, qualitative tests prior to analysis and sample preparation. Part II presents chemical and modern instrumental methods: emission spectroscopy (flame, plasma), absorption spectroscopy [molecular and atomic (flame and electrothermal)], ionometry, gas and liquid chromatography, and elemental analysis (C, H, N, O, S). It concludes with two interesting chapters, one on automation in instrumental analysis and the other on quality control of analytical data.

Although other books on soil analysis or on specific methods of physicochemical analysis have been published, none has such a comprehensive approach, none treats the subject in such depth. The authors of this book have presented their material in a logical and lucid manner that will prove invaluable not only to the chemist faced with problems of soil analysis, but also to analysts who should generally be initiated into modern methods of analytical chemistry.

Although soil analysis is valuable primarily to the pedological researcher, fundamental geologist or agronomist, it plays a major role in many other areas as well. Knowledge of the environment calls for soil analysis with investigation of toxic elements or compounds liable to contaminate the environment, such as industrial emissions through the atmosphere or liquid effluents. The mining industry exploits certain specific constituents of the soil such as clays, carbonates, oxides, silicates etc., and the applied research services of chemical, biotechnological and pharmaceutical industries are also concerned with certain organic constituents of the soil.

Extensive bibliographical references and an Index complete the utilitarian value of this monograph. Overall, the book is impressive and will prove very useful not only to the chemist responsible for analysis, but also the scientist and the researcher requiring such data.

We wish this new book the greatest success.

Maurice Pinta and Adrien Herbillon

Acknowledgements

This book was prepared at IRD, the French Research Institute Development (formerly ORSTOM). It synthesizes the experience of over 50 years of this organization in establishment, operation and collaborative work of analytical laboratories in France and various countries of the world: Algeria, Bolivia, Brazil, Burkina-Faso, Cameroon, Congo, Ivory Coast, Equatorial Africa, Lebanon, Morocco, Mexico, Niger, Senegal, Syria, Togo and Tunisia.

We are grateful to Mr Gérard Winter, former Director-General of IRD, Mr Georges Pedro and Jean-Yves Martin, former Presidents of the Scientific Commissions on Hydrology-Pedology and 'Engineering and Communication Sciences' as well as to Michel Rieu, Antoine Cornet, Guy Hainnaux and Claude Paycheng for their support of this project.

We are further grateful to Mr Maurice Pinta, author of many books and papers on analytical chemistry, for his interest in this undertaking and critical reading of the manuscript. We also express our recognition Prof. Adrien Herbillon, former Director of the Pedobiology centre of CNRS at Nancy for his constant support of this endeavour and constructive criticism.

The manuscript has been enriched by many corrections and comments from specialists in physicochemical analysis and soil analysis. First of all, we thank Mr Francis Sondag, Director of the Laboratory for Surface Formations of IRD in Bondy and Ms Florence Le Cornec, spectrographer at IRD for corrections in Chapter 11 and elsewhere. Our gratitude to Mr Thibault Sterckeman, Director of the Soil Analysis Laboratory at INRA at Arras and President of the Commission for Standardization 'Soil Quality—Chemical Methods' at AFNOR, for his warm appraisal of this work and pertinent comments. We thank the analytical research specialists of CPB-CNRS Ms T. Chone, Mr E. Jeanroy, J. M. Portal, J. Rouiller, G. Belgy and D. Merlet for their corrections and remarks. Thanks are also due to Mr J. P. Legros (soil scientist at INRA, Montpellier) for corrections on sampling and quality of analysis, F. Laloe, statistician at IRD, for corrections in Chapter 18, C. Riandey, spectrographer at IRD, for corrections in Chapters 10 and 11 and M. Misset, soil scientist at IRD for the material in Appendix 3.

Numerous specialists added elements to the preparation of this monograph, among whom special thanks go to Ms A. Bouleau, S. Doulbeau, M. Gautheyrou, E. Gavinelli, M. Gérard, M. C. Larre-Larrouy, G. Millot, M. L. Richard, I. Rochette and P. Talamond as well as Mr G. Bellier, J. Bertaux, J. C. Brion, B. Dabin, J. Fardoux, C. Hanrion, M. Marly, G. de Noni, P. Pelloux, J. Pétard, A. Plenecassagne, E. Roose, M. Sourdat, M. Sicot, J. C. Talineau, J. F. Vizier, P. Zante (IRD); A. Aventurier and P. Fallavier (CIRAD, Montpellier); P. Bottner (CNRS, Montpellier); and Ms M. Dosso (CNEARC, Montpellier). Finally, we extend our warm gratitude to Ms S. Rigollet for secretarial help, Ms C. Dieulin and Mr P. Raous for their timely help with computer problems and Mr E. Elguero for formatting some of the statistical tables.

The Authors

In Memoriam, Jean Susini

This book was initiated by Jean Susini, who himself participated in its writing until his sudden death in July 1994. Jean Susini, chemist, joined IRD formerly ORSTOM, after World War II following a difficult ordeal of deportation. Among the pioneers of this young institute he actively took part in the establishment of research structures and followed their development. His specialty led him to collaborate in the setting up and operation of laboratories for physicochemical analysis of soils, waters and plants in several countries, in particular France, Cameroon, Algeria and Tunisia.

He was a principled person who believed in man and his capacity to control his future. He appreciated progress but was prudent regarding new technologies whose functions he felt had to be understood in terms of sustainable development. He also sought simplicity, for example, the least expensive techniques to solve a given analytical problem, without, however, excluding more complex techniques when the need arose. Above all, he championed the notion that persons in the field and in the laboratory not lose their initiative by becoming mere button pushers on elaborate machines. Concomitantly, he quite often declared that many instruments were underutilized or even unutilized without human inputs.

For all these reasons he wanted this book to be written and we hope we have been true to his ideas.

The Authors

Contents

PART I

Sampling

Sampling

1. Introduction

Pedology a natural and observational science, has long used the empirical data accumulated by generations of field workers, botanists, agricultural scientists, foresters, cattle-breeders etc. These persons were the only source of knowledge and soil was considered simply a support of natural vegetation or agricultural production. Their observations led to the definition and categorization of soil conditions according to their 'natural' production potential.

Use of fertilizers for maintaining land quality, increasing yields, correcting certain defects and sometimes remedying steep falls in production, quickly necessitated more precise diagnosis of the land; it became imperative 'to know the soils better in order to better utilize them'.

Various methods of study and characterization of the soil in the field were proposed, all of which required taking samples of the plough layer for essential 'agricultural-chemical' analyses. The results of these analyses constituted a basis for refining predictions and determining application of fertilizers.

Initiated for agricultural purposes, these analyses were gradually extended to more fundamental problems as the 'physical, chemical and mineralogical basis of soil science' became better known (Chamayou and Legros, 1989). Today they display a panoply of the simplest to the most sophisticated methods, often borrowed from other disciplines (geology, geophysics, medical sciences etc.) and adapted to soil science.

The range of analytical possibilities today is such that one must first define the purpose of these determinations, that is, the research objective; the manner of sampling depends on the type of analysis desired. It is thus essential that on-site soil sampling correspond to the problem to be resolved and, above all, be properly executed. Proper interpretation of the results depends on this. Thus the major sources of error to be mastered are the why and how of sampling.

The various classic statistical or geostatistical treatments of analytical data to be applied later to the sampling results presume a precise protocol for sampling as well as a map of sampling sites and indication of number of samples to be taken. Geostatistical treatment requires, for example, consideration of a minimum number of field measurements and analytical determinations (of the order of 50) to enable useful interpretation of the scatter diagram. This expresses the spatial variation and the 'range' of the parameter measured, and should allow mapping of trends in the form of isovalue curves, with assumption of spatial stability of the type of variability in the parameter measured. If temporal variations are to be studied, obviously the number of observations will increase.

These three components—**research objective, sampling methodology** and **type of treatment** of the results—should be understood *a priori* in all soil analyses.

2. Research Objectives

The sampling protocol for soil in the field may differ greatly according to the objectives. These are so diverse that a complete delineation has not been attempted. The principal themes of research, be they fundamental or applied, can be grouped under several broad headings, however, to each of which one method of sampling corresponds.

2.1 Study of Soil Genesis

This usually involves geological, geographical, palaeogeographical, geomorphological, etc. approaches. The spatial unit considered may vary dimensionally from a simple pedological profile showing sequence and catena of soils to the drainage basin. This approach, earlier done by point studies, relying on a vertical concept and aimed at morphological characterization and subtle physical chemistry of the pedon[1] or the solum[2], was later gradually extended to spatially larger observation units.

Soil characterization was thereafter completed by studies of the microstructure and sequential organization, having recourse to other methods of analysis, micromorphological for example.

The latest trend in pedogenetic studies is associated with understanding the functional mechanisms that direct the evolution of soils. To be specific, new methods of measurement, certain parameters of evolution, sampling and analysis of soil solutions have been employed.

A major problem in working towards pedogenetic conclusions is that of representativeness of field samples, the results of which are to form the basis of interpretation. Rarely, if ever, done in a systematic manner, scientists have to rely on a judicious choice of sampling sites. Given the variability in certain soil characteristics, a study of the landscape, experience and knowledge of the terrain enable determination of the best observation sites for obtaining representative samples.

2.2 Mapping

Samples are as varied as scales of observation are possible; they also serve different purposes—soil inventory, cartographic synthesis, maps of agricultural land improvement, forest maps, grassland maps and geotechnical survey maps.

---

[1]   Pedon: elementary volume necessary and sufficient to define at a given instant all the structural characteristics and materials composing the soil. The lateral dimensions should be sufficiently large (1 to 10 m^2 depending on the soil) to enable study of the nature of all the horizons (after J. Lozet and C. Mathieu, 1993, *Dictionnaire de Science du Sol*, Lavoisier, France).

[2]   Solum: vertical section through a pedological cover observable in a profile pit or trench (Référentiel pédologique, 1992 INRA, France).

In the first case, the work is accomplished on a small or medium scale and directed towards preparing soil maps and simple inventories designed for regional planning. The survey and sampling method is then based on reconnaissance surveys of homogeneous soil units, most often defined from preinterpreted documents (base maps, aerial photographs, satellite images) or sometimes directly from the field. These homogeneous units should be characterized by one or several profiles representative of the **soil type**, the description and physicochemical and mineralogical analytical data of which usually serve as the **reference profile**.

Two considerations should always be borne in mind (Legros, 1996):

(1) Field sampling is essential because variability in the field is considerable; sampling errors may vitiate the representativeness of the sample taken.

(2) In most cases, sampling is done neither randomly nor systematically (Fig. 1.1), but in the best possible way according to the sampler's knowledge of the terrain, in arbitrary fashion.

2.3 Diagnosis and Monitoring of Fertility or Pollution

A direct agronomic application is often the purpose of this exercise: rainfed agriculture (cereals, trees, grassland), forestry and various irrigated crops. Its aim is a broad characterization of a given geographic area, sometimes with natural boundaries, but often artificially delimited (fenced-in, plots). Furthermore, subsequent to the preliminary diagnosis, it is distinctly useful, even indispensable, to follow up changes in each parameter (N, P, K, organic matter, various pollutants etc.) over time under the impact of land improvement, returning to the same sites for fresh sampling. Sampling sites should be precisely located from the beginning of the exercise. There are several kinds of sampling schemes:

— random, for example random selection from a numbered square grid (Fig. 1.1a);

— stratified random, following predefined areas of subpopulations (Fig. 1.1b);

— systematic, along one or more lines drawn at regular intervals (transect);

— systematic, at the centre or at intersections of a regular bidirectional grid (square grid, Fig. 1.1c), more rarely tridirectional (triangular grid).

In the zigzag pattern, sample locations are defined from an original random starting point in a small starting area (Sabbe and Mary, 1987).

The second and third types of sampling are easy to execute and provide comparable results when the land is relatively homogeneous, without disturbance or apparent trend or gradient. Errors may be introduced if the perimeter has been prepared or treated along a straight line (deep-trenching, subsoiling, broadcasting of chemicals), in which case random sampling is preferable.

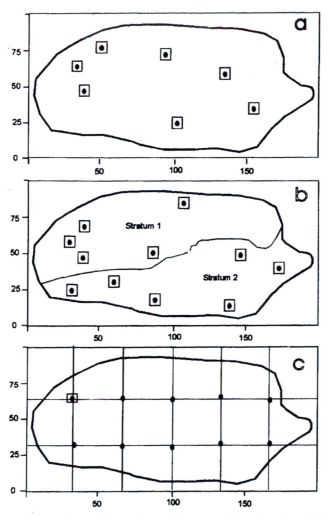

Fig. 1.1 Three types of sampling schemes for a site (after Crépin and Johnson, 1993). a) Random sampling; b) Stratified random sampling; c) Systematic sampling (grid point).

3. Sampling Methods

Except for systematic sampling schemes, the precise choice of sampling site should in every instance take into account certain factors that could have a direct influence on the characteristics of the underlying soil: microrelief and land usage (natural tree vegetation or crop, for example) and also proximity to drainage channels, buried drains or various outlets.

Once the site has been decided, its location characterized (particularly site conditions) and its environment described (PrISO11259, 1997;

PrISO10381-1, 1997), sampling is effected directly using manual or motorized augers, or after digging a soil profile pit using a pick or shovel (Fig. 1.2). Breaks-in-slope, quarries, banks of streams and wells used prudently, add valuable supplementary information but are generally avoided for representative sampling.

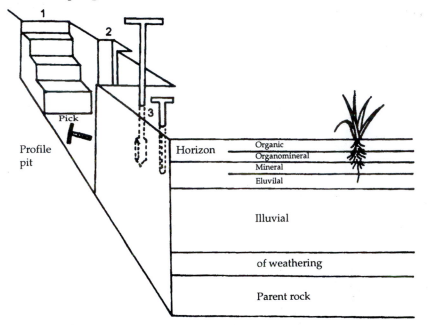

Fig. 1.2 Types of sampling in a schematic profile (presumed to be mature and clay-leached in this example). 1) along steps; 2) vertically along a face; 3) by auger.

According to the objective, it is possible to distinguish:

—pedological samples: samples enabling analysis and comparison of the different horizons in vertical succession and some preliminary interpretations regarding the soil (clay-leaching and accumulation of certain elements, for example eluviation and illuviation);

—samples for agronomic purposes: these are often bulk samples taken from the plough layer for evaluation of soil fertility or research on pollutants.

The principal methods of taking these soil samples are presented below.

3.1 Sampling with Manual Auger

Study and description of the soil are limited in this case because of the disturbance caused during boring: colour, texture, stoniness and roots may, however, be estimated *in situ* (see Chapter 2). Various implements are used according to the nature of the soil, its friability, plasticity and hardness.

3.1.1 Helical augers

These are the most common augers (Fig. 1.3a) and can be used for most soils if not too hard, gravelly or cemented (by gypsum, calcium carbonate or iron oxides). The diameter of the helix varies from a few centimetres (4 cm) to a few tens of centimetres (30 cm). The most popular are of the order of 7 to 8 cm in diameter; those for sampling of sandy soils have a more closed helix (Fig. 1.3b) while others are adapted to penetration of gravelly soils (Fig. 1.3c). Special shapes can be made to order but the steel should be hard enough to resist bending under pressure.

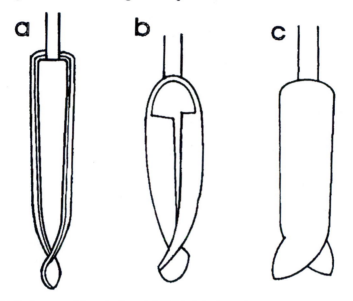

Fig. 1.3   Helical augers a) Standard model; b) Model adapted to sandy soils; c) Model adapted to gravelly soils.

The implement is welded to a rigid steel shank or screwed onto it and adjustable, so that the wrench, with extension rods, allows access to greater depths, say about 10 metres. In such cases, a guiding platform is useful for maintaining verticality and to preclude wobbling of the set of extension rods. In the easiest cases (coarse-textured soils, plough layer etc.), the handle of the auger may be lighter, split or even foldable.

Soil samples are then drawn from different depths, usually marked on the shank.

3.1.2 Australian auger

This implement is a cylinder 10 cm long in which the base is set with cutting teeth (Fig. 1.4a). It enables penetration of rather hard material that cannot be penetrated by the helical auger, but withdrawal of the sample is a tricky operation, more so if the soil is dry and greatly disaggregated. Sampling

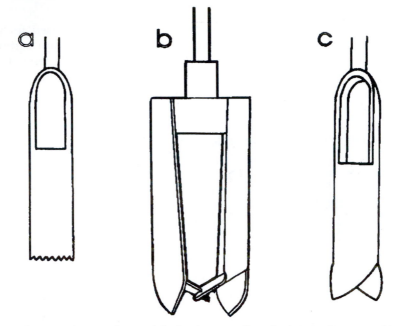

Fig. 1.4 Some implements designed for hard or gravelly soils a) Australian auger; b) Auger for very hard soils; c) Riverside auger (USA).

depth is thus often limited to the length of the bracket that retains the sample at the bottom of the cavity of the auger. This type of auger is nevertheless very useful for crossing a shallow, compacted or cemented horizon.

3.1.3 Riverside auger
This auger is obviously built on the same principle as a cylinder but has just two teeth; depth of sampling is greater than with the Australian auger—15 to 20 cm (Fig. 1.4c). The standard diameter is 7 cm; it is used for penetration of gravelly soils.

3.1.4 Semicylindrical augers
Also called 'gouge augers', these implements are part cylinder (more than half) with sharp edges that enable drawing of an undisturbed core (ELE International[1]) from soft, non-gravelly soils. Those used in agronomy are light, rather fragile and very small in diameter (30 mm or sometimes less, say 19 mm). The sample is thus limited in volume but is continuous and gives a good idea of the morphology of the plough layer and the succession of horizons (Fig. 1.5b).

In mangrove, the 'mud scoop' is used (Fig. 1.5a). Built on the same semicylindrical model but about 10 cm wide and about 120 cm long, it

---

[1] *See* Appendix 6 for addresses.

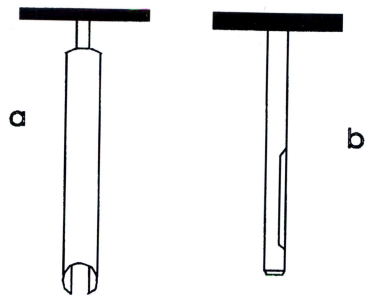

Fig. 1.5 Semicylindrical augers (for soft and non-gravelly soils, mangroves etc.) a) Mud scoop; b) Auger for undisturbed samples (0.6 to 1.2 m long).

enables taking a large soil core. Its use is limited strictly to organic oozes saturated with water. It is very useful in mangrove swamps being easy to handle and fast, and enables removal of a large number of continuous cores quickly for immediate measurements on undisturbed samples (pH, for example, to determine potential acidity).

3.1.5 Special augers

For more specific requirements, or for use in particular terrains, special implements are available with certain manufacturers or are fabricated by the users themselves (Fig. 1.6): chisel augers, stone-breaking augers (plain or flanged), bell-and-screw augers for boring and sampling in fine sand, and various valved augers usable in water-saturated media.

3.2 Sampling with Machines

3.2.1 Motorized auger

The implement consists of a spiral screw about a metre long (the screw threads at the bottom may be limited to a few tens of centimetres). The implement is rotated by a portable petrol engine provided with a handle for two persons to hold it above the soil. The soil sample is continuously raised to the surface by rotation of the screw; it is totally disturbed and identification of its depth is well-nigh impossible. The implement is quickly installed but its use is limited to digging a borehole for different purposes or for

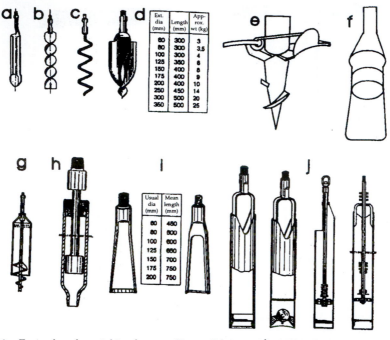

Fig. 1.6   Examples of special implements (Bonne-Espérances[1]) a) Chisel auger; b) Strip auger;
c) Corkscrew auger; d) Auger for boring by rotation in cultivated soils, clays and
sandy clays; e) Auger for gravelly soils; f) Spoon auger; g) Bell-and-screw auger for
boring by rotation and for sampling in fine sands; h) Hammering apparatus; i) Stone-
breaking augers (flat and flanged); j) Valved augers for water-saturated media.

determining the depth to a hard layer for example; it is totally inadequate
for soil sampling even if the auger shank is rotated down to a shallow
depth.

3.2.2 Drills and corers
These are very cumbersome machines requiring much manpower and a
large work area. Some are movable and can be pulled along the land or
used from a vehicle.

Percussion systems are sometimes used: they are extended by 'gouge
augers' or reinforced cylinders with cutting head that enable taking undisturbed samples from several metres depth (Fig. 1.7).

Others, such as the stationary drill with piston, function automatically
once installed. Such a drill was used by Marius (1978) for example, in the
clayey oozes of the mangroves of Senegal. It enabled drawing 1 m long
cores to a depth of 10 to 12 metres successively. Utilization is laborious,
however, and limited by the nature of the ooze, which should be exclusively

---

[1] *See* appendix 6 for addresses.

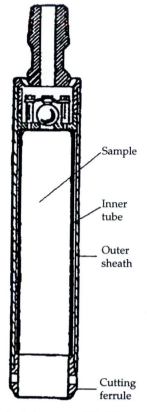

Sample

Inner tube

Outer sheath

Cutting ferrule

Fig. 1.7 Apparatus approved by the Laboratory of Buildings and Public Works (France) for taking intact samples. This apparatus is fitted with a split inner tube, which enables extraction of a second sheath of thin sheet steel containing the sample. The implement is driven into the soil by hammering (in this case it is usually surmounted by a hammering apparatus and a hammering head linked to the shank) or by pressure from a screw-jack supported on a special mounting.

clay since the presence of a thin sandy layer blocks coring. Such samples provide valuable information about the characteristics and physicochemical environment of deep oozes.

3.3 Sampling from a Profile Pit

The operation is carried out after digging a pit or trench with vertical walls (Fig. 1.2) to a certain depth, depending on the objective (agricultural, pedological or weathering profile).

It is a reliable method because it is normally used after a minute examination and preliminary description of the pedological profile. The horizon boundaries, irregularities and anomalies are identified and it is then easy to do systematic or selective sampling.

3.3.1 Bulk sampling
This is simple to do but requires certain precautions:

—Sampling should be done with a proper tool (pick, knife) from the bottom of the pit upward; it is generally done in the previously identified horizons but may be done systematically every 10 or 20 cm.

—A larger sample than the one required for analyses must be taken so that proper homogenization can be done in the laboratory; the sample quantity varies—generally 250 g, 500 g or 1 kg, sometimes more if the soil is gravelly, and still more if required (to preserve a standard sample for example).

—Bulk samples are stored in bags (biodegradable cloth or plastic) or even in boxes of different kinds.

—The label should be clear, legible, indelible and precise to pose no ambiguity in later operations.

3.3.2 Selective sampling
According to the heterogeneities identified by profile examination, it is often useful to take small specific samples, distinct from yet complementary to the bulk sample: salt concentrations, crystals, mottles, nodules, concretions, weathering rings, coatings, etc. These are meticulously collected, preferably directly from the field if conditions permit, otherwise in the laboratory from the original bulk sample.

Undisturbed samples, with structure intact, may be required: small aggregates, pads, clods etc. Although sampling poses no particular problem, transportation requires certain precautions (rigid packing and cotton wool).

3.4 Sampling for Agronomic Purposes

Whatever the objective, these samples concerning the many facets of agricultural improvement of soil should be taken with consideration of spatial as well as temporal aspects.

In a cultivated area or one prepared for a crop, therefore disturbed, it is essential to establish a three-dimensional sampling design.

In prepared soils, the depth to which the soil has been worked should be reached or in the case of meadow and even forest soils, the top few centimetres in which most of the nutrient elements are concentrated. In any case, the depth of sampling should be uniform and reported identically to enable comparison of data from the same thickness of soil (organic matter content for example).

The sampling design (Fig. 1.1) should be predetermined and adapted to the type of endeavour or to the surface morphology of the land. Although it is always preferable to take a large number of samples, in practice on average about fifteen sites per hectare are maintained, which often serve for making a composite sample.

Sampling, at least from the surface layer, is done with a spade, generally better suited than the auger for heterogeneous sites.

Controversy continues regarding significant temporal variations in nutrient elements in soil during the crop season. It is therefore recommended that the growing period be considered for sampling. Sampling should be done either before or several months after application of organic manure, fertilizer or amendments.

3.5 Reference Samples

It is useful, if not essential, to have one to three different standard samples in the laboratory in order to obtain a wide range of certain properties (texture, soluble salts, organic substances etc.). These samples are included in the usual series of analyses to verify the precision of results and to preclude certain deviations which might otherwise escape notice. The reference result is obtained by a preliminary analysis done by three analysts. It is therefore necessary to hold in reserve a large amount of a standard sample. The standard is obtained by taking a large volume of soil, which should be carefully homogenized after drying and sorting out coarse particles, for example by using a drum of 200 litres or a concrete mixer, or by turning over soil heaps.

There are international exchange networks for soil analysis such as that based in the Netherlands, I.S.E. (International Soil Analytical Exchange, Department of Soils and Plant Nutrition, Wageningen Agricultural University). This Exchange, which has about 300 member laboratories distributed in 69 countries, provides data (means, standard deviations, medians etc.) for several determinations. Analysis can be done of samples provided by the network and the data confidentially compared with that obtained in the laboratory of the network (Houba *et al.*, 1992).

3.6 Special Samples

These are samples other than those taken at random or selectively, which are withdrawn in known volumes with special equipment; they are taken for determination of moisture content, bulk density, permeability and micromorphological studies or experimental studies on soil monoliths.

3.6.1 Samples for determining moisture content

Such samples are taken with an auger. The only problem is avoidance of handling as that disturbs the original moisture condition; the sample should be stored and transported in a container that can go directly into a hot-air oven, such as a cylindrical metal box with a hermetically sealed lid or, better still, a glass bottle with a screw-type lid of Bakelite. Condensation of moisture on the inside of the lid is insignificant if precaution is exercised to completely fill the receptacle. It is also recommended that wooden boxes with compartments, very useful for transportation and safe storage of samples, be fabricated.

3.6.2 Core sampling of soil

Principle
The purpose is to take a known volume of undisturbed soil for determination of dry weight and bulk density or certain other moisture characteristics. The method, termed 'by cylinder' (Audry *et al.*, 1973), is the subject of the AFNOR standard NF X31-501 (1992). The aim is to determine the bulk specific gravity, which is numerically equal to the bulk density of a known volume of soil sampled *in situ*:

$$da = \frac{P}{Va}$$

The volume (*Va*) is sampled with a cylinder directly pressed into the soil and the sample evenly trimmed at both ends. The sample is dried and weighed (*P*). The equipment for sampling consists of a cylinder that may have various features; it is driven into the soil by hammer blows or by pressure applied with a jack or screw jack. Once the cylinder has been pressed in, the soil block is cut out with a spade and trimmed at both ends.

Equipment
The cylinder should be sharp-edged and suitably bevelled to avoid compaction of soil when it is driven in (Fig. 1.8); its size and shape (height:diameter ratio) depend on the research objective and the soil properties:
—with a small cylinder, the bulk density of thin horizons and small soil volumes can be determined; but it will give results greatly influenced by differences in detail that use of a larger cylinder would more or less mask;
—compaction while driving in the sampler should always be avoided; the soil will tend to compress more, the more compact and stickier it is and the longer and narrower (and less smooth) the cylinder; but the shorter the cylinder in relation to its diameter, the easier to remove a good unbroken soil sample; an average practical value is thus arrived at within the range $1/2 < H/D < 2$;
—the total volume of the cylinder should be between 100 and 1000 cm^3; however, up to 2000 cm^3 is allowed.
The anvil (headpiece) transmits the force of the blows or continuous pressure applied along the axis of the cylinder without damaging it. The equipment should have an opening large enough to allow observation of the interior of the cylinder as it is pushed in to detect any possible compression (Fig. 1.8).

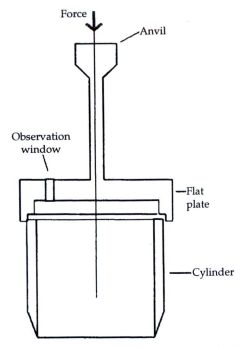

Force

Anvil

Observation
window

Flat
plate

Cylinder

Fig. 1.8 Sectional view of the implement for taking soil cores (Audry *et al.*, 1973).

The jack or screw jack has no special features except that it is provided with a series of chocks and pins that ensure its convenient placement and straight driving. It is generally operated on one vertical wall of a pit taking the support of the opposite wall.

Tools for extraction and trimming: spade or shovel without handle, knife, machete or any sharp straight-edged tool.

Container for storage and drying: according to the precision sought in determination of moisture content, an ordinary tightly closed plastic bag or an airtight moisture can (plastic bag soldered shut or sealed with adhesive, or metal or plastic box sealed with adhesive tape).

Sampling technique

The cylinder may be driven into the soil in any direction; in practice, it is generally pushed in vertically at the surface and on steps on the side of a pit or horizontally into the wall of a pit. The cylinder should be driven in true, without internal compaction; the best method for avoiding compaction (hammer blows, manual pressure or pressure applied with a jack) is selected according to the soil and its condition.

The cylinder should be forced in flush with the soil surface or, better still, slightly deeper. The soil block in it is removed with a shovel, with sufficiently deep digging. In some soils it is necessary to actually carve out the block containing the cylinder.

It is convenient to then place the block on a plate to enable clean trimming of the ends and cleaning of the outside of the cylinder. Removal and storing of the sample is preferably done on a sheet of plastic to avoid any loss of soil.

The principal problems related to this type of sampling are:
— forcing in the cylinder (especially if coarse fragments are present);
— danger of soil compaction in the cylinder;
— trimming the core, which should be level;
— transportation under hermetically sealed conditions for determination of soil moisture content.

This is a relatively rapid technique that, applied with minimal equipment and a few precautions, enables precise determination of bulk density.

3.6.3 Taking soil monoliths

This type of sampling, trickier in execution, especially in certain soils, has the advantage of providing an undisturbed volume of soil on which various laboratory studies can be done. The monolith is variable in shape and size: cubical, cuboidal or cylindrical.

3.6.3.1 THE VERGIÈRE METHOD

Principle

This is a well-standardized method (Bourrier, 1965) in which the objective is laboratory determination of the hydraulic conductivity or bulk density (d) of monolithic soil samples (cubes of $10 \times 10 \times 10$ cm), in which the moisture content at sampling is also determined (Audry *et al.*, 1973).

The cubical sample is cut from the soil (Fig. 1.9a), enclosed in an open-ended metal box of side 12 cm, in which watertightness at the base is ensured by a clay seal applied externally, to enable pouring paraffin wax into the free space in-between the sample and the box. The sample is detached and trimmed and the box hermetically sealed with paraffin wax (Fig. 1.9b).

In the laboratory, the following steps are executed:

—determination of the total volume V_1 (soil + box + paraffin wax) by liquid displacement and weighing the sample at the same moisture content as at sampling;

—determination of hydraulic conductivity at a constant head of 10 cm imposed on the sample; measurement of the outflow Q enables calculation of K by application of Darcy's law (S is the cross-section area in cm^2, H the hydraulic head in cm and L the side of the cube in cm)

$$Q = KS \frac{H}{L}$$

—determination of the volume of the paraffin wax and the box by simple weighing, for calculating the volume V of the soil, which is weighed dry, giving P_2 then

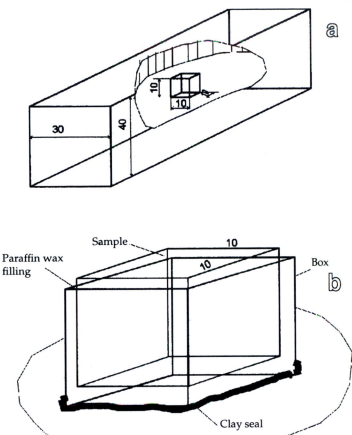

Fig. 1.9 Taking of soil monoliths (Vergière method: Bourrier, 1965; Audry *et al.*, 1973) a) Cube cut from a step in a trench; b) Paraffin wax filling (sides in cm).

$$d = \frac{P_2}{V}$$

Equipment
—trenching tools for digging the pit;
—equipment for taking the monolith (after Bourrier, 1965):
  gas stove,
  2.5-litre saucepan with insulated handle and spouts on opposite sides,
  5-litre plastic bucket for storing the clay for sealing,
  box for storing small tools,
  strong soil knife with 25-cm blade,
  10-cm soil knife,
  U-shaped steel frame with 12-cm opening for shaping the cube,
  mason's level,

2 m metal ruler,
finger trowel,
triangular painter's scraper 8 cm,
short pickaxe or equivalent,
American shovel,
narrow shovel with flat bottom, termed coal shovel,
pair of electrician's scissors or secateur (for cutting roots), paintbrush,
rags, etc.
—equipment for drying and storage; and consumables:
open-ended boxes 12 × 12 × 10 cm high, with two end covers, preferably
of zinc,
paraffin wax,
electrical tape to ensure airtightness of box when closed,
wooden box with pigeonholes for transporting sampling boxes.

Sampling procedure
Vergière method (Bourrier, 1965)
—Dig a pit with steps along one wall, corresponding to the different horizons to be sampled.
—On each step (starting from the bottom to avoid trampling), shape a 10 × 10 × 10 cm cube of soil using level, knives, set-squares, trowels and scraper (see Fig. 1.9a).
—Emplace an open-ended box with an airtight seal (Fig. 1.9b).
—Fill the interspace with paraffin wax: caution, very hot wax, very fluid, tends to penetrate the soil pores; also wax with a high coefficient of expansion is likely to undergo greater contraction the hotter it is. It is necessary to work when a solid scum starts to appear on the surface of the liquid, several times by partial filling if need be (Tobias, 1968).
—Remove the cube (pickaxe, shovels), trimming of the lower face, labelling the sample and marking its orientation.
—Fix the lids: directly (normal Vergière method) or after applying on the inner and outer faces a thin coating of a molten mixture of paraffin wax and beeswax, using a paintbrush. It is useful to line the insides of the lids with very thin foam rubber or synthetic foam that holds the sample by its elasticity without compressing it. The lids are held in place with electrician's tape applied criss-cross, then made airtight with the same tape.
Vergière modification: The paraffin wax is replaced by foam rubber. The cube is fully shaped, detached and placed on a lid. The open-ended box is replaced by four separate sheets held together by the two lids.
Inconvenient: there is risk of rupture of the cube during handling before coating with wax in the laboratory; risk of compaction of the cube; difficulty in sealing the sample hermetically.

Tobias modification (1968):
—a large monolith (40 × 40 cm × variable height) is cut on the profile face, the soil removed in 15 cm slices with a saw, and the slices transported to the laboratory packed in a box with foam rubber, cotton, etc.;

—the cubes are cut in the laboratory with a handsaw or hacksaw (cutting is very quick with either) and a specially sized joiner's-type box;

—then the material is impregnated with paraffin wax in a Vergière box.

Disadvantages: the method is only applicable to soils with sufficient cohesion; it does not allow determination of field moisture content other than on a nearby sample; there is always risk of the cubes being damaged in the laboratory. Use of the saw and the cutting box should be confined to the same soil.

Many adaptations are possible in accordance with nature of the soil and its cohesion (Audry *et al.*, 1973), but the principle remains the same.

In the laboratory various determinations can be done on the sample, which should be oriented properly. However, this type of sampling is essentially designed for determination of permeability under strict Darcy conditions and of drainage under standard controlled conditions (hydraulic head, initial moisture content, orientation of the sample). The major problems of the method are its limitation to soils with adequate cohesion, time taken for sampling and the meticulousness required.

3.6.3.2 OTHER MONOLITHS

It is possible to take much more voluminous monoliths for various purposes:

—Parallelepipeds several tens of centimetres in size (Loyer and Susini, 1978), held between four plates of transparent plastic; they are sampled and used in the laboratory for testing the response of various specific properties under controlled moisture contents (temperature, pH, redox potential, pNa etc.);

—Soil columns precut and inserted in large PVC tubes.

3.6.4 Samples for micromorphological studies

The soil is cut into thin sections for micromorphological studies after resin impregnation. Unassembled boxes are used, composed of stainless steel sheets, precut and shaped into a parallelepiped. They are forced directly into the soil and closed with two metal lids; the assembly is held together with adhesive tape. The box size depends on the type of thin section desired; the most common are $12 \times 7 \times 5$ cm. Impregnation is accomplished in the laboratory in special workshops.

3.6.5 Lyophilized samples

Aeration and drying after sampling of soils saturated with water can lead to great physical changes due to shrinkage (structure, microstructure, porosity etc.), or chemical changes (redox potential, oxidation of elements such as iron, manganese and sulphur, pH etc.). Application of sampling methods and preparative procedures specific to this type of sample is obviously essential.

Attempts at direct observation without impregnation in glass tubes sealed immediately after filling, have shown that the method is most useful for studying living animal or plant matter.

Attempts at direct impregnation have been made with a hydrophobic resin (Carbowax 6000, Union Carbide, New York); the moist soil is directly immersed in liquid resin at 60°C; after three days the sample is removed and allowed to solidify in air (Mackenzie and Dawson, 1961).

As these simple methods are not always satisfactory, lyophilization is obviously the preferred method for certain soils, for example mangrove soils (Vieillefon, 1970). The method has been successfully applied to mangrove soils containing more than 200% water (dry soil basis) and more than 30% organic matter, in which the dense fibres are obviously a factor favourable for cohesiveness of the sample. Similarly, in samples rich in salts there is no expansion, as seen when soils are air-dried.

To summarize, lyophilization has several advantages for chemical study of the forms of various elements, in particular sulphur compounds, because it enables much better sampling than determinations on a fresh sample. It also provides a satisfactory solution for micromorphological investigation of soils saturated with water. It is also very useful for studies pertaining to environmental protection, especially determination of degradable pesticides.

Principle

One of the major advantages with lyophilization lies in the possibility of very long storage of simple and complex organic substances, solid or volatile. The principal steps are:

—freezing at very low temperature;
—sublimation under a vacuum of less than 0.5 mbar of the water crystallized into ice;
—desorption of bound water.

The first two operations leave in the lyophilized material only a very low percentage of water (1-3%), termed residual water, which varies according to the material. This dehydration is generally not taken too far as residual water plays a protective role for oxygen.

In all cases, it is essential that the stored material be protected from moisture and oxygen. Lyophilized materials have low bulk density and very high porosity. Fine crystallization of ice actually produces multiple cavities more or less interconnected, giving a large internal surface area that may reach tens of m^2 per gram. This condition is favourable for impregnation of the sample with resin.

Procedure

Oriented samples are taken in waxed cuboidal boxes (see Sec. 3.6.4) or in large plastic tubes (length 1 m, diameter 7 cm). Lyophilization is accomplished in a tray lyophilizer of large capacity, directly for the first kind of sample, or after sectioning the cylindrical sample. According to the type of equipment used, samples up to 4 cm thick are allowable; obviously lyophilization is faster if the thickness is reduced. On average, 4 to 5 hours suffice.

A similar technique has been used for marine sediments by placing the samples immediately (at the site) in Dewar flasks containing liquid nitrogen for rapid freezing before lyophilization (Alix and Ottman, 1969).

After lyophilization the samples may be stored indefinitely in a desiccator. Impregnation can then be done later with an appropriate resin. The lyophilized sample fully retains the dimensions of the original sample, whereas normal drying leads to considerable shrinkage and large cracks. Contrarily, crystallization of ice tends to create a micropolyhedral structure that does not, however, impede study; the natural shapes (pores, plant remains, sand grains, various segregations) remain undisturbed in place.

4. Sampling Soil Solutions

For these studies, begun in the 1960s, two types of methodolgy are possible depending on whether free soil water, i.e., water corresponding to a moisture content greater than its retentive capacity, or bound water is to be sampled. For the former a lysimetric apparatus is used and for the latter a suction system that allows extraction of the soil solution (Briggs and McCall, 1904).

4.1 Sampling Free Soil Water

The apparatus differ considerably according to the situation and the designer: gutters for collecting oblique subsurface drainage water (ERLO pits), partly perforated plastic pipes inserted vertically in the soil (piezometers for sampling water with a valved tube, Fig. 1.6) or horizontally in the soil (drains) etc.

At a given site, seasonal qualitative (sometimes quantitative) monitoring of the soil solution is possible if inert materials are used, especially in dilute media. The apparatus have the disadvantage of drawing in free air, which limits later analyses and interpretations. One technique for extraction of free water (Maitre, 1991), on the other hand, allows sampling operations and also storage and filtration in the absence of air and light. To study dissolved elements, sometimes in low concentrations, the materials used are chemically inert. They are perforated, airtight polypropylene flasks (Fig. 1.10), which are inserted in the soil by means of a post-hole auger. These small lysimeters have two tubes that protrude out of the soil, one for applying light suction and the other for collecting the solutions under controlled atmosphere.

4.2 Sampling Free and Bound Soil Water

The principle is to extract the solutions by applying suction to a microporous plug; the plug is connected through a PVC pipe of varying length to a vacuum pump (Fig. 1.11).

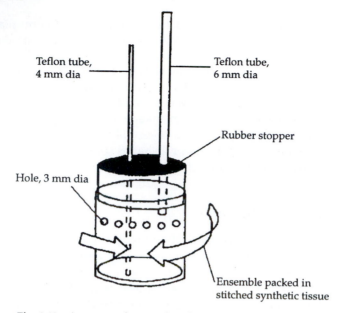

Teflon tube,
4 mm dia

Teflon tube,
6 mm dia

Rubber stopper

Hole, 3 mm dia

Ensemble packed in
stitched synthetic tissue

Fig. 1.10 Apparatus for sampling free soil water (Maitre, 1991).

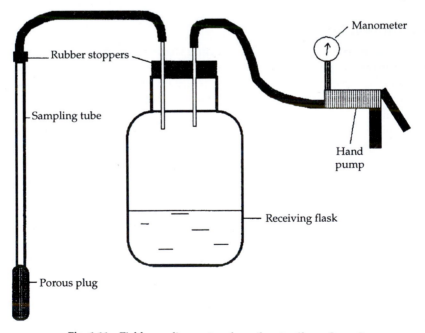

Manometer

Rubber stoppers

Sampling tube

Hand
pump

Receiving flask

Porous plug

Fig. 1.11 Field sampling system for soil water (free + bound).

Although the principle is simple, several technical problems relating to the equipment and its mode of use have been raised (Cheverry, 1983; Litaor, 1988; Debyle *et al.*, 1988).

—The choice of material for the porous plug (ceramic, fritted glass, teflon) is crucial because of its possible chemical reaction with the solution extracted: release of elements (calcium, magnesium, sodium and aluminium salts, silica etc.), or adsorption (phosphates, potassium). Washing the plug in acid and rinsing it thoroughly is a mandatory preliminary. Teflon, which *a priori* appears suitable, is hydrophobic. To compensate this defect, some manufacturers slyly add a small amount of ceramic, which releases undesirable compounds.

—The hazard of clogging by certain soil clay minerals or oxides.

—Disturbance of the natural circulation in the soil by the plug.

—The force and duration of applied suction, which influence the sphere of extraction and the more or less bound nature of the water extracted, and therefore its concentration.

Among the most commonly used plugs are the Nardeux Humisol[1] which vary in diameter (30, 50 and 63 mm), pore size (1-10 μm), length, thickness and area of porous wall; they are available at tube lengths of 15-180 cm and have manual vacuum pumps with adjustable suction. They are inserted in the soil using augers of suitable diameter which, of course, somewhat disturb the porosity of the soil.

If all precautions are taken and reservation exercised with respect to their use for analysis of certain elements in dilute solutions, these solution collectors seem to be the most efficient and simplest means for collection of soil solution *in situ* but require modifications in the nature and quality of porous material used.

Their areas of application are various: extraction of soil solution in general and of nutrient, saline or polluting solutions.

Use of a water-retentive gel such as 'water beads' (S. A. Beck, Post Box 2, 67037 Strasbourg Cédex, France) may be considered for trapping soil solutions. These extremely hydrophilic polyacrylamide beads are able to retain up to 400 times their weight of water. Inserted in the soil, they can without doubt extract the solutions. One problem remains, however: the mode of extraction of retained salts.

5. Sampling Soil Gases

Field extraction of soil gases for analysis is a tricky operation that has challenged the ingenuity of pedologists, agronomists and microbiologists. Soil air, collected with or without a suction pump, is generally adsorbed on

---

[1] *See* Appendix 6 for addressess

activated charcoal, then transported to the laboratory in airtight flasks for analysis.

Because of difficulties in sampling, handling and drying these free gases representing soil air, the current trend is immediate execution of qualitative or semiquantitative analyses. Some of the techniques available (tests *in situ*) are described in Chapter 2.

References

Alix Y. and Ottman F. 1969. Emploi de la lyophilisation pour l'imprégnation d'un sédiment meuble saturé d'eau. *C.R. Soc. Géol. France*, 63-69.

Audry P., Combeau A., Humbel F.X., Roose E. and Vizier J.F. 1973. *Bulletin du groupe de travail sur la dynamique actuelle des sols*. ORSTOM, Paris, **2**: 126 pp. + annexes.

Bourrier J. 1965. La mesure des caractéristiques hydrodynamiques des sols par la méthode Vergière. *Bull. Techn. du Gén. Rur.*, n° 73, 96 pp.

Briggs L.J. and Mc Call A.G. 1904. An artifical root for including capillary movement of soil moisture. *Science*, **20**: 566-569.

Chamayou H. and Legros J.P. 1989. *Les bases physiques, chimiques et minéralogiques de la science du sol*. Presses Univ. de France, 593 pp.

Cheverry C. 1983. L'extraction de la solution du sol par le biais de bougies poreuses: une synthèse bibiliographique des problèmes méthodologiques posés par ces dispositifs. *Bull. GFHN*, **14**: 47-71.

Crépin J. and Jonhson R.L. 1993. Soil sampling for environmental assessment. In *Soil Sampling and Methods of Analysis*, Carter M.R. (ed.) Lewis Publishers, pp. 5-18.

Debyle N.V., Hennes R.W. and Hart G.E. 1988. Evaluation of ceramic cups for determining soil solution chemistry. *Soil Science*, **146**: 30-36.

Houba V.J.G., Uittenboggard J. and Pellen P.J. 1992. *Report of International Soil Analytical Exchange (I.S.E)*. Wageningen Agricultural University, 148 pp.

Legros J.P. 1996. *Cartographies des sols. De l'analyse spatiale à la gestion des territoires*. Presses polytechniques et universitaires romandes, Lausanne, 321 pp.

Litaor M.I. 1988. Rewiew of soil solution samplers. *Water Ressources Research*, **24**: 727-733.

Loyer J.Y. and Susini J. 1978. Réalisation et utilisation d'un ensemble automatique pour la mesure en continu et *in situ* du pH, Eh et du pNa du sol. *Cah. ORSTOM, sér Pédol.*, **16**: 425-437.

Maitre V. 1991. Protocole d'extraction, de conservation et de filtration des eaux libres du sol. *Sc. du sol*, **29**: 71-76.

Makenzie A.F. and Dawson J.E. 1961. Preparation and study of thin material section of wet organic soil materials. *J. Soil Sci.*, **12**: 142-144.

Marius C. 1978. *Etude pédologique des carottages profonds dans les mangroves du Sénégal et de Gambie*, missions 1976-1977, A.T.P. «Mangroves et Vasières», ORSTOM, Dakar, 46 pp. multigr.

NF X31-501, 1992. Mesure de la masse volumique apparente d'un échantillon de sol non remanié- Méthode au cylindre. *In Qualité des sols*, 1994, AFNOR.

Pr ISO 11259, 1997. Description simplifiée du sol (X31-001). Evaluation des critères, terminologie et codification, vocabulaire. AFNOR.

Pr ISO 10381-1, 1997. Echantillonnage—Etablissement de programmes d'échantillonnage. AFNOR.

Sobee W.E. and Marx D.B. 1987. Soil sampling: Spatial and temporal variability. In soil testing: sampling, correlation, calibration and interpretation, Brown J.R. ed., Special Publication No. 21, SSSA, Madison, USA.

Tobias C. 1988. Mesure au laboratoire de la perméabilité d'échantillons de sols non remaniés. *Cah. ORSTOM Sér. Pédol.*, **6**: 251-257.

Vieillefon J. 1970. Intérêt de la lyophilisation pour l'imprégnation des sols gorgés d'eau. In *Séminaire de microscopie des sols*, INRA, Versailles, 3 pp.

Bibliography

Baize D. and Jabiol B. 1995. *Guide pour la description des sols.* INRA, Versailles.

Byrnes M.E. 1994. *Field Sampling Methods for Remedial Investigations.* Lewis Publishers, Inc.

Gomez A., Leschber R. and L'Hermite P. (eds.). 1986. *Sampling Procedures for the Chemical Analysis of Sludge, Soils and Plants.* Elsevier Applied Science.

Gras R. 1988. *Physique du sol pour l'aménagement.* Masson, Paris, 608 pp.

Hodgson J.M. 1978. *Soil Sampling and Soil Description,* Clarendon Press, Oxford.

Maignien R. 1969. *Manuel de prospection pédologique,* ORSTOM, Paris, 132 pp.

Mathieu C. and Pieltain F. 1998. *Analyse phsique des sols. Méthodes choisies.* Lavoisier Tec et Doc., Paris.

NF X31-071. 1983. Matériaux types. Définitions, prélèvements. In *Qualité des sols,* AFNOR (1994).

NF X31-100. 1992. Échantillonage. Méthode de prélèvement d'échantillons de sol. In *Qualité des sols,* AFNOR (1994).

Pr ISO 10381-4, 1997. Échantillonage. Procédure d'investigation des sols naturels et cultivés (X 31-008-4), AFNOR.

Pr ISO 10381-5, 1997. Échantillonage. Procédure d'investigation concernant la dépollution des sites urbains et industriels (X31-006-5), AFNOR.

Robert M. 1996. *Le sol. Interface dans l'environnement, ressource pour le développement.* Masson, Paris, 276 pp.

Russell-Boulding J. 1994. *Description and Sampling of Contaminated Soils: A Field Guide.* Lewis Publishers, Inc.

Saxena S.K., Gill S.A. and Lukas R.L. 1998. *Subsurface Exploration and Soil Sampling.* American Society of Civil Engineers.

Tan K.H. 1996. *Soil Sampling, Preparation and Analysis.* Marcel Dekker.

Wilson N. 1995. *Soil Water and Ground Water Sampling,* Lewis Publishers, Inc., 208 pp.

2

Preliminary Field Tests

1. Use

Evaluation directly in the field of the most important soil characteristics is often necessary to refine the profile description, sequence the field work and chronicle the laboratory analyses to be done later.

But these rapid field tests, mostly qualitative or semiquantitative, cannot be taken as proper assessments. Furthermore, they do not include field determinations related to hydrophysical behaviour or structure of the soil, and those that more strictly belong to the pedological domain than to the analytical domain: morphological description of the structure or resistance to the penetrometer for example. Some of these determinations, such as soil moisture content, should preferably be done in the laboratory but can sometimes be accomplished *in situ* using special methods.

The actual tests can be grouped according to their mode of operation, that is, whether they work on sensory perception or chemical examination.

2. Sensory Tests

2.1 Visual Tests

2.1.1 Determination of soil colour

This is essential for the description of the soil profile (Maignien, 1969).

It enables a preliminary insight into the abundance or absence of certain compounds (organic matter, iron oxides, etc.) and into the level of development of certain processes (oxidation, reduction, rubefaction, etc.), it also permits comparison between soils of different origin.

Colours are determined by reference to a code. Direct evaluation is not advisable because precision varies with the individual and the terms are subjective.

Two systems can be used:
—the polar co-ordinate code of Cailleux and Taylor;
—the Munsell Soil Color Chart (Fig. 2.1).

Experience has shown that the second, inspite of its high price, is more useful because:
—it is used today by the majority of pedologists worldwide, which enables simpler correlation;
—the stages are rational and correspond to soil characteristics;
—it is an easy mnemonic system.

Its nomenclature consists of two complementary systems (Fig. 2.1):
—a simple or compound colour name;
—a Munsell notation (numerals and letters).

The terms are used in scientific papers as well as in common parlance. They have
—a precise object;

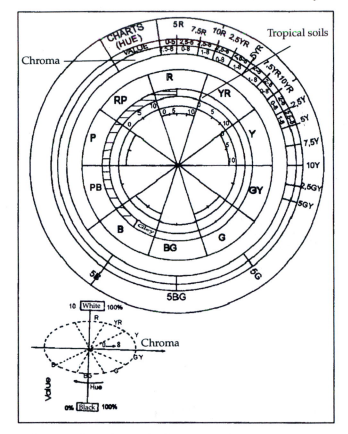

Fig. 2.1   Munsell system for coding soil colours: R = red, Y = yellow, G = green, B = blue,
P = purple; the valid zone is from 5 R (some tropical soils) to 5 B (some gley soils);
hatched zone not used for soils. Munsell Soil Color Chart, 2441 North Calvert Street,
Baltimore Md 21218, USA.

—an abbreviation for field descriptions;
—a specific relationship between the colours observed and the possibility
of statistical treatment.

The Munsell Soil Color Chart usually comprises 196 coloured chips. These
chips are systematically grouped according to their Munsell notation on
plates with a grey background, assembled in a folder with removable pages.
It is possible to add supplementary plates of 28 colours, in particular for
yellow and red colours. Further, special composite plates have been created
specifically for gley soils (values 4 to 7, hues 5 Y, 5 GY, 5 G, 5 GB and 5 B).
The notations consist of the sequence hue, value and *chroma*.

The hue is the predominant spectral colour. It is related to the dominant
wavelength of light. All the colours of one plate have the hue represented
by a symbol in the upper right-hand corner. The symbol for hue is the
upper case initial of the colour:

—R for red
—YR for yellowish-red (orange)
—Y for yellow
—N for neutral
—G for green
—B for blue,

preceded by a number from 0 to 10. For YR hues the plates become more yellow and less red as the number increases. The number 5 corresponds to a mixture of equal portions of yellow and red; 2.5 corresponds to 2.5 parts yellow for 7.5 parts red; the point 0 coincides with the point 10 of the hue following red. Thus the YR hues extend from 10 R (zero YR) to 10 YR (zero Y).

The value is linked to the relative brightness of the colour. It is a function (approximately the square root) of the amount of light.

In practice, this corresponds to addition of a known and variable amount of grey to the colour of the hue. Different greys are obtained by mixing white and black in defined proportions, each of which characterizes one value. The notation of values thus ranges from 0 for absolute black to 10 for white. The value 1 corresponds to a grey composed of nine parts black and one part white, the value 2 to 8 parts black and two parts white, and so forth.

The symbols for value are Arabic numerals that appear on the left of each plate opposite each row of chips. Vertically, from bottom to top, the colours become successively brighter by visually equal steps. Their value increases.

The intensity (*chroma*), often termed saturation, is the relative purity or strength of the spectral colour. It corresponds to the colour of the hue to which a certain amount of grey or known value has been added. This amount of grey decreases from left to right.

The notation of chroma appears at the bottom of each plate below the vertical column of chips. It is symbolized by Arabic numerals from 0 to 8. It increases from left to right with decrease in grey tone.

To summarize, the colours form three scales in their arrangement:
—from one plate to the next in hue;
—on the same plate, vertically in value, horizontally in chroma.

Thus in each plate the darkest colours are at the bottom left, the colour progressively lightening upwards and to the right.

The sequence to be followed is hue, value, chroma. The figures for value and chroma are separated by a solidus, for example, 2.5YR 4/6 = red.

If the colour notation requires use of a subdivision of an integer, decimals are used, never fractions. For example, the colour midway between 10YR 4/3 and 10YR 5/3 is 10YR 4.5/3; if the colour is located between 7.5YR 5/6 and 5YR 5/6, it is written 6.25YR 5/6, and similarly for intensity.

To determine the colour of a horizon, a piece of the sample is compared with the colour chips. As only rarely is the colour of the sample exactly that of a chip, the nearest colour is taken. Determination of colour should be done in daylight.

When the colour is recorded, it is essential to note the moisture state of the sample. As a matter of fact, the colour changes with water content. This phenomenon is striking in certain soils. Sometimes it can be characteristic of the soil type. Thus 'Leached Tropical Ferruginous' soils differ by 2 to 3 units of value and chroma between the dry and moist state.

In practice, colour determination is done either on an air-dried sample or on samples with moisture content close to field capacity. In most descriptions the colour given is that of a moist sample. In dry regions it is necessary to give the colour when dry, then moisten the samples with water from a spray bottle and record the colour when the water film has disappeared.

2.1.2 Coarse fragments

It is important to rapidly estimate in the field the proportion of coarse fragments in a soil in relation to its fine earth. This ratio is used, for example, to determine an index of the weight of fine earth per hectare, which correlates closely with the yield of cereals on certain soils (Baize, 1988). This determination is done gravimetrically.

For more general pedological purposes, estimation of gravel and stones can be done visually by comparing the vertical face of the horizon or profile with schematic drawings showing volumetric percentages.

2.1.3 Porosity

It is necessary to distinguish true porosity, which is diffuse and a quantitative characteristic determined in the laboratory, from a description of pores, which is a more qualitative than quantitative characteristic.

In the field the abundance of visible pores should be estimated with the naked eye in a section across the bulk of the horizon or in a section through the structural units (Maignien, 1969). A distinction should be made between this porosity within the aggregates (intra-aggregate porosity) and that corresponding to the spaces that separate the structural units (structural or interaggregate porosity). Thus micropores (the limit of visibility of which, $\simeq 0.1$ mm, still relates to the macropores of hydraulic scientists) are distinguished from macropores, by taking into account the shrinkage cracks that could have considerable influence on the drainage of a dry clayey soil when water comes from rain or irrigation.

With respect to the porosity within the aggregates, the following standards were proposed after observation under a hand-lens (Duchaufour, 1977):
—compact: less than 10 pores per cm^2;
—moderately porous: 10 to 25 per cm^2;
—porous: 25 to 50 pores per cm^2;
—very porous: more than 50 pores per cm^2.

2.2 Tactile Tests

2.2.1 Texture

Until the results of particle-size analyses are available, which give precise information on the texture of a sample, it is often useful to obtain a quick evaluation of it directly in the field. This helps in characterizing different profiles according to their origin or different reference or diagnostic horizons, to highlight textural differences or even to define certain pedogenetic processes (clay-leaching, accumulation of clay, or textural degradation).

The texture of a horizon is determined by crushing and rolling a dry sample between the fingers when possible, then doing the same with a moist one. With a bit of experience, a quick assessment of the texture can be arrived at from the feel by applying the following criteria:

—when moist, clay forms a paste that adheres to the fingers, is easily deformed by finger pressure and takes the desired shape;

—silt, when dry, is soft to the touch like talc; when moist it forms a paste that does not stick to the fingers and quickly becomes friable;

—fine sand, in which the particles are not readily visible, sounds gritty and when rubbed gives a more or less distinct impression of rugosity according to particle size;

—coarse sand is visually identifiable by its size.

For certain clayey soils the following test may be applied (E. Roose, IRD, pers. comm.): (1) moisten and crush the soil with the fingers; (2) roll it between the hands into a rod 5 cm long and 5 mm in diameter; if the rod does not break, the clay content may be estimated as greater than or equal to 25%; (3) if step 2 is possible, bend the rod into a ring; if this is possible, the clay content is greater than 35%.

Some soils, such as Andosols with allophane or Salisols with sodium carbonate (A.F.E.S., 1990), do not always obey these criteria, the gels and certain salts giving a rather specific feel (soapy).

Another difficulty is estimation of the mixture of different proportions of textural components, which according to their percentage give different tactile sensations. Clays definitely feel different according to their mineralogical composition. Kaolinite for example, sticks much less than montmorillonite. An abundance of organic matter invites overestimation of the actual proportion of silt.

So the value of the test is very relative and demands considerable field experience. A trained surveyor can estimate texture to within 5%. Restandardization is necessary, however, after absence from the field for some time.

Note: Some laboratory suppliers (ELE International[1]) offer small portable texture-estimation kits for use in the field. They consist of graduated tubes, a dispersing agent and a flocculating agent, and enable quick estimation of the three fractions—clay, silt and sand.

---

[1] *See* Appendix 6 for addresses.

2.2.2 Consistence

This property of aggregates is highly dependent on the moisture state of the material. It is difficult to estimate in wet or moist material. When dry, the behaviour of an aggregate pressed between the fingers allows definition of different classes (soft, more or less coherent, hard).

In Thiosols and Sulfatosols of mangroves or polders, consistence is an essential physical characteristic (Marius, 1990). It has been defined by the index n, related to the water content, texture and organic matter by the formula

$$n = \frac{H - 0.2\,R}{C + 3\,OM}$$

where H is the gravimetric water content (%) of the soil in the field (calculated on dry soil basis), $R = 100 - C - OM =$ silt + sand, C is clay content (%), OM is organic matter (%) (= organic carbon $\times$ 1.724).

The higher the n, the less developed (or mature) the soil. Estimation of consistence enables determination *in situ* of the degree of physical development of a solum. Five classes of development corresponding to five classes of consistence have thus been defined (Table 2.1).

Table 2.1 n value and consistence classes of acid sulphate soils (A.F.E.S., 1990)

n value	Consistence class	Classification of development	Description
> 2	1	Undeveloped	Fluid, soft, cannot be retained in the hand
1.4-2	2	Slightly developed	Without consistence, very plastic; passes through the fingers
1-1.4	3	Half-developed	Very malleable, plastic, sticks to the hand but escapes between the fingers
0.7-1	4	Nearly developed	Malleable, slightly plastic, sticks to the hand. Has to be forced to pass through the fingers
< 0.7	5	Developed	Very firm, resists manual pressure

2.3 Olfactory Tests

When an observation pit is opened it is useful to smell certain samples, particularly those from depth, that emit an odour of gas emanating from anaerobic fermentation: H_2S or methane in mangrove soils. This odour indicates a particular physicochemical condition that can be qualitatively assessed.

Reagent ampoules are also available that give a colour reaction with certain gases (see Sec. 3.5 below).

2.4 Taste Tests

In saline media, determination of the presence and sometimes the nature of salts is useful. Salts may be expressed as efflorescences, powder or crusts, or simply be present in the soil solution. Very soluble salts have a very pronounced taste, which enables distinction of sodium chloride, sulphate and carbonate. Aluminium salts (alums) have a very specific bitter (acrid) taste, very different from those of the aforesaid.

With some experience it is possible to at least recognize the presence or absence of salts, if not estimation of the degree of salinity of a sample; this capability varies from person to person.

3. Chemical Tests

3.1 Calcium Carbonate Content

Some carbonates, such as calcium carbonate, exert a strong influence on the morphology of profiles. Calcium carbonate specifically directs the evolution of organic matter and slows down clay leaching. It is thus important and easy to point out the presence or absence of carbonates directly in the field.

The presence of calcium carbonate is identified by spraying dilute hydrochloric acid on the sample, using a spray bottle; the acid produces a more or less vigorous effervescence by releasing carbon dioxide. For a relative estimate of its content according to horizons, samples are taken from successive horizons and juxtaposed in small heaps; a few drops of acid added to each heap enable better observation and comparison of the speed and intensity of the reaction. It must be remembered that a calcareous sand reacts more vigorously than a marl with comparable carbonate content, and that the effervescence could also be related to the presence of sodium or magnesium carbonate. The reaction of the latter is slower and manifests only when heated, and thus may pass unnoticed in the field. Siderite ($FeCO_3$), present in some soils, hardly reacts. It is possible to roughly distinguish the following classes (Maignien, 1969):

—non-calcareous: no effervescence;

—slightly calcareous: slight effervescence, barely visible, but clearly audible;

—calcareous: visible effervescence;

—highly calcareous: presence of calcium carbonate is generally visible (mycelia, small pinhead-size nodules, accumulations etc.).

Percentage classes are often proposed from these observations. They are difficult to formulate and it is often preferable to wait for the results of analysis.

3.2 Soil pH

Soil pH is the chief chemical parameter used to define the soils of mangroves and polders, especially Thiosols[1]. Actually when measured in the field and sometimes recorded by special probes (Loyer and Susini, 1978), it is generally close to neutrality or very slightly acid, between 6 and 7. When measured on an air-dried sample, it may drop to values lower than 4, say 3.5. The acidity that develops when the samples are dried is mostly caused by oxidation of pyrite, with production of sulphuric acid. It is termed *potential acidity* and corresponds to 'pH *in situ* minus pH dry'.

3.3 Test for Amorphous Material

This technique (Fields and Perrot, 1966) is used for demonstrating the presence of amorphous material in soils, particularly aluminium hydroxide in the form of $Al(OH)_3$. The principle of the test is based on the reaction

$$Al(OH)_3 + 6\,NaF \rightarrow AlF_6\,Na_3 + 3\,NaOH$$

Release of OH^- leads to a great rise in pH that is indicative of a more or less large amount of amorphous material (pH > 9.4, in 2 minutes).

In the field, an ashless filter paper impregnated with phenolphthalein and air-dried or oven-dried is used for the test. A small clod or a small quantity of crushed soil is placed on the filter paper and wetted with a $1.0\,M$ NaF solution (pH 7.5) using a very fine dropper. The reaction produced in the soil releases OH^- to the solution, which wets the filter paper. This increase in pH causes the phenolphthalein to change colour (pH ~ 8.5), the colour appearing red towards pH 9.4 and becoming redder, even purple or violet around pH 10-11, according to the abundance of aluminium in the material containing allophanes, non-crystalline aluminium polymers or chelated Al^{3+}.

The presence of organic matter may mask the reaction. If observation is difficult, the filter paper is turned over to observe the more visible coloration on the under side.

3.4 Degree of Decomposition of Hygrophilic Plant Debris

The degree of decomposition of hygrophilic plant debris (present in Histosols for example) is estimated in the field (Von Post, *in* Laplace-Dolonde, 1992).

The test is done on the fibrous material by manually squeezing the sample and estimating the colour of the liquid that flows out and the nature of the fibrous residue. The scale of humification comprises ten degrees (Table 2.2).

---

[1] *See* soil classification and soil reference base in Appendix 5.

Table 2.2 Scale of humification of histic horizons according to Von Post (A.F.E.S., 1990)

	Degree of humification	Decomposition	Plant structures before test	Content of amorphous material	Material obtained by manual pressure	Nature of residue left behind on the palm
R_1*	h1	None	Perfectly identifiable	Zero	Clear water	Undecomposed plant materials
O_i	h2	Insignificant	Easily identifiable	Zero	Yellow to brown solution	Very slightly decomposed plant materials
	h3	Very slight	Identifiable	Very low	Brown or black solution	Slightly decomposed plant material, slightly moist fibrous mass
	h4	Slight	Identifiable with difficulty	Low	Turbid solution	Residue (moist) has slightly granular consistence
R_2	h5	Moderate	Recognizable but not identifiable	Medium	Turbid solution with a little peat	Slaked pasty residue, plant structures still visible to the naked eye
O_c	h6	Moderate to strong	Not recognizable	High	Muddy water; less than 1/3 of the peat passes through the fingers	Granular and soft residue with some visible plant structures
	h7	Strong	Indistinct	Very high	Muddy water; almost half of the peat passes through the fingers	Slaked residue with visible plant structures
R_3	h8	Very strong	Very indistinct	Very high	Mud: 2/3 of the peat passes through the fingers	Soft and slaked residue, at times with undecomposed ligneous residues
O_a	h9	Almost total	Practically not discernible	Very high	Almost all the homogeneous soil-water mixture passes through the fingers	The small quantity of plant structure included in the residue is rarely recognizable
	h10	Total	Not discernible	Very high	The entire homogeneous peaty mass passes through the fingers	No residue

*Equivalents:
R_1 - R_2 - R_3 = indices of the International Peat Society
O_a - O_c - O_i = indices of the US Department of Agriculture-Natural Resources Conservation Service

3.5 Detection of Soil Gases

The difficulty in sampling free gases in soil prompted investigators to measure gas release directly *in situ*. Certain apparatus (Bachelier, 1966) are designed for precise determinations, while others belong to the level of preliminary field tests.

One such apparatus is the Dräger gas detector, used for determining the carbon dioxide content in the soil air (Bachelier, 1968).

This apparatus uses graduated reaction tubes through which a known volume of air is passed using a calibrated hand-pump. Carbon dioxide in the air thus pumped reacts by combining with hydrazine in the reaction tube; transformation of hydrazine is shown by a colour reagent.

For taking samples, probes of different length are installed permanently in the soil. These probes consist of copper tubes of 12-16 mm diameter pushed into previously drilled 16 mm holes. The holes should be 2-3 cm deeper than the probes to provide a small air chamber at the base of each probe. The probes protrude 5 cm above the soil and are provided with a wide soldered collar. These collars are attached to the soil with plaster of Paris and ensure airtightness on the surface. The probes are closed at the top by a lightly greased stopcock or a simple rubber stopper.

Reading of the CO_2 of the soil air is taken by attaching the reaction tube to the stopcock of the probe or by intoducing it into the probe itself through an extension tube provided with a rubber ring that closes the top opening of the probe. The air aspirated by the hand-pump passes through the graduated reaction tube which, for a given volume of air, gets coloured over a length corresponding to the CO_2 present. Five strokes of the pump are required for CO_2 reaction tubes graduated from 0.1 to 1.0%. Ten strokes are necessary for reaction tubes graduated from 0.02 to 0.3%.

The manufacture in the 1950s of the first reaction tubes designed for analysis of air and gases, slowly replaced by other reaction tubes, opened up possibilities in the area of detection of soil gases. It is possible, for example, to directly attach a reaction tube to the suction pump that forms a gas-detecting assembly. Knowing that today there are more than 200 types of selective reaction tubes, the possibilities of determination of several gases and vapours are greatly enhanced (Dräger, 1989).

The problems of their use in soils have been solved by the invention of a probe apparatus—a reaction tube that enables direct analysis of oxygen, carbon dioxide, hydrogen sulphide, methane etc. The method is quick: a small hollow drilling probe of adjustable length, carrying a detachable drill head (Fig. 2.2) is pushed into the soil with the drill head attached. When the test depth has been reached, a gentle upward pull of the probe detaches the head. The appropriate reaction tube is then introduced into the body of the pumping probe and the gases aspirated by a manual bellows are fixed by the reactive compound; the change in colour allows direct reading and its

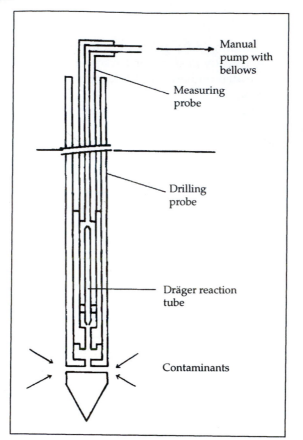

Manual
pump with
bellows

Measuring
probe

Drilling
probe

Dräger reaction
tube

Contaminants

Fig. 2.2 The Dräger drilling and measuring probe.

translation to concentration, as a function of the number of strokes of the pump. This may be programmed for automatic computer calculation.

3.6 Chemical Test Kits

Progress in electronics and computer technology has enabled miniaturization of measuring instruments and improvement of their viability and their programmes. Through reduction in size of the equipment, its energy consumption and price, it is now possible to supply users with field instruments capable of providing perfectly reproducible results and data storage (price ranging from FF 200 to about FF 25,000 or more, according to quality and features).

These instruments are designed for chemical or physical analysis. They are used for:

—analysis *in situ* of unstable samples that cannot be analyzed in the laboratory;

—obtaining results immediately that enable planning field strategy;

—selecting the most representative samples before sending them to the laboratory (homogenetic etc.);

—doing the indispensable pretreatments of sterilization, concentration and reduction of weight before sending them to the laboratory;

—continuous determinations *in situ*, for observing long-term phenomena with storage of data and cable transmission later.

Many determinations priorly done in the laboratory can now be done in the field with easily transportable self-contained instruments (determinations on a small quantity by immersion, circulation, penetration etc.). The choice of determinations and tests depend on the absolute or relative precision required, mode of transport and field requirements.

It is possible (for liquids) to measure conductivity and pH and to do colorimetry with instruments weighing less than 1 kg, or through recourse to more sophisticated instruments (with data entry) interfaced with a computer, weighing around 6 kg, to measure pH, Eh, dissolved oxygen, conductivity (salinity), moisture content, temperature and ion content.

3.6.1 Analytical kits for soils and waters

—classic assemblies: Barbier-Morgan[1];

—modern assemblies: Hach, Lamotte, Palintest[1] etc.

These consist of disc comparators or paired comparison cuvettes, or a battery-powered electronic colorimeter with filters for 410, 490, 520, 570 and 640 nm.

Various reagents are supplied, enabling determination of: aluminium, ammonium, boron, cyanide, bromine, chlorine, chloride, copper, soluble iron, fluorine, molybdate, nitrate, nitrite, phosphorus, potassium, silicon, sulphate, sulphide, zinc, water hardness etc.

Some assemblies also include a micro-pH-meter and conductivity meter operating on 1.5V or 9V batteries.

The kits (in a case) weigh from 1 to 5 kg and both qualitative and quantitative analyses are possible. The reagents are supplied as tablets for a single analysis or in automatically filling vacuum-sealed ampoules. The most common analyses are for:

alkalinity	10-100 ppm
chloride	2-100 ppm
CO_2	10-1000 ppm
Ca^{2+} hardness	50-500 ppm
total hardness	2-200 ppm

Ampoules are also available which enable other analyses or preservation of samples in an acid, alkaline or complexing, etc. medium (stabilization, sterilization).

---

[1] *See* Appendix 6 for addresses.

Merck[1] supplies a very large set of qualitative and quantitative tests: for example, nitrates can be determined with the *Merckoquant* strip test using a 9 V reflectometer in the range of 0 to 500 mg L^{-1} with a precision of ± 10%.

3.6.2 Useful small portable instruments

These generally operate on 1.5 V or 9 V batteries.

A. *Demineralized water* is prepared in the field using plastic bags that contain a mixed-bed resin. The bag is filled with on-site water, the contents shaken and the bag inverted. About five litres of demineralized water with conductivity of 1-2 mS cm^{-1} can be obtained from one bag.

B. *The samples can be reduced* in the field using mixers spinning at 5000-25,000 rpm that run at 12 V, weighing about 1.5 kg.

C. *For liquid samples*, portable, battery-operated, self-contained sampling equipment with carrying case are available, consisting of pump, filter, collector and programming for 24 hours.

D. *Weighing in the field* is accomplished with top-loading electronic microbalances (*see* Chapter 6) that can weigh up to 100 g to the nearest mg; 1.5 V operation; weight 0.2 to 2 kg.

E. *pH and Eh* can be determined by several electronic instruments that allow work on just a single drop.

pH standards are supplied in small capsules for dissolving on site in resin-treated water or in sealed glass ampoules (25 ml) of NBS quality. Calibration is done at two points, usually at pH 4 and pH 7 or at pH 7 and pH 9.

A range of instruments is available, from a tiny one with flat sensor for work if necessary on one drop (credit-card size, weight 40 g, digital display, 3 V lithium battery, runs for 500 hours) to larger, heavier hand-held airtight models with temperature compensation, automatic calibration, RS232 interface etc.

F. *Various conductometers:* 25 models powered by 1.5 V or 9 V batteries, weighing from 40 g to about 700 g, range 0.0 to 1999.9 S cm^{-1}, 0 to 199.9 S cm^{-1} or 0.00 to 19.99 S cm^{-1} with a precision of 0.3 to 2%, that do measurements on small volumes (microcell with integral or separate electrode, temperature control, RS232 interface etc.).

G. *Refractometers and salinometers*, range 0-10% in steps of 0.1% NaCl, 0-28% in steps of 0.2%; weight 150 g.

H. *Cuvette colorimeters or Lovibond-type comparators* (*see* Chapter 9).

I. *Moisture meters* for determination of moisture content by polarization of the test sample under the effect of an electromagnetic field at 50 MHz; sample weight 20-180 g.

---

[1] *See* Appendix 6 for addresses.

J. *Climatic measurements*: atmospheric humidity or moisture content of a solid (by measuring capacitance between two electrodes).

Integral or separate probe, airtight; penetration probes for solids or flat 'dagger' probes: range 10 to 95%; precision ± 2%; 1.5 V or 9 V batteries; weight 100 to 800 g.

Electronic thermometers in various temperature ranges with penetration probe (–20°C to 70°C, in 1°C or 0.1°C) multichannel temperature recorders, printer interface etc.; weight 50-100 g, 1.5 V batteries.

K. *Oxygen meters for biochemical oxygen demand (BOD)*, with RS232 interface: range 0-20 mg L^{-1}; salinity 0-40%; temperature –5°C to 400°C; 9 V batteries; weight 180 g.

Bar code pen and self-sticking labels numbered from 1 to 100 for labelling samples in the field.

L. *Chlorometer*: range to 5 mg L^{-1}; resolution 1%; 9 V batteries; weight 320 g.

M. *Other instruments*:

Gas chromatographs are sometimes used in the field for determination of complex gases in a volcanic environment (for doing analysis before any contact with the air) or in the zone of permanent hydromorphy (methane, H_2S etc.).

Mass spectrometers are available that can be transported in light all-terrain vehicles but these costly instruments are used only for special studies or projects with high economic returns.

Gamma-ray spectrometers, with germanium detectors for example, are used in studies of the environment and pollution. A compact computer allows collection and computation of data directly in the field.

Miniaturized portable optical microscopes can attain magnifications of 1000; for geologists they can be equipped with polarizers and equipment for photomicrography (they have been improved by addition of a fibre-optic cable network).

Portable X-ray fluorescence spectrometers (about 10 kg) enable detection of trace elements during ore genesis surveys for example and their quantification. Fully automated, they can be used without special configuration. The data after acquisition are directly entered and stored in memory. Six trace elements can be analysed in about one minute. The batteries allow operation for ten hours; these instruments are supplied with several measuring heads (sealed ^{109}Cd, ^{241}Am etc. sources).

References

A.F.E.S. 1990. *Association française pour l'étude du sol: référentiel pédologique Kyoto 90*. A.F.E.S.-I.N.R.A., 203 pp.

Bachelier G. 1966. Mesure *in situ* du dégagement de gaz carbonique des sols à l'aide de l'ampoule de Koepf. *Cah. Orstom Sér. Pédol.* IV: 93-97.

Bachelier G. 1968. Problèmes relatifs à l'atmosphère du sol et utilisation possible d'un détecteur de gaz pour la mesure de sa teneur en gaz carbonique. *Cah. Orstom sér. Pédol.* **VI**: 95-104.

Baize D. 1988. *Guide des Analyses courantes en Pédologie.* INRA, 172 pp.

Cailleux A. and Taylor G. 1950. *Code expolaire.* Boubée et Cie.

Dräger R. 1989. *Livre de poche concernant les tubes réactifs.* Kurt Leichnitz, 7e éd., 315 pp.

Duchaufour Ph. 1977. *Pédologie. V. I. Pédogenèse et classification.* Masson, Paris, 477 pp.

Fields M. and Perrot K.W. 1966. The nature of allophane in soils. III. Rapid field in laboratory test for allophane. *NZ. J. Sc.* 9: 623-629.

Laplace-Dolonde A. 1992. Histosols. In *Référentiel Pédologique. Principaux sols d'Europe.* A.F.E.S.-I.N.R.A., pp 119-128.

Loyer J.-Y. and Susini J. 1978. Réalisation et utilisation d'un ensemble automatique pour la mesure en continu et *in situ* du pH, Eh et pNa du sol. *Cah. ORSTOM sér. Pédol.*, XVI: 425-437.

Maignien R. 1969. *Manuel de Prospection Pédologique.* Orstom Paris, 132 pp.

Marius C. 1990. Sols sulfatés acides. In *Référentiel pédologique.* A.F.E.S.-I.N.R.A., pp. 199-202.

Bibliography

ASTM. 1998. *ASTM Standards Related to the Phase II Environmental Site Assessment Process.* Amer. Assoc. for Testing Materials.

Brown J. R. (ed.). 1987. *Soil Testing: Sampling, Correlation, Calibration and Interpretation.* Soil Sci. Soc. Amer., Spec. Publ. no. 21.

Furcola N. C. 1995. *Annual Book of ASTM Standards*, Sec. 4: Construction, vol. 04.09: Soil and Rock (II: D 4943—Latest Geosynthetics). Amer. Soc. for Testing Materials.

Hans Likke H. and van Gestel C.A.M. (eds.). 1998. *Handbook of Soil Invertebrate Toxicity Tests.* John Wiley and Sons, New York.

Havlin J. and Jacobson J.S. (eds.). 1994. Soil testing: Prospects for improving nutrient recommendations. *Proc. Symp. S-3, S-4, S-8, S-9, Soil Sci. Soc. Amer.* Soil Sci. Soc. Amer., Inc.

Liu C and Evett J.B. 1996. *Soil Properties: Testing, Measurement and Evaluation.* Prentice-Hall, New Jersey.

Lunne T, Robertson P.K. and Powell J.M. 1997. *Cone Penetration Testing in Geotechnical Practice.* E. & F. Spon.

Mandal J.N. and Divshikar D.G. 1995. *Soil Testing in Civil Engineering.* Balkema Publ.

Menon R.G., Chien S.H. and Hammond L. Leroy. 1989. *The Pi Soil Phosphorus Test: a New Approach to Testing for Soil Phosphorus.* Int. Fertilizer Dev. Centre.

Nagaraj T.S. 1993. *Principles of Testing Soils, Rocks and Concrete.* Developments in Geotechnical Engineering, no. 66. Elsevier Science, Ltd.

Sanglerat G., Olivari G. and Cambou B. 1983. *Problemes pratiques de mécanique des sols et de fondations*, vol. 1: *Généralities. Plasticité. Calcul des tassements. Interprétation des essais in situ.* Dunod, 352 pp.

Tarradelias J., Bitton G. and Rossel D. (eds.). 1996. *Soil Ecotoxicology.* Lewis Publ., Inc.

Wasemiller M.A. and Hoddinott K.B. (eds.). 1997. *Testing Soil Mixed with Waste or Recycled Materials.* Amer. Soc. for Testing Materials, Spec. Tech. Publ. 1275.

Westerman R.L. 1990. *Soil Testing and Plant Analysis.* Soil Sci. Soc. Amer., Inc.

Sample Preparation

1. Concepts and Organization

1.1 Introduction

The laboratory generally does not interfere in mixing of samples in the field (see Chapter 1), the first step in their preparation for analysis. This operation is basic to ensuring the representativeness of the analytical data and is liable to error because of the heterogeneity of the materials (destruction of structure, grinding of coarse fragments to 'fine earth', loss of fine particles etc.), or addition of contaminants during the preparatory sequence.

Receipt and preparation represent the first link in ensuring quality. The laboratory considers the samples it receives representative of the study area and establishes preparation and homogenization protocols. Profile descriptions and notes pertaining to the presence of concretions, nodules, fragile materials (for example, there is risk of recarbonation of the fine earth by coarse fragments during grinding of some soils on marly limestones) or substances that can cause interference in the analyses enable a better choice of analytical methods (samples to be stored moist etc.), better understanding of the meaning of the analyses required, the first data check and verification of the apparent coherence of the results at the profile level.

The preparative laboratory should include areas for receipt of samples (inventory, provisional grouping etc.), contamination-free areas for preparation (drying, grinding etc.) and areas for long-term preservation (storerooms).

The grinding zone, which generates dust if closed-system equipment are not used, is isolated from drying and storage zones, and has dust traps at the sources. The grinding equipment is attached to the floor with *Silentblocs* to reduce noise and preclude transmission of vibrations to the building.

The areas for preparation and storage are designed according to the programmes, nature and volume of samples, climatic constraints and determinations required to be done (representative volume, fineness of grinding, absence of contamination etc.). Administrative, budgetary etc. considerations are also taken into account.

The preparative steps are thus manifold. However, it is possible to define a certain number of fundamental options on which a preparative unit should be based.

1.2 Mixing and Labelling

The samples arriving at the laboratory are often in many forms: plastic bottles or bags, cardboard boxes, metal tubes and so on. Their field identification is often only by a number that may already be in use in the laboratory, which can cause confusion. Hence it is usually necessary to complete this identification and to remix the samples after preparation in order to achieve uniform presentation and eliminate errors in labelling.

The preliminaries are many, irrespective of whether receipt of the samples is entirely manual or computer-checked. For example, each package could receive a coded record number, the record comprising several identified items:

—sender, address, date of arrival, geographic origin, scientific programme etc.;

—finance memo;

—complete list of samples sent and checkmarks on receipt;

—list of analyses required and constraints in preparation and analysis.

The samples can be listed under their original number with a check key (depth for example). They are prepared according to standard or personalized protocols.

After drying and reduction to 2 mm, the 'fine earth' is sometimes passed through the sample-splitter and stored in plastic bottles (about 1 kg when the range of analyses is wide) or boxes, with the record number and original number legibly marked (composing stick, occasionally with a bar code added).

When a rotary sample-splitter is used, eight subsamples of about 125 g each are obtained. If necessary they may be distributed according to the requirements of different analyses. For example:

—one bottle could be used for the sample that will be ground to 0.2 mm;

—two bottles may be preserved closed, protected from all contamination; they might possibly serve as reference and reserve samples for checking or for later determinations;

—five bottles could be simultaneously distributed to the specialized analysis units according to the analyses desired.

Mixing is adapted to all forms of subsamples, in order to allow sample management through centralizing tubes for the small samples. Lack of standardization of containers and components for storage or manipulation is a constant annoyance. Nowadays one may expect storage bottles that might possibly be handled by robotic arms (*see* Chapter 17).

It is recommended that the structures for storage and manipulation be of modular design (Fig. 3.1) and likewise the receptacles for appropriate storage of test samples (Fig. 3.2) in order to obtain standardization, which starts at the stage of sample receipt and preparation and continues up to dispatch of the analytical results.

The common denominator for all these components should be found, which is sometimes difficult in view of the standard equipment available in the market.

1.3 Storage and Preservation of Samples

Soil samples are analyzed after some delay that may sometimes extend to even almost one year. For samples preserved moist, these delays must necessarily be shorter. If analyses are required to take into account microbial

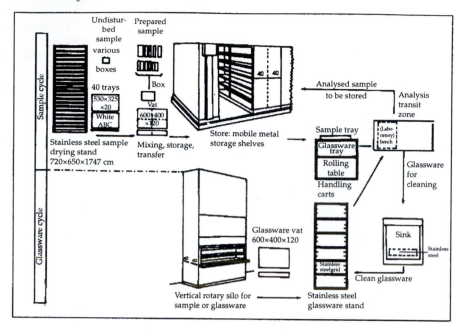

Fig. 3.1 Modular laboratory system for mixing and transfer of samples.

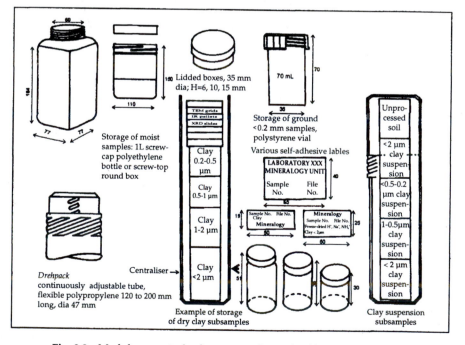

Fig. 3.2 Modular receptacles for storage of samples (dimensions in mm).

activity (mineral nitrogen for example), determinations should be done immediately—the same day—or on samples rapidly frozen at about −40°C. Moist samples should be kept only for a short period (in double-walled bags or airtight bottles) in darkness at a temperature lower than 15°C if possible. They tend to lose part of their original moisture, condensation appearing on the inner wall of the bags. If they are exposed to light, proliferation of green algae can be seen. Redox and recrystallization reactions occur.

For air-dried samples, in equilibrium with the atmospheric humidity, the changes are generally limited except for particular cases (Andosols, Histosols, mangrove soils etc.). Conservation can be done for several years and samples can be provided for supplementary determinations at a later date or for checking after three or four years. It must be ensured in such cases, however, that the samples are protected from possible contamination by dust, gases (NH_3 etc.), radioactive particles etc. Low temperature, good ventilation and low humidity are mandatory components of good preservation.

Some research laboratories restrict storage of samples to about five years; samples beyond this age are returned to the original sender or destroyed with his permission. True standard samples of the laboratory and/or certified reference samples should be preserved for checks on each series of analysis.

1.4 Receipt of Samples

The samples identified by their original number are grouped according to the demand memo into homogeneous lots.

Standard preparation
Air-drying—careful mechanical grinding—sieving through 2 mm round-hole screen—determination of percentage of fine earth—sorting of mineral and organic coarse fragments. Possibly, subsampling with rotary sample splitter—grinding an aliquot to 0.2 mm or 0.1 mm in agate mortar may be required for total chemical analysis.

Personalized preparation
For example,
—preparation and preservation at original moisture content (Andosols, Histosols, Spodosols, Andic soils, peats etc.);
—preservation in the original condition until analysis, desiccation by lyophilization of an undisturbed sample, impregnation for thin-section studies;
—drying and preparation with selective sorting of phases for quantification (nodules, concretions, crystals etc.);
—sieving through square-mesh screen (2 mm or more) for structural analysis;
—uncontaminated grinding for determination of trace elements;

—preservation at −18°C and in some cases at −40°C, or stabilization by chloroform vapour for determination of mineral nitrogen.

These personalized preparative procedures should be done in close harmony with the research programme and should be adapted to very varied situations.

2. Preparation Protocols

2.1 Preparation of Moist Samples

2.1.1 Principle

Certain analyses can only be done on soil preserved in its original moisture condition (NH_4^+, NO_3^-, pesticides etc.). Besides, some types of soils have properties that may be irreversibly modified during drying (Andosols, Histosols, etc.)

Homogenization of these samples and elimination of coarse fragments without significantly altering the moisture content is therefore imperative. The moisture content may reach 200 to 300% (expressed by the water-retention capacity, see Sec. 2.1.3) due to the presence of aluminosilicates, gels, amorphous materials, various oxides and hydroxides, and humified organic substances.

Sampling with a screen with 2 mm round holes is generally impossible. What is used, preferably in the field itself, is a sieve with a 4 mm square-hole mesh, whereby a representative moist sample without modification is obtained. Soils can lose moisture by simple pressure (a phenomenon similar to thixotropy but without return to the original phase). Mechanization is therefore ruled out as pumice and organic debris are particularly fragile when moist. Further, given the high microbial activity of these soils, samples have to be preserved at sufficiently low temperature in darkness to avoid condensation and growth of green algae.

Determination of pF is done on these moist samples, as are irreversibility thresholds (remoistening), various charge determinations (exchange complex), forms of organic matter, aggregate analysis, extractable manganese and so forth.

The results are expressed in relation to the soil dried at 105°C and analysis done on a fixed '105°C-dried soil' equivalent weight, in order to obtain samples that have an initial surface area identical to that in the field, and which are comparable irrespective of changes in their moisture content.

It should be noted that samples taken to study the products of microbial processes should be analysed as soon as possible after sampling or, if not, preserved moist at −40°C (weighing of an aliquot before freezing enables direct thawing of test sample portions in the extraction reagent without renewal of moisture content or microbial growth).

2.1.2 Equipment
—stainless steel sieves, 4 mm square-hole mesh, 300 mm diameter, with stainless steel receiver;
    —double-walled plastic bags or airtight bottles with screw cap;
    —hard wooden spoons.

2.1.3 Procedure
—Rub the moist sample with the hand or a hard wooden spoon to pass it through the sieve, retaining the coarse fragments (stones and roots = mineral and organic waste of 4 mm size); avoid drying of the 'fine earth' so obtained.
    —Weigh the moist fine earth (P_0) with a tared sieve receiver.
    —Keep aside stones and roots for drying at 105°C (after a quick rinsing in water); weigh (stones P_1; roots P_2).
    —Quickly homogenize the moist sieved soil and put approximately half in the double-walled plastic bag.
    —Preserve in darkness at a temperature lower than 20°C if possible.
    —Take an aliquot in a tared box to determine the moisture content (at 105°C) of the moist sample (NF X31-102, 1981).
    —Keep the other half of the sample exposed to air for about 15 days, then grind and sieve through a 2 mm round-hole sieve. Pass the sample through the sieve to make it equivalent to the moist sample (sieved through 4 mm mesh); the gravel sieved between 2 mm and 4 mm enables correction of the results.

Calculation (Fig. 3.3)

  I. Total weight of moist soil P
        Weight of stones (dried at 105°C) P_1

Stone waste in per cent of moist soil $C_1 = 100 \dfrac{P_1}{P}$

Weight of organic debris (dried at 105°C) P_2

Organic waste in per cent of moist soil $R_1 = 100 \dfrac{P_2}{P}$

  II. Total weight of soil sieved through 4 mm P_0
       Aliquot taken in tared box for moisture content P_3
       Aliquot dried at 105° C P_4
       Water lost $H = P_3 - P_4$
       Moisture content of soil *in situ* expressed by
       water-retention capacity

(pedology: may exceed 100%) $Z_1 = 100 \dfrac{H}{P_4}$

Moisture content of soil *in situ* expressed for
later correction of results (chemistry: NF X31-102) $Z_2 = 100 \dfrac{H}{P_3}$
Weight of 105°C-dried soil in per cent of
moist soil (calculation of equivalent weight for
working with moist soil) $Z_3 = 100 - Z_2$

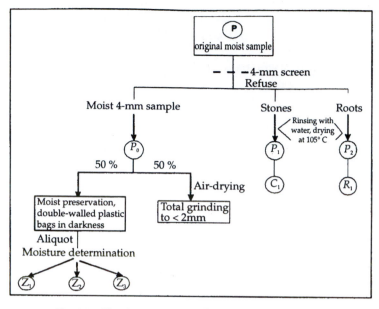

Fig. 3.3 Classic preparation scheme for moist samples.

2.2 Drying Samples

2.2.1 Principle

This operation is necessary to restrict microbial growth, to ensure medium- and long-term preservation of samples, to enable better separation of aggregates and better homogeneity for representativeness of test portions. The reference basis is a thin soil layer of soil air-dried at room temperature and protected from direct sunlight.

However, the concept of 'dry soil' is subjective given the nature of soils and the mineral and organic substances constituting them. The water extracted during drying represents 'free' water plus a small amount of hygroscopic water, which can be distinguished from structural water. Equilibrium depends on temperature, particle size, porosity, more or less hygroscopic salts, if present, and lastly, the relative humidity of the air. Errors in the 'dry state' are of various kinds.

Other techniques may be used to speed up drying: raising the temperature, forced ventilation without excessive draft, radiation, etc.

2.2.2 Air-drying

Equipment

—wooden tray with protective plastic film or plastic tray about 50 × 25 × 5 cm (1-2 kg sample);

—mobile tiered shelves for 40 plastic trays (530 × 325 × 20 cm).

Procedure
—Allow to dry in air at room temperature ($< 40°C$) in a well-ventilated area without excessive draft until the sample has lost its plasticity, crumbles in the fingers and remains in equilibrium with the surroundings (24 to 72 hours for most soils of temperate climates, about a week or more for certain tropical soils).
—Obtain the total weight of air-dried soil.

Remarks
The samples dried in air retain a hygroscopic water content of 0.5% for sandy soils to about 8% for clayey and organic soils. In the case of Andosols this 'equilibrium' water may reach 10 to 30%, rarely 50%, which necessitates systematic conversion of the analytical results to 105°C dry basis.

The changes effected by drying affect analysis of all the physical and chemical properties of soils. Some changes are drastic in certain soils, others scarcely significant. It may be considered that drying can alter:
—organic matter, mineralizable nitrogen, organic phosphorus and redox processes;
—charge distribution and pH (by oxidation of sulphur and proton transfer), exchangeable potassium, exchangeable manganese, cation exchange capacity, oxidation of iron and manganese etc.;
—physical properties of aggregation, water retention etc.

2.2.3 Oven-drying at 40°C

Principle
Acceleration of drying by moderate heating (50°C) with ventilation. Drying should be as short in duration as possible and resorted to only if analytical results are urgently required or if the local humidity is such that air-drying is impossible.

There are only a few modifications of properties compared to air-drying for sandy or clayey soils but it may be noted that:
—halloysite tetrahydrate is rapidly transformed to halloysite dihydrate;
—organic matter can be slightly altered;
—Fe^{++} can be oxidized to Fe^{+++};
—exchangeable manganese is modified;
—extractable P_2O_5 can increase slightly if drying is excessively prolonged;
—pH values may be very slightly modified (lowered) if the temperature does not exceed 50°C; the same is true for charges and the value of CEC;
—mineral nitrogen is modified, especially in calcareous soil (loss of NH_4^+, oxidation of NO_2^- etc.);
—gypsum ($CaSO_4\,2H_2O$) converts to anhydrite ($CaSO_4$ or $CaSO_4\,0.5H_2O$).
Some effects of heating are still not known for every type of soil.

Equipment
—stainless steel or enamelled trays;
—forced-draft air oven, 0-100°C.

Procedure
   —Spread the sample in a thin layer on a stainless steel tray and keep in a forced-draft air oven for about 8 hours at 50°C.
   —Weigh the tray + soil and put back in the oven for one more hour to check whether drying is complete (weight remains constant).

2.2.4 Drying in a microwave oven

Principle
This technique is applied to small samples and allows some analyses to run without delay but cannot be used on a large scale.
   Speedier drying is achieved by microwaves at low temperature. Microwaves are part of the domain of Hertzian radiation between far infrared and radio frequencies (30 cm to 1 mm). They are used in soil science for drying and also for wetdigestion.
   The wavelength of 12.25 cm (frequency 2450 Hz) is the only one approved by French law. Microwaves are produced by a Magnetron type oscillator composed of a cylindrical diode under high vacuum subjected to a magnetic field parallel to the cylinder axis. An inner metallic ring acts as the anode. When heated to high temperature it produces thermoionic emission. The assembly is placed in a metal enclosure that reflects the microwaves onto the dielectric substances placed on the tray and prevents their dispersion outwards. Part of the energy of this electromagnetic field is converted to heat by dielectric loss.
   Water, a substance with permanent dipole moment, has the property of getting polarized. The molecular vibrations cause a rapid conversion to the vapour state. Drying soils is quick and since they are 'transparent' to microwaves (like ceramics, glass and plastics), only slight heating occurs. The depth of penetration of radiation is a function of the frequency. To moderate the penetration a turntable or rotating antenna is used, which allows the energy to be dissipated in all directions. By using a moderate regulation of energy (about 50%), heating can be lower than 50-60°C in a cycle of 10-30 minutes. As metals reflect 100% of the radiation, a metal receptacle cannot be used since the reflected radiation may damage the emitting source.

Equipment
   —microwave oven with turntable or rotating antenna;
   —glass or plastic trays (100 mm diameter).

Procedure
   —Spread the sample in a thin layer on a glass or plastic tray (caution: never use metal trays and never operate the apparatus when empty).
   —Regulate the energy between 0 and 100% (generally at 50% to avoid abnormal heating of the sample).

—After 10 minutes, weigh the sample, then replace it in the oven for 10 minutes and again reweigh to verify weight constancy; if the latter has not been attained, put back in the oven for another 10 minutes.

Remarks

The dehydration achieved is not always uniform and depends chiefly on sample size, initial moisture content, nature of the water retained as well as of the mineral and organic substances, thickness of soil layer and uniformity of the microwave field, besides the impossibility of measuring the effective temperature at any instant during treatment.

In the case of allophanic soils, this drying has enabled attainment of constant weight in 35 minutes in samples containing 250% water (pedological moisture, Z_1) without noticeable heating.

Safety

Caution: the oven can function only with the door closed. Microwave radiations can cause very severe internal burns.

2.2.5 Drying by lyophilization

Principle

Drying by lyophilization is achieved by rapid freezing of the sample at a temperature lower than –80°C followed by sublimation of the water under strong vacuum.

This type of drying is useful for soils with a high shrinkage coefficient containing clay minerals that become hard and compact, soils with high porosity to preserve the existing structure, and mangrove soils with rapid chemical evolution. It also precludes oxidation and possible deterioration of unstable molecules (pesticides).

Freezing should be accomplished very rapidly so that there is no fragmentation or coagulation of the aggregates. Water sublimation is quite rapid and the usable volume of the trap should be calculated on this basis. Desorption of vigorously retained water (more or less bound water) is very slow and incomplete and hence the operations should be standardized to avoid very prolonged heating of the sample at the end of the cycle. Optimization of lyophilization should take into consideration the nature of the sample and its resistance to thermal degradation. The temperature should be as low as possible during sublimation, as otherwise the cycle is prolonged and the unit cost increased.

Equipment

—lyophilizer with apparatus for rotating the samples;
—wide-mouth 500-mL Pyrex flasks or stainless steel trays.

Procedure

The moist samples (4 mm or undisturbed samples) are placed in wide-mouth Pyrex flasks (stopper raised). These flasks are rotated in the freezing chamber to avoid possible formation of surface crusts with low diffusion coefficient (for undisturbed samples the rotation apparatus is not used).

At the instrumental level, the procedure is adapted to the type of equipment used.

After lyophilization each sample is stored in its Pyrex flask tightly closed to avoid regaining of hygroscopic moisture, which is always significant with this type of drying.

2.3 Crushing of Aggregates, Grinding, Sieving

2.3.1 Objectives

Essentially the following:

—Preliminary concentration and purification of the fine soil phase by eliminating the coarse mineral and organic fragments larger than 2 mm, without breaking them.

—Analyses under standardized conditions by homogenizing the physicochemically most active fine particles and providing better representativeness of the test samples.

—Increasing the immediate reactive surface area and expediting solid-liquid contact (extractions etc.).

2.3.2 Principle

Selection by careful crushing (on a calibrated separating device—screen with 2 mm round holes or 2 mm square-mesh sieve) of:

—the most active components originating from the structural units of a soil with a more or less stable cohesive force;

—stable elementary particles with defined mechanical properties (particles > 2 mm, unweathered primary rock fragments, pedological features, undecomposed organic matter, etc.).

The heterogeneous population of particles without uniform shape but of similar dimensions (isogranulometric) that is passed through the separatory sieve contains assemblages of different sizes with edges, faces and volumes related to the micromorphology of the soil. (This preparation may be considered a preliminary test of the stability of the aggregates when subjected to a controlled external force capable of breaking their coherence.).

The concepts held in 1926-1930 for standardizing the preparation of soils were based on a naturalist view and agronomic conclusions at an elementary level of observation (aggregates visible to the naked eye). The idea was to characterize the structure (micro-and macrostructure) by evaluating the amount of aggregates obtained by mechanical separation of dry soil, the sequence going from an elementary aggregate of 50 μm, a veritable 'soil molecule', to aggregates of 2000 μm. The assemblage of particles represents the 'fine earth' comprising the active phase of the soil, closely related to pedological development and soil fertility. It suffices then to do a simple mechanical separation by passing the clods and aggregates through a 2 mm round-hole sieve with moderate force. This norm is still used as the basis for reference by most laboratories (NF X31-101, 1992).

However, albeit the procedures have not been modified for 60 years, the concepts have changed considerably. In 1956, Hénin, considering that the structure could be better defined by measurement of its stability when moist, adopted the 2 mm square-hole sieve. The aggregates can enter the square hole diagonally and thus are slightly larger but still correspond to the 'size of soil grains that can favour germination of seeds and have a structure we would like to preserve'. This standard is now sometimes internationally adopted as the basis for separation. The conventional size 2 mm is arbitrary and has no scientific significance because of the multiplicity of aggregate shapes that cannot be defined by a single dimension (diameter of a spherical particle).

Studies conducted during the last decade on thin sections with image analysers at different scales of magnification clearly show that the original concept is liable to criticism. The cyclical dynamics of structure (degradation and recombination related to anthropic and climatic factors) also underscores the difficulty in measuring the *in-situ* structure by laboratory means.

However, the perenniality of the procedure described here is easily explained by the fact that crushing the soil is based on the application of a sufficient, moderate and progressive force of disruption that enables:

—rupture of the cohesive forces binding the fine particles participating in the structure and having various high activities;

—separation of the coarse particles, elementary particles with better defined volume, representing the textural components.

The principle of crushing may thus be summarized in general as: 'crushing structural elements without modifying the original texture'. This definition is apt for samples with particulate structure that are very easily crushed, with continuous or massive structure rich in fine silt with low cohesion, with fragmentary structure or with high cohesion, particularly if the structure is built up rather than mechanical.

For personalized preparations, it is necessary to add to the preparation of particles larger than 2 mm, that of pedological features such as well-differentiated glaebules[1], nodules and concretions that should be accounted for in the balance ($CaSO_4$, Fe + Mn etc.).

Sieves with holes < 2 mm are necessary together with manual picking. Separation of the 'textural clay' smaller than 2 μm should be done in an aqueous medium.

2.3.3 Manual crushing—fine earth < 2 mm

Equipment
   —hard wooden roller and tray;
   —mortar and pestle (agate, hard porcelain or glass according to dangers of contamination);

---

[1] Glaebules: in soil micromorphology, three-dimensional units in the s-matrix of the soil material, usually elongate to equidimensional (from Lozet and Mathieu, 1986).

—sieve with 2 mm round holes, AFNOR NF33, 200 mm diameter, tall (77 mm) form, with stainless steel lid and receiver;
  —stainless steel 2 mm square-mesh sieve, AFNOR NF34, 300 mm, low form with stainless steel lid and receiver.

Protocol for crushing aggregates with mortar and pestle or with hard wooden roller
The choice of the nature and size of mortar will depend on the hardness of the sample, risk of contamination and the volume to be treated. Crushing aggregates with a hard wooden roller (*see* Chapter 4) is much more gradual:
  —Break the aggregates by controlled pressure of the pestle, then work with a rotary motion applying light pressure to avoid breaking unweathered rock fragments.
  —Sieve through a screen with 2 mm round holes to separate as much as possible the particles passing through the sieve while avoiding excessive grinding.
  The fine earth is weighed as are the residual coarse organic and mineral materials.
  Note: for structural analysis the round-hole screen is replaced by a 2 mm square-hole sieve. Then manual crumbling is done with gentle forcing on the sieve to avoid modifying the size and shape of the aggregates.
  Manual crushing and sieving may lead to sampling errors because of the great heterogeneity of soils and the manner in which the aggregates are made to pass through the separatory surfaces or sieves.
  Passage through the sieve should be as uniform as possible; for this a to-and-fro-horizontal rotary motion must be applied to the soil along with regular vertical shaking. The screen of the sieve must not be overloaded as this could deform it. To avoid loss of fine or light particles, a lid should always be used.
  Sieving should be done sequentially with crushing of the aggregates to separate as completely as possible the particles that have reached the desired size without excessive breakage. It must be noted that the particles passing through a sieve will be 10-20% smaller than the dimension of the mesh of the sieve.

Calculation: determination of 'fine earth'
Weigh separately
  —the fine earth P_1
  —coarse fragments P_2
  —plant debris P_3
  ($P_1 + P_2 + P_3 = P_4$, the original weight of the sample)

Fine earth, % $100 \dfrac{P_1}{P_4}$

Coarse fragments, % $100 \dfrac{P_2}{P_4}$

Plant debris, % $100 \dfrac{P_3}{P_4}$

Occasionally the product of selective hand-picking is also weighed (P_x).

Nodules, concretions, % $100 \dfrac{P_x}{P_4}$

General remarks
The coarse fragments that represent significant features of pedological origin (lime nodules, iron-manganese concretions, gypsum crystals etc.) should be separated without breaking. They should be quantified at two levels:
   —as coarse fragments or residue from the 'fine earth';
   —as pedological features for establishing the contents of calcium carbonate, gypsum, iron and manganese immobilized in the horizon.
   The presence of friable, highly weathered rock poses for the sample preparer the greatest of insoluble problems if the soil and the purpose of the studies are not known. Only the research scientist indenting analysis can decide whether the more or less indurated accumulations should be combined with the 'fine earth'.
   Expression of the results is generally done on a weight basis but may also be done by volume if the 'coarse fragments' fraction is very large. According to their nature and size, these large particles can affect the physical, hydric and agronomic properties of the soil (mechanical properties and workability, porosity, water retention etc.). It is advisable to briefly include in the analytical data records the following useful information:
   —rounded shapes, pebbles (alluvial deposits);
   —presence of charcoal (ancient fires, human settlement etc.);
   —weathering stage, presence of flint, concretions, coarse calcareous particles, gypsum crystals, etc.

2.3.4 Mechanical crushing—fine earth < 2 mm

Principle
The term grinding may be retained here rather than crushing as the machines distinguish only imperfectly the differences in hardness of the particles and their mechanical resistance.
   Grinders with a rotating drum with 2 mm perforations and containing a free central roller of variable weight (discontinuous operation) or those with retractable pounders (continuous open operation) enable very rapid grinding of most soils (see Chapter 4). They constitute the basic equipment of a preparative laboratory for soil survey and agronomic samples or samples containing scarcely any fragile minerals and concretions. They enable a large daily output with representativeness satisfactory for the usual analyses.
   Grinding is isogranulometric, the particles being progressively sieved, and therefore pollution is practically negligible if stainless steel components are used.

Remarks on use of rotating-drum grinder

This kind of grinding is suitable for very dry soils with medium or light texture. On the other hand, with very clayey and sticky soils (Vertisols) it is sometimes not possible to halt grinding at the correct moment[1]; the particles then harden and cannot pass through the screen, which gradually gets clogged.

In the case of Andosols that retain a large residual moisture content at equilibrium, the pressure of the drum releases part of the water, setting in motion agglutination of the soil which makes grinding impossible.

Remarks on grinding with flails

Grinding is very quick and the fractions are isogranulometric. Vertisols are ground without difficulty, so too air-dried Andosols, but some geological weathering products (schists, micas etc.) and fragile concretions may partly pass into the 'fine earth'.

Contamination is very slight with equipment made of stainless steel, as the time of contact with the grinding tools is greatly reduced.

2.3.5 Manual or mechanical total grinding of soils

Principle

From a soil sample crushed to 2 mm, an aliquot is taken and totally ground to 0.5 mm (NF28 sieve), 0.2 mm (NF24) or 0.1 mm. These particle sizes allow representative test samples (see Chapter 18) when sizes smaller than 2 g are required.

Equipment

—agate mortar of 150 mm diameter (about 180 mL) and pestle;
—discontinuous mechanical grinder (see different types in Chapter 4);
—0.5 mm square-hole sieve (NF28) with lid and receiver;
—0.2 mm square-hole sieve (NF24) with lid and receiver;
—0.1 mm square-hole sieve with lid and receiver;
—plastic vial 70 mm × 35 mm diameter.

Procedure

—Grind by hand with a rotary motion of the pestle, applying slight pressure.
—Sieve frequently to obtain uniform particle size and to avoid overgrinding.
—Pass the entire sample through the sieve.
—Transfer to the storage vial and label.

---

[1] In humid tropical climate with a constant atmospheric relative humidity close to 90%, a Vertisol containing 2:1 clay minerals retains 8% moisture and cannot be ground by this method.

Remarks

Samples ground in this manner are used for total dissolution and analyses requiring easy diffusion of ions, such as determination of the point of zero charge or certain extractable anions, provided overgrinding that could lead to redistribution of charges is avoided.

Mechanical grinding

All these systems are discontinuous and function in open or closed mode (*see* Chapter 4).

Equipment is selected on the basis of volume, hardness, contamination, possible recovery of the entire sample and the time needed for cleaning the apparatus.

In closed systems there is no loss during grinding but clays tend to stick to the grinding balls or rings and it is very difficult to recover them quantitatively, which enriches the sample with hard, non-sticky products (error of omission).

Grinders with oscillating flails of agate, without rings, are the most suitable as recovery and cleaning are satisfactory, as is the speed of grinding.

2.4 Laboratory Sampling of Soils

Principle

After preparation of the soils and crushing and grinding to 2 mm, it is necessary to mix the fractions obtained to arrive at as homogeneous a 'fine earth' population as possible. The reproducibility of results and representativeness of the soils will then be improved.

In a laboratory with various specialized units, record management is facilitated by preparation of 2 mm subsamples, of which one or two are preserved in the store (free of all potential contamination during the analytical chain) for checks if required; some of these subsamples are reground to 0.5 mm, 0.2 or 0.1 mm according to the analyses required.

Equipment

— manual riffle sampler (or Jones sampler) in stainless steel;
— rotary sample-splitter with vibration and feeding hopper (Fig. 3.4);
— enamelled iron, stainless steel or plastic tray for manual sampling;
— stainless steel spoon.

Microsamples with additives

In some cases, homogenization of microsamples for total analysis demands a grinding-mixing in the presence of a matrix corrector (X Ray fluorescence, etc.), and in mineralogy, thorough mixing with an additive in required proportion using microvibrators with grinding balls or flails. The mixing protocol is as follows:

— Place a weight x of sample ground to 0.2 mm in an agate mortar.

— Add an additive (KBr for pelleting, the non-absorbing oil Nujol or matrix corrector, etc.).

Fig. 3.4   Rotary sample-splitter.

—Pass through the homogenizing tube with suitable grinding ball (agate, methacrylate etc.); for substances readily decomposed by heat, the tube can be cooled cryogenically.

—Place in the electromagnetic mixer for three minutes.

3. Representativeness of Samples

Sampling errors are almost always the greatest in physicochemical analyses, especially for soil substrates. In decreasing order, the first will be the sampling error linked to field sampling (*see* Chapter 1), then the error in laboratory subsampling described here, and lastly the error properly termed error of measurement.

Statistical checks (*see* Chapter 18) should integrate all the quantifiable factors and should be applied most often to the 'unique sample' that reaches the laboratory, the sample that presents some difficulties as the population distribution is unknown and bases for comparison non-existent. Quality control requires consideration of chemical, physicochemical, physical and biological aspects.

A study conducted in the laboratory[1] (unpublished data) showed that carbon is one element that gives occasion for large errors because of its great heterogeneity in the soil. In 108 samples prepared by poorly executed

[1]   'Organic Matter' Laboratory, IRD Bondy, France 1986.

grinding and sieving (discarding of hard-to-grind organic or mineral components) aberrant results could be shown in 30-100% of the samples, depending on the origin of the batches (range of error 10-160% compared to the lowest value).

The errors rest on the differences in preparation of the 2 and 0.2 mm or 0.5 and 0.2 mm fractions. However, even with stricter standardization of the grinding conditions, some errors originating from the grinding and sieving were still observed, greater than the instrumental error in carbon determination, showing the importance of preparation.

It is therefore essential to periodically check the quality of sample preparation. Nevertheless, after the indispensable calibration before starting laboratory analysis, these checks will not be applied to all the samples (because of the expense involved) but should be done starting from the general choice pertaining to quality of the analyses and reliability of the laboratory (introduction of standard samples, of soil matrices, of random repetition of the same sample in the series etc.).

Statistical study at the level of preparation is most often simplified by using equipment of known and constant performance (no loss of light, fine particles, no loss in the sieve screens, no chemical contamination by grinder components or inadequate cleaning, total recovery of ground product, perfect homogenization of ground material without overgrinding etc.), and employing trained personnel for sample preparation or having the indenters themselves prepare the samples.

Remarks

—If the population of original 'fine earth' components is very high and sieving adequately selective, errors will be negligible with respect to particle-size analysis, analyses of structural units in aggregates and also physicochemical properties.

—If the population of 'non-fine earth' components of particle size less than 2 mm is high (because of excessive grinding of rocks, concretions, nodules, glaebules etc.), electroionic and chemical properties will be significantly modified, especially if the sample comes from horizons rich in fragile weathered rock (soft limestone, clayey schist etc.).

If drying is a limiting factor (as in Andosols, Histosols etc.), chemical, physicochemical and biological properties will be modified irreversibly and sometimes even drastically (the moisture content Z_1 (Fig. 3.3) at pF 3.0 can go from 300% to less than 50%).

—If contamination of the sample by the grinders is high, determination of trace elements will be to no avail.

It can be seen that the 'preparation' stage, often neglected and confined to untrained personnel, actually influences the precision of the entire chain of analyses. No aspect that can improve the quality of analysis should be neglected; the weakest link decides the strength and thus the reliability of the laboratory.

Bibliography

Standardization

Lozet J. and Mathieu C. 1986. *Dictionnaire de Science du sol*. Technique et Documentation, Lavoisier, Paris.

NF X31-101. 1992. Préparation d'un échantillon de sol pour analyse physico-chimique. Séchage, émottage et tamisage à 2 mm. In *Qualité des sols*. AFNOR, pp. 15-21.

NF X31-102. 1981. Détermination de l'humidité résiduelle d'échantillons de sols préparés pour analyse. In *Qualité des sols*, AFNOR, pp. 22-23.

NF ISO 11465, 1994. Détermination de la teneur pondérale en matière sèche et en eau—Méthode gravimétrique. In *Qualité des sols*, AFNOR, pp. 517-524.

Pr ISO 11074-2, 1997. Termes et définitions relatifs à l'échantillonnage.

Soil Drying: General

Allen S.E. and Grimshaw H.M. 1962. Effect of low temperature storage on the extractable nutrient ions in soils. *J. Sci. Food Agr.*, **13**: 525-529.

Bartlett R.J. and James R. 1980. Studying dried, stored samples—some pitfalls. *Soil Sci. Soc. Am. J.*, **44**: 721-724.

Cunningham R.K. 1962. Mineral nitrogen in tropical forest soil. *J. Agr. Sci.*, **59**: 257-262.

Davey B.G. and Conyers M.K. 1988. Determining the pH of acid soils. *Soil Sci.*, **146**: 141-150.

Fujimoto C.K. and Scherman G.D. 1945. The effect of drying, heating and wetting on the level of exchangeable manganese in Hawaiian soil. *Soil Sci. Soc. Am. Proc.*, **10**: 107-112.

Gardner W.A. 1986. Water content. In *Methods of soil analysis (part I)*. Klute A. (ed.) A.S.A.-S.S.S.A., pp. 493-544.

Goloberg J.P. and Smith K.A. 1984. Soil maganese: values, distribution of manganese 54 among soil fractions and effects of drying. *Soil Sci. Soc. Am. J.*, **48**: 559-564.

Keog J.L. and Mapples R. 1973. Evaluating methods of drying soils for testing. *Bull. 783. Agric. Exp. Station*, Univ. Arkansas.

Kubota T. 1972. Aggregate formation of allophanic soils: effect of drying on the dispersion of the soils. *Soil Sci. Plant Nutr.*, **18**: 79-87.

Leggett G.E. and Argyle D.P. 1983. The DTPA extractable iron, manganese, copper and zinc from neutral and calcareous soils dried under different conditions. *Soil Sci. Soc. Am. J.*, **47**: 518-522.

Maynard D.G., Kalra Y.P. and Radford F.G. 1987. Extraction and determination of sulphur in organic horizons of forest soils. *Soils Sci. Soc. Am. J.*, **51**: 801-805.

Peverill K.I., Briner G.P. and Douglas L.A. 1975. Changes in extractable sulphur and potassium levels in soil due to oven drying and storage. *Austr. J. Soil Res.*, **13**: 69-75.

Philips I.R., Black A.S. and Cameron K.C. 1986. Effects of drying on the ion exchange capacity and adsorption properties of some New Zealand Soils. *Comm. Soil Sci. Plant Anal.*, **17**: 1243-1256.

Raveh A. and Avnimelech Y. 1978. The effect of drying on the colloidal properties and stability of humic compounds. *Plant and Soils*, **50**: 545-552.

Schalscha E.B., Gonzalez C., Vegara I., Galindo G. and Schatz A. 1965. Effect of drying on volcanic ash soils in Chile. *Soil Sci. Soc. Am. Proc.*, **29**: 481-482.

Searle P.L. and Sparling G.P. 1987. The effect of air drying and storage conditions on the amounts of sulphate and phosphate extracted from a range of New Zealand top soils. *Commun. Soil Sci. Plant Anal.*, **18**: 725-734.

Warden R.T. 1991. Manganese extracted from different chemical fractions of bulk and rhizosphere soil as affected by method of sample preparation. *Commun. Soil Sci. Pl. Anal.*, **22**: 169-176.

Microwave Drying

Beary E.S. 1988. Comparison of microwave drying and conventional drying technics for reference materials. *Anal. Chem.*, **60**: 742-746.

Gee G.W. and Dodson M.E. 1981. Soil water content by microwave drying, a routine procedure. *Soil. Sci. Am. J.*, **5**: 1234-1237.

Hankin L. and Sawhney B.L. 1978. Soil moisture determination using microwave radiation. *Soil Science*, **126**: 313-315.

Gautheyrou J. and Gautheyrou M. 1987. Le point sure séchage micro-onde. Orstom. *Courrier des laboratoires*. Mars N°1, pp. 1-21.

Miller R.J., Smith R.B. and Biggar J.W. 1974. Soil water content microwave oven method. *Soil Sci. Soc. Am. Proc.*, **38**: 535-537.

Moody J.R. and Beary E.S. 1987. Standard Association of Australia 1986. Determination of the moisture content of a soil: microwave-oven drying method. A.S. 1289. B 1.4.

Other Methods of Drying
Barthakur N.N. and Tomar J.S. 1988. A novel technique for drying soil samples. *Commun. Soil Sci. Pl. Anal.* **19**: 1871-1886.

Crushing Aggregates for Soil Analysis
International Society of Soil Science. 1928. The study of soil mechanics and physics. *Proc. 1st Int. Cong. Soil Sci.* II: 359-404.

Lebret P. and Levant M. 1985. Étude du raccord entre les techniques classiques de granulométrie et modernes de microgranulométrie. *Bull. Centre Géomorph.*, Caen, C.N.R.S. **30**: 7-22.

Mandelbrot B. B. 1983. *The Fractal Geometry of Nature*. W. H. Freeman and Co., New York.

NOVAK Report. 1930 Finesse de tamisage 2 mm des échantillons de sol en vue de l'analyse. *C. R. Assoc. Int. Sci. Sol*, **5**: 15.

Oden S. 1925. The size distribution of particles in soils and the experimental methods of obtaining them. *Soil Sci.* **19**: 1-36.

Ralph B. 1984. Application of image analysis to the measurement of particle size and shape. *Anal. Proc.* **21**: 506-508.

Rieu M. and Sposito G. 1991. Relation pression capillaire-teneur en eau dans les milieux poreux fragmentés et identification du caractère fractal de la structure des sols. *C. R. Acad. Sci.*, Paris, vol. 312, ser. II, pp. 1483-1489.

Tanner C.B. and J. J. Bourget. 1952. Particle shape discrimination of round and square holed sieves. *Proc. Soil Sci. Soc. Amer.* **16**: 88.

Young I.M. and Crawford J. W. 1991. The fractal structure of soil aggregates: its measurement and interpretation. *J. Soil Sci.* **42**: 187-192.

Sieving
AFNOR X 11-501. 1938. Passoire à trous ronds 2 mm NF 33. Tamis à mailles carrées 2 mm NF 34.

AFNOR NF X 11-501. 1970. Tamis et tamisage—dimensions nominales des ouvertures.

AFNOR NF X 11-504. 1970. Exigences techniques et verification.

AFNOR NF X 11-507. 1970. Analyse granulométrique—tamisage et contrôle.

AFNOR NF X 18-304. 1970. Granulométrie des agrégats.

ASTM. 1998. *Manual of Test Sieving Methods* (ASTM Committee E-29 on Particle and Spray Characterization). ASTM Manual 32, Amer. Soc. Testing and Materials.

ISO 565-1972. Tamis et contrôle.

Kennedy S. K., Meloy T. P. and Durney T. E. 1985. Sieve data. Size and shape information. *J. Sedim. Petrol.* **55**: 356-614.

Metz R. 1985. The importance of maintaining horizontal sieve screens when using a Ro-Tap. *Sedimentology*, **32**: 613-614.

Rotary Sieve-grinder
Chepil W. and Bisal F. 1943. A rotary sieve method for determining the size distribution of soil clods. *Soil Sci.* **56**: 95-100.

Chepil W. S. 1952. Improved rotary sieve for measuring state and stability of dry soil structure. *Proc. Soil Sci. Soc. Am.* **16**: 113-117.

Effect of Grinding
Soltanpour P. N., Khan A. and Schwab A. P. 1979. Effect of grinding variables on the NH_4OCO_3-DPTA soil values for Fe, Zn, Mn, Cu, P and K. *Commun. Soil Sci. Pl. Anal.* **30**: 903-909.

Sampling
Baiulescu G., Dumitrescu P. and Zugravescu P. 1991. *Sampling*. Lavoisier, n°312314, 184 pp.
Castanho N. 1987. Methods of soil sampling. In *Comments on Modern Biology*, Part B. (1)3-4: 221-227.
Cline M.G. 1944. Principles of soil sampling. *Soil Sci.*, **58**: 275-288.
Commissariat à l'Énergie Atomique. 1986. *Statistique appliquée à l'exploitation des mesures* (2e éd.) Masson, 569 pp.
Etchevers J.D. 1987. *Chemical Soil Analysis—The reasons for their drawbacks*. Labex Work shop-ISRIC.
Gy P. 1988. *Hétérogénéité, échantillonnage, homogénéisation. Ensemble cohérent de théories*. Masson, 624 pp.
Gy P. 1992. *Sampling of Heterogeneous and Dynamic Material Systems: Theories of Heterogeneity, Sampling and Homogenizing*. Elsevier Science Ltd.
Gy P. 1998. *Sampling for Analytical Purposes*. John Wiley and Sons, New York, 153 pp.
Lacroix Y. 1962. *Analyse chimique. Interprétation des résultats par le calcul statistique*. Masson, 68 pp.
Pelletier J.P. 1917. *Techniques numériques appliquées au calcul scientifique*. Masson, 366 pp.
Ruelle P., Ben Salah D. and Vauclin M. 1970. Méthodologie d'analyse de la variabilité spatiale d'une parcelle agronomique-Application à l'échantillonnage. *Agronomie* **6**: 529-539.
Tan K. H. 1996. *Soil Sampling, Preparation and Analysis*. Marcel Dekker.
Walker W.M., Siemens J. C. and Peck T.R. 1970. Effect of tillage treatments upon soil test for soil acidity, soil phosphorus and soil potassium at three sites. *Commun. Soil Sci. Plant Anal.*, pp. 367-375.

4

Grinding and Sieving Equipment

1. Grinders

1.1 Grinder Selection

Grinding of soils, rock and plants can be done only in grinders adapted to the nature of the samples as these are very variable in consistence and hardness:

—rock samples are generally brittle and can be conveniently prepared by pressure, crushing and percussion;

—samples rich in clay, soft minerals or amorphous materials are often sticky and require grinding with an oscillating disc, avoiding grinding balls that lead to large losses since recovery of the sample is difficult;

—fibrous samples (litter, plant debris etc.) should be prepared with shearing or cutting grinders;

—fossil gums and soft plant debris can be ground rapidly with the addition of volatile freezing agents (liquid nitrogen or dry ice) in airtight percussion tubes.

Selection of the material used in grinding equipment (Table 4.1) is dictated by the hardness (Tables 4.2 and 4.3) and the chemical nature of the samples. It is necessary to know the magnitude and nature of contamination in the entire grinding chain (storage containers, grinder, sieve, sampling etc.). Therefore, according to the elements to be analyzed, various grinding procedures are adopted for several subsamples.

In general, hard materials can be prepared in equipment made of agate, various steels, or fritted carbides or oxides—each with a specific contamination (to preclude cross-contamination, the same type of equipment is used for the entire preparation).

Hard porcelain, inexpensive, should not be used if trace elements are to be analyzed.

Agate can now be considered the least contaminating material, its SiO_2 content being as high as 99.9% (Table 4.4).

Sample size should also be considered.

Rocks, stones and clods should first be mechanically reduced to a size compatible with the grinding components (by grinding-pressing or percussion).

The sample volume controls the type of grinder to be utilized for treating macrosamples larger than one kilogram or adapted to microsamples (grams or fractions of a gram).

Specific areas of application necessitate a choice between:

—discontinuous-feed systems, allowing grinding in airtight conditions with possible addition of liquids or solids under controlled atmosphere and temperature;

—faster 'open' continuous systems adapted to large sample volumes, but susceptible to loss or contamination of samples.

One could also take into consideration, whenever possible, the length of the grinding cycle, the ease and speed of cleaning (some equipment require

Table 4.1 Materials used in grinding equipment

Property	Sintered or ceramic products[1]						Steel		
	Agate	Hard porcelain	Boron carbide	Tungsten carbide	Fritted corundum	Zirconium oxide	Chrome steel	Stainless steel 18/8	Manganese steel
Chemical composition	SiO_2 99.91%	SiO_2, Al_2O_3, K_2O	Boron carbon	W-C-Co	Al_2O_3 99.72%	ZrO_2 97%	Fe-Cr-C	Fe-Ni-Cr	Fe-Mn-C
Specific gravity	2.6	2.46	2.51	14.75	3.9	5.75	7.85	7.85	7.85
Hardness (Mohs)	6.5-7.0	8.0	9.5	8.5	9.0	8.5	5.5-6.0	5.0-5.5	5.5-6.0
Shock resistance	Low	Low	Medium	Medium	Low	Medium	High	High	Very high
Abrasion resistance	Fairly high	Low	High	High	Medium	Fairly high	Medium	Medium	Fairly low
Structure (crystallinity)	Crypto-crystalline	Crystalline	Crystalline	Crysalline	Crystalline	crystalline			
Chemical resistance	Attacked by HF and NaOH			Attacked by HCl and HNO_3					
Contamination	Very low (Si, Al, Na, K, Mn, Ca, Mg)	Si, Al, K	B, C, Si Fe	W, C, Co	Al, Si	Zr, Mg, Ca, Fe	Fe, C, Si, S, Mn, Cr, V	Fe, C, Si, Mn, Cr, Ni	Fe, Mn, C, Si, P, S, Ti
Specific use	XRD[2], trace elements, IR[2], XRF[2], NAA[2]	Cheap, agronomic purposes	To be tested for XRD[2] and for trace elements		Bauxite		Unusable for trace elements		Percussion grinding

[1] Sintering is done by agglomeration of very finely ground (1μm) materials, enabling maximum density and almost zero porosity to be attained. The material is moulded under pressure, then baked. In some cases an organic binder is injected at 1500 bars in a mould. After removal of the material from the mould, the binder is eliminated by slow heating. The mortars and pestles are then machined and shaped.

[2] XRD = X ray diffraction, IR = infrared spectrometry, XRF = X ray fluorescence, NAA = neutron activation analysis.

Table 4.2 Hardness scales of various minerals

Reference mineral	Mohs	Vickers	Old Knoop	Modified Knoop	Modified Mohs		
Talc	1	2.4			1	Talc	$3MgO \cdot 4Si \cdot O_2 \cdot H_2O$
Gypsum	2	36	32	32	2	Gypsum	$CaSO_4 \cdot 2H_2O$
Calcite	3	110	135	135	3	Calcite	$CaCO_3$
Fluorite	4	190	163	163	4	Fluorite	CaF_2
Apatite	5	540	430	430	5	Apatite	$CaF_2 \cdot 3Ca_3(PO_4)_2$
Feldspar	6	800	560	560	6	Orthoclase	$K_2O \cdot Al_2O_3 \cdot 6SiO_2$
Quartz	7	1120	820		7	Vitreous silica	SiO_2
Topaz	8	1430	1340	820	8	Quartz stallite	SiO_2
Corundum	9	2000		1340	9	Topaz	$(AlF_2) SiO_4$
Diamond	10	10,000	7000	1360	10	Garnet	$Al_2O_3 \cdot 3FeO \cdot 3SiO_2$
				(1860)	11	Fused zircon	ZrO_2
				2100	12	Fused alumina	Al_2O_3
				2480	13	Silicon carbide	SiC
				2750	14	Boron carbide	B_4C
				7000	15	Diamond	C

Table 4.3 Hardness of various substances

Mohs hardness	Substances
Very hard, brittle 6.5-8.5	Basalt, diabase, granite, magnesia, porphyry, quartz, rutile, sillimanite, corundum...
Hard 4.5-6.5	Cinders, lead ore, copper sulphide, iron chromate, siderite, feldspar, haematite, calcite, magnetite, ferromanganese, marble, various ores...
Moderately hard 2.5-4.5	Bauxite, pumice stone, brown haematite, refractory fired clay dolomite, feldspar, gneiss, Iceland spar, magnesia, barite
Soft 1.5-2.5	Pumice stone, brown haematite, gypsum, mica, graphite, lignite, potash salts, rock salt, talc
Fibrous	Cellulose, peats, plants...

more time for decontamination than for the grinding per se), the cost and life of the grinding components, and the comfort factors (noise, vibrations etc.).

1.2 Manual Mortars

Mortars (Fig. 4.1, Table 4.5) are often chosen to prepare samples of small volume, whose grinding can be adapted to certain tasks. One may speak of reduction rather than grinding if just one particle size is desired (for example extraction-concentration of soft rocks like calcareous modules).

Table 4.4   Composition (%) of the first choice agate

SiO_2	99.91
Al_2O_3	0.02
Na_2O	0.02
Fe_2O_3	0.01
K_2O	0.01
MnO	0.01
CaO	0.01
MgO	0.01
Sp. gravity	2.6
Hardness (Mohs)	6.5-7

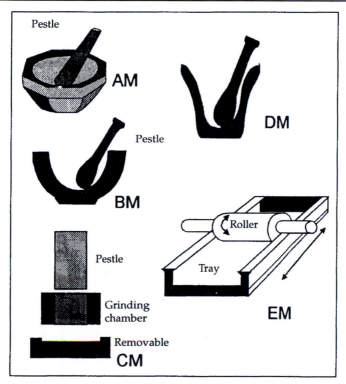

Fig. 4.1   Manual mortars. The symbols correspond to those in Table 4.5. AM: Agate mortar BM: Hard porcelain mortar CM: Abich mortar DM: Steel mortar EM: Mortar with hard wooden roller

Low-form mortars are very suitable for grinding by rotary motion of the pestle and pressure on the sample. However, for a given diameter their capacity is less than that of tall-form mortars. Risk of loss by dust and external contamination are high.

Abich mortars, made of steel, enable rapid crushing of small rock fragments by breaking. Residual contamination is low at this level and the sample may then be transferred to a grinder of another type, or the grinding

Table 4.5 Types of manual mortars.

Symbol (Fig. 4.1)	Grinder type	Grinding principle (system)	Nature of grinding tools		Particle size		
			Usable	Unsuitable	Permissible	Final	
AM	Low-form mortar	Crushing (controlled pressure)	1, 3, 5*	—	Reduction/ grinding of soil to 2 mm	1μm soil	3-5 mL (rapid cleaning); diameter 30-200 mm (9-600 mL)
BM	Mortar	Crushing	2, 10*	—	Reduction of soil to 2 mm	2 mm, 1 mm, 0.5 mm, 0.2 mm soil	75-300 mm diameter (rapid cleaning)
CM (pestle, grinding chamber, removable base)	Abich mortar	Percussion crushing pressure	9*	1, 2, 3, 4, 5, 6,	About 20 mm 12*	Variable (rock crushing)	Maximum diameter 30 mm (rapid cleaning; small samples)
DM	Mortar	Percussion	11*	1, 2, 3, 4, 5, 12*	About 20-50 mm	Rock crushing	155 mm diameter (rapid cleaning), 2000 mL
EM	Tray	Controlled crushing	Veinless hard wood 12*	—	Reduction of soil to 2 mm (controlled), clods	2 mm	100-500 g (rapid cleaning)

* 1 Agate
2 Hard porcelain
3 Boron carbide
4 Tungsten carbide
5 Corundum
6 Zr oxide
7 Chrome steel
8 Stainless steel 18/8
9 Manganese steel
10 Pyrex glass
11 Turned cast iron
12 'Ironwood' of Guyana, sp. gr. 1.35 (Guaiacum, Ixora or *Krugiodendron ferreum*)

may be continued until the desired particle size is attained. For larger samples a cast-iron mortar may be used.

Mortars with hard wooden roller and tray allow gradual crushing and accommodate relatively large soil samples. Contamination by 'ironwood', which is very hard and non-fibrous, is negligible and insignificant.

These manual grinders are the most suitable for reduction of samples to 2 mm and are superior to tall-form mortars using a pestle, which are more suitable for pulverizing fragile rocks, concretions, nodules etc.

1.3 Principal Types of Mechanical Grinders

Mechanical grinders (Figs. 4.2 and 4.3, Table 4.6) are indispensable for routine preparation of soil samples if their volume and number are large.

They are definitely efficient when the grinding is undifferentiated and ought to give a uniform final particle size within selected limits (sieve system).

Their selection should take into consideration the criteria given in Sec. 1.1, rapidity of cleaning and decontamination being major advantages.

Closed-system grinders eliminate all external contamination and losses. Recovery is close to 100% except for grinding balls, which are always difficult to clean. Grinders with oscillating disc are the easiest to clean because the surfaces of the disc or hammer are flat or only slightly convex.

2. Sieves and Strainers

2.1 Sieving Equipment: General Comments

Sieves are fabricated from various materials. To restrict contamination, sieves of stainless steel 18/8 should be used, which give negligible contamination under normal conditions of use.

In some cases, recourse can be had to a wooden body and nylon mesh; manual dry-sieving below 50 μm is not possible.

Mechanical sieves with gyratory/shaking or electromagnetic motion for sieves of 100 mm to 400 mm diameter enable working with a set to about 60 μm. Dry-sieving is better suited to poorly structured materials (Fig. 4.4).

Mechanical wet-sieving is also possible by attaching a special spraying head and airtight sealing rings to the apparatus (Fig. 4.5). Very sticky samples or electrostatically charged samples pass through without difficulty albeit other constraints do exist, such as floating of light organic materials or pumice with closed pores and loss of soluble endogenous substances.

Special electroformed sieves of nickel can be used with screen size of 15, 10 and even 5 μm for dry or wet sieving. Stirring is achieved in an ultrasonic field of 20-50 kHz frequency. Low-amplitude vibrations eliminate screen clogging. Since the sieve sets are airtight, there is no loss of fine particles.

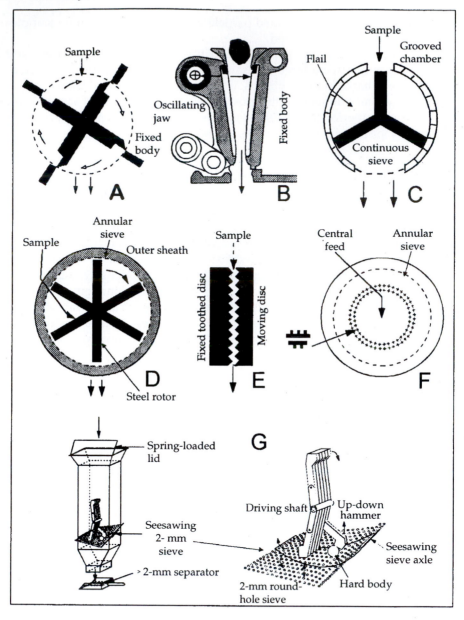

Fig. 4.2 Types of grinders (open system). The identifying letters correspond to those in Table 4.6.
A: Grinder with knives B: Jaw crusher C: Grinder with flails D: Rotary percussion grinder (central feed, viewed from above) E: Disc grinder F: Centrifugal grinder with toothed or corrugated rotor (viewed from above) G: Grinder with retractable hammers.

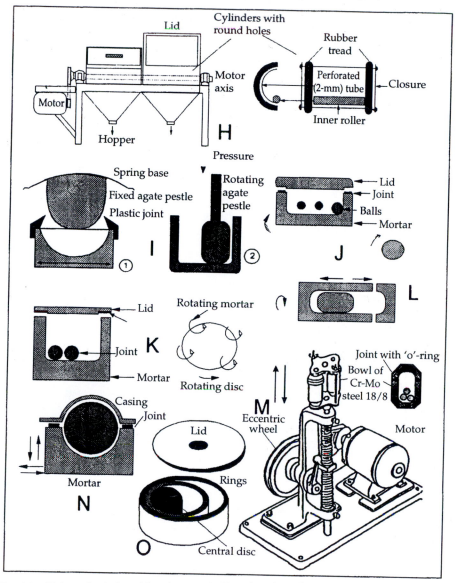

Fig. 4.3 Types of grinders (closed system). The identifying letters correspond to those in Table 4.6.
H: Grinder with roller I: Grinder with agate mortar: 1-rotating mortar, fixed pestle; 2-rotating mortar and pestle: eccentric inverse rotation J: Centrifugal ball grinder K: Planetary mortar with grinding balls L: Horizontal vibratory rotary grinder M: Vertical percussion grinder with steel balls N: Oscillating electro-magnetic ball grinder O: Grinder with oscillating disc and rings.

Table 4.6 Types and characteristics of mechanical grinders useful in analysis of soils, rocks, plant products (the identification letters correspond to those in Figs. 4.3 and 4.4; the numbers identifying the nature of the grinding tool are those in Table 4.5).

	Grinder type	Grinding principle (system)	Nature of grinding tool		Particle size		Minimum permissible quantity and nature of samples
			Available or usable	Not usable	Permissible mm	Final after sieving°	
Open system (continuous feed)							
Plant litter	Cutting grinder (A)	Cutting, shearing (wet or dry grinding)	7, 8	1, 2, 3, 4, 5, 6	5	0.2 mm isogranular or assorted	1 kg to 1 g dry or moist material, root litter
Rock crushing	Jaw crusher (B)	Breaking (dry grinding, dust-proof chamber)	9, 8, 4	1, 2, 3, 5, 6	60	1 mm	> 1 kg (long cleaning time), brittle products, rock, lime-stone, ores, silicates, etc.
Undiffe-rentiated grinding	Grinders with flails (C)	Percussion, shearing (dry grinding, dust-proof chamber)	7, 8	1, 2, 3, 4, 5, 6	20	0.1 mm isogranular°	> 1 kg (long cleaning time), Mohs hardness < 6.0, brittle fibrous products, soils without rock fragments
	Rotary percussion grinders (D)	Shearing, percussion (dry grinding, dust-proof chamber)	8, 7	1, 2, 3, 4, 5, 6	15 (according to hardness)	0.1 mm isogra-nular°	< 1 kg (long cleaning time); Mohs hardness < 4.0; brittle products, soils, peats, usable with dry ice
	Disc grinders (E)	Shearing; pressure (dry grinding, dust-proof chamber)	7, 8	1, 2, 3, 4, 5, 6,	20	2 to 0.2 mm	kg (long cleaning time); soils without rock fragments
	Centrifugal grinders with toothed or corru-gated rotors (F)	Percussion effect of shearing force (dry grinding, dust-proof chamber)	7, 8	1, 2, 3, 4, 5, 6	10	0.1 mm isogranular°	kg (difficult to clean); Mohs hardness < 5.0; brittle, fibrous products, soils, peats
Controlled grinding to 2 mm	Grinders with retractable hammers (G)	Controlled percussion (dry grinding, dustproof chamber)	7, 8, 9	1, 2, 3, 4, 5, 6	10	2.0 mm° isogranular	100 g (rapid cleaning); soils without fragile concretions or nodules

(Contd.)

Table 4.6 (*Contd.*)

	Grinder type	Grinding principle (system)	Nature of grinding tool		Particle size		Minimum permissible quantity and nature of samples
			Available or usable	Not usable	Permissible mm	Final after sieving°	
Closed system (discontinuous feed) — Controlled grinding to 2mm	Grinders with rollers (H)	Pressure, controlled crushing (dry grinding, dust-proof chamber)	7, 8, 9	1, 2, 3, 4, 6	10	2.0 mm° isogranular	300-500 g (long cleaning time for highly clayey Vertisols); non-clayey, non-Andic soils
Various fine-grinders	Mortar-grinder (I)	Crushing pressure on dry or moist materials (dust-proof chamber)	1, 2, 3, 4, 5, 6, 7, 8	—	5 mm	1μm	0.5-50 mL; 50-500 mL; cleaning rapid for soils and sands
	Centrifugal ball-grinder (J)*	Dry or moist crushing (airtight chamber)	1, 2, 3, 4, 5, 6, 7, 8	—	< 8 mm	1 μm	50-100 mL; soils, sands, etc.; recovery of clay from balls difficult
	Planetary ball-grinders (K)*	Friction-percussion crushing, dry or moist (airtight chamber)	1, 2, 3, 4, 5, 6, 7, 8	—	< 8 mm	< 1 μm	30-300 mL; cleaning laborious for soils, sands, etc.; recovery of clay difficult
	Horizontal vibratory grinder (L)	Percussion crushing (airtight chamber)	1, 2, 3, 4, 5, 6, 7, 8, 9, Teflon	—	4 mm	< 1 μm	1.5-5 mL; grinding or mixing for mineralogy, KBr pellets (cryogeny possible)
	Ball-grinder with vertical percussion (M)	Percussion dry or moist (airtight chamber)	8, 9	1, 2, 3, 4, 5, 6	4 mm	< 1 μm	65-150 mL clods; long cleaning time for soil and sands; recovery of clay difficult
	Oscillating electromagnetic ball-grinder (N)	Percussion crushing; moist or dry (airtight chamber)	1, 2, 3, 4, 5, 6, 7, 8	—	< 5 mm	< 1 μm	0.1-10 mL; rapid cleaning for soils and sands
	Grinder with oscillating disc and rings (O)	Friction; crushing by shocks; moist or dry (airtight chamber)	1, 2, 3, 4, 5, 6, 7, 8	—	4 mm	< 1 μm	50-250 mL; fairly rapid cleaning with disc (soils, sands)

* For grinding: small number of large balls; For mixing: small balls (about 6 to 8).

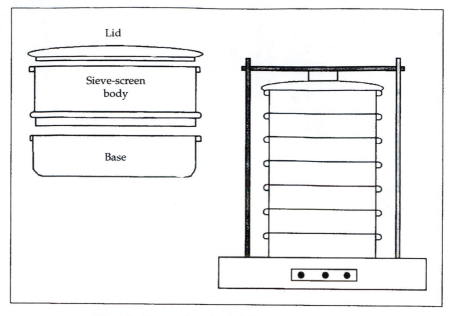

Fig. 4.4   Sieve and mechanical dry-sieving apparatus.

This type of equipment is used for separation of coarse silts or for testing the uniformity of secondary grinding.

The size of particles is referred, for simplification, to the equivalent sphere, which is determined by a single dimension, the diameter. Use of a square-mesh sieve introduces two dimensions: that of the two parallel sides and the diagonal. Use of rectangular mesh for fibrous materials (organic litter etc.) introduces three dimensions: the two sides of the rectangle and the diagnonal.

2.2 Correspondence Tables of Principal French and World Standards

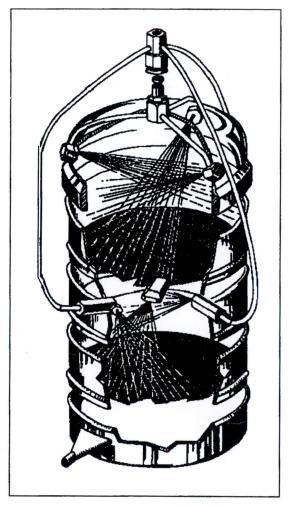

Fig. 4.5 Wet-sieving apparatus (Fritsch[1])

---

[1] *See* Appendix 6 for addresses.

80 *Soil Analysis*

Table 4.7 Principal standards for sieves.

ISO — International Standards Organization, ISO-R-565 1967 Principal opening, μm	FRANCE — AFNOR French Standardization Society, NF-X11-504 Dec 1970 Principal opening, μm	Series designation	UNITED STATES — ASTM-E-11-70 1970 Number designation	Opening, μm	The W.S. Tyler Co. Tyler Standard screen Scale Sieves Mesh designation	Opening, μm	GERMANY — German Standards DIN-4188 1969 Opening, μm	BRITAIN — British Standards Institution BS-410 1969 Opening, μm
	20						20	
	25						25	
	32						32	
			400	38	400	38		
	40	17					40	
45			325	45	325	45		45
	50	18					50	
			270	53	270	53		
63	63	19	230	63	250	63	63	63
			200	75	200	75		
	80	20					80	
90			170	90	170	90		90
	100	21					100	
			140	106	150	106		
125	125	22	120	125	115	125	125	125
			100	150	100	150		
	160	23					160	
180			80	180	80	180		180
	200	24					200	
			70	212	65	212		

(Contd.)

Table 4.7 (Contd.)

ISO	FRANCE		UNITED STATES				GERMANY	BRITAIN
International Standards Organization	AFNOR French Standardization Society		American Society for Testing and Materials		The W.S. Tyler Co. Cleveland 14, Ohio		German Standards	British Standards Institution
ISO-R-565 1967	NF-X11-504 Dec 1970		ASTM-E-11-70 1970		The Tyler Standard screen Scale Sieves		DIN-4188 1969	BS-410 1969
Principal opening, μm	Principal opening, μm	Series designation	Number designation	Opening, μm	Mesh designation	Opening, μm	Opening, μm	Opening, μm
250	250	25	60	250	60	250	250	250
			50	300	48	300		
	315	26					315	
355			45	355	42	355		355
	400	27					400	
			40	425	35	425		
500	500	28	35	500	32	500	500	500
			30	600	28	600		
	630	29					630	
710			25	710	24	710		710
	800	30					800	
			20	850	20	850		
mm 1	mm 1	31	18	mm 1	15	mm 1	mm 1	mm 1
			16	1.18	14	1.18		
	1.25	32					1.25	
1.4			14	1.4	12	1.4		1.4
	1.6	33					1.6	
			12	1.7	10	1.7		

(Contd.)

Table 4.7 (*Contd.*)

ISO International Standards Organization ISO-R-565 1967 Principal opening, μm	FRANCE AFNOR French Standardization Society NF-X11-504 Dec 1970 Principal opening, μm	Series designation	UNITED STATES — American Society for Testing and Materials ASTM-E-11-70 1970 Number designation	Opening, μm	UNITED STATES — The W.S. Tyler Co. Cleveland 14, Ohio · The Tyler Standard screen Scale Sieves Mesh designation	Opening, μm	GERMANY German Standards DIN-4188 1969 Opening, μm	BRITAIN British Standards Institution BS-410 1969 Opening, μm
2	2	34	10	2	9	2	2	2
			8	2.36	8	2.36		
	2.5	35					2.5	
2.8	3.15	36	7	2.8	7	2.8	3.15	2.8
			6	3.35	6	3.35		
4	4	37	5	4	5	4	4	4
	5	38	4	4.75	4	4.75	5	
5.0	6.3		3.5	5.6	3.5	5.6	6.3	5.6
				6.3	3	6.7		
				6.7	2.5	8		
8	8			8		9.5	8	8
	10			9.5			10	
11.2	12.5			11.2		11.2	12.5	11.2
				12.5		13.2		
16	16			13.2		16	16	16
	20			16		19	20	
				19				

(*Contd.*)

Table 4.7 (*Contd.*)

ISO	FRANCE	UNITED STATES				GERMANY	BRITAIN
International Standards Organization	AFNOR French Standardization Society	American Society for Testing and Materials		The W.S. Tyler Co. Cleveland 14, Ohio		German Standards	British Standards Institution
ISO-R-565 1967	NF-X11-504 Dec 1970	ASTM-E-11-70 1970		The Tyler Standard screen Scale Sieves		DIN-4188 1969	BS-410 1969
Principal opening, μm	Principal opening, μm / Series designation	Number designation	Opening, μm	Mesh designation	Opening, μm	Opening, μm	Opening, μm
22.4			22.4		22.4		
	25		25			25	
			26.5		26.5		
31.5	31.5		31.5			31.5	
			37.5				
	40					40	
45		45					
	50	50				50	
		53					
63	63	63				63	
		75					
	80					80	
90		90					
	100	100				100	
		106					
125	125	125				125	

Table 4.8 Principal standards for strainers

Perforated mild steel sheets of thickness 5/10 for dia 1-1.6, 8/10 for dia 2-6.3, 10/10 for dia 8-12.5, 15/10 for dia 16-40, 25/10 for dia 50-125 Round holes	PRINCIPAL STANDARDS			
	ISO International Standards Organization ISO-R-566, 1967 Round holes	FRANCE AFNOR French Standardization Society NF-X11-504, Dec, 1970 Round holes	UNITED STSTES American Society for Testing and Materials ASTM-E-323, 1970 Round holes	
Opening, mm	Principal opening, mm	Principal opening, mm	Opening, mm	Designation, inches
1	1	1	1	0.039
			1.18	0.045
1.25		1.25		
	1.4		1.4	0.055
1.6		1.6		
			1.7	0.066
2	1	2	2	0.078
			2.36	3/32
2.5		2.5		
	2.8		2.8	7/64
3.15		3.15		
			3.35	0.127
4	4	4	4	5/32
			4.75	3/16
5		5		
	5.6		5.6	1/32
6.3		6.3	6.3	1/4
			6.7	17/64
8	8	8	8	5/16
			9.5	3/8
10		10		

(Contd.)

Table 4.8 (*Contd.*)

Perforated mild steel sheets of thickness 5/10 for dia 1-1.6 8/10 for dia 2-6.3 10/10 for dia 8-12.5 15/10 for dia 16-40 25/10 for dia 50-125	PRINCIPAL STANDARDS			
	ISO International Standards Organization ISO-R-566, 1967	FRANCE AFNOR French Standardization Society NF-X11-504, Dec, 1970	UNITED STSTES American Society for Testing and Materials ASTM-E-323, 1970	
Round holes	Round holes	Round holes	Round holes	
Opening, mm	Principal opening, mm	Principal opening, mm	Opening, mm	Designation, inches
12.5	11.2	12.5	11.2	1/16
			12.5	1/2
			13.2	17/32
16		16	16	5/8
20		20	19	3/4
25	22.4	25	22.4	1/8
			25	1
			26.5	1 1/16
31.5	31.5	31.5	31.5	1 1/4
			37.5	1 1/2
40		40	45	1 1/2
50	45	50	50	2
			53	2 1/8
63	63	63	63	2 1/2
80		80	75	3
100	90	100	90	3 1/2
			100	4
125	125	125	106	4 1/4
150			125	5
200			132	6
			203	8

5

Preliminary Qualitative
Laboratory Tests

1. Introduction

Rapid qualitative tests are often performed in the field during the survey phase (*see* Chapter 2) because they lead to change in field strategy and to the taking of special-purpose supplementary samples.

So far as the laboratory is concerned, good knowledge of the samples is imperative (especially if the information memos are incomplete) for selecting the methods most suited to and compatible with the nature of the soils. Tests are done on raw samples when they reach the laboratory to preclude any irreversible change. The tests should be simple, specific or selective, and sensitive enough to detect the elements that could interfere with determinations because of their relative concentration. Excessive sensitivity is not always an advantage in the complex systems that soils represent.

It is preferable to select reactions that take place in a wide range of pH and result in gas release, clearly noticeable coloration or abundant precipitate of an insoluble salt. These reactions should be possible with a small volume of the reagent and small quantity of soil, without separation of phases if rapid decantation is possible.

They are done using drops in test-tubes, on a white porcelain test-plate or on filter paper.

2. Anions, Cations and Various Tests

2.1 Apparatus

—test-tubes, 10 mL,
—graduated pipette 5 mL in mL,
—test-plate of white porcelain,
—glass bubbler.

2.2 Reagents

—acetic acid, concentrated;
—concentrated hydrochloric acid diluted about ten-fold (about 1 mol L^{-1} HCl);
—sulphuric acid (about 4 mol L^{-1});
—nitric acid (about 1 mol L^{-1});
—lime water;
—barium chloride solution, 10% in water;
—lead chloride solution, 1%;
—ammonium oxalate solution, 10%;
—silver nitrate solution, 10%;
—ammonium persulphate, crystalline;
—sodium fluoride solution, saturated (about 40 g L^{-1});

—phenolphthalein solution, 1% in ethanol (for impregnation of 30 mm diameter filter paper and drying).

2.3 Reactions in Acid Medium

2.3.1 Carbonates (CO_3^{2-})

Principle

$$CO_3^- + 2H^+ \xrightarrow{\text{HCl}} CO_2 \uparrow + H_2O + \text{salt}$$

The CO_2 produced may be confirmed by the following reactions (lime water in a bubbler):

$$CO_2 + Ca(OH)_2 \longrightarrow \frac{CaCO_3}{\downarrow \text{white}} + H_2O$$

$$2CO_2 + Ca(OH)_2 \longrightarrow Ca(HCO_3)_2 \,_{\text{soluble, colourless}}$$

The CO_2 when bubbled through the lime water first causes precipitation of insoluble calcium carbonate (cloudiness), then excess of CO_2 causes formation of calcium bicarbonate and the cloudiness disappears.

Procedure
—Take in a test-plate:
  a pinch of soil + about 5 drops of 1 mol L^{-1} HCl.
  Formation of bubbles indicates the presence of carbonates. Record the intensity of the effervescence, speed of appearance of the bubbles and the possible localized sources of gas release (nodules). If the reaction is very slow, repeat the reaction in a test-tube with 0.5 g soil + 4 mL water + 1 mL HCl and heat to 70°C to release CO_2. The presence of dolomite is probable (always check for the smell of hydrogen sulphide—indicating the presence of S^{2-}—when HCl is added).

2.3.2 Sulphates (SO_4^{2-})

Principle

$$SO_4^- + BaCl_2 \xrightarrow{\text{HCl}} \frac{BaSO_4}{\downarrow \text{white}} + 2Cl^-$$

Procedure
  —0.5 g soil + 5 mL H_2O + 2 drops HCl;
  —shake, decant;
  —add 2 drops of 10% $BaCl_2$ to the clear liquid;
  —a white precipitate (or a white cloudiness) indicates the presence of sulphate;
  —the two reactions for carbonate and sulphate may be joined in one single operation;

HCl addition:
  release of gas: presence of carbonates,
  no release of gas: no carbonates,
  decantation, addition of $BaCl_2$,
  cloudiness or precipitate: presence of SO_4^{2-}

2.3.3 Sulphides (S^{2-})

Principle

$$S^{--} + 2HCl \longrightarrow H_2S\uparrow + 2\,Cl^-$$

In the bubbler

$$H_2S + PbCl_2 \longrightarrow \underset{\downarrow black}{PbS} + 2HCl$$

Procedure
  —0.5 g soil + 1 mL dilute HCl;
  —quickly connect to the bubbler containing a soluble lead salt;
  —release of H_2S (nauseating odour of rotten eggs) causes precipitation of insoluble black lead sulphide in the bubbler.

2.3.4 Chlorides (Cl^-)

Principle

$$Cl^- + AgNO_3 \xrightarrow{\ HNO_3\ } \underset{\downarrow white\ curd\ turning\ dark\ in\ light}{AgCl} + NO_3^-$$

Procedure
  —0.5 g soil + 2 drops dilute HNO_3 + 5 mL H_2O, shake, decant;
  —add 2 drops of 10% $AgNO_3$ to the clear liquid;
  —presence of chloride indicated by appearance of a white curd-like precipitate if chlorides are abundant, and by a white cloudiness if present in small amount.

2.3.5 Calcium (Ca^{2+})

Principle

$$Ca^{++} + (NH_4)_2C_2O_4 \xrightarrow{\ CH_3COOH\ } \underset{\downarrow white}{CaC_2O_4} + 2NH_4^+$$

Procedure
  —0.5 g soil + 1 drop acetic acid + 5 mL H_2O;
  —shake;

—decant;
—add 3 drops ammonium oxalate to the clear liquid;
—a white precipitate indicates the presence of calcium.

2.3.6 Manganese (Mn^{2+})

Principle
Oxidation in dilute acid medium (other than HCl) to give violet-coloured permanganate ions:

$$Mn^{2+} \xrightarrow{\text{oxidation} + CH_3COOH} Mn^{7+}$$

Chlorides interfere with the reaction as follows:

$$2\ KMnO_4 + 8HCl \rightarrow 2MnO_2 + 3Cl_2 + 2KCl + 4H_2O$$

Procedure
—0.5 g soil + 3 drops HNO_3 + 5 mL H_2O;
—shake;
—decant;
—add $AgNO_3$ to the clear liquid to precipitate any chlorides present, + a pinch of ammonium persulphate, then heat over a Bunsen burner;
—a mauve-violet colour appears (the white curds of silver chloride do not interfere).

2.4 Various Reactions and Rapid Instrumental Measurements

2.4.1 Allophane

Principle
Test based on rise in pH when NaF is added to soils containing allophane (see Chapter 2, 'Field Tests'):

$$Al(OH)_3 + 6NaF \rightarrow AlF_6Na_3 + 3NaOH$$

Procedure
A pinch of soil is placed on a 30 mm diameter filter paper impregnated with phenolphthalein and 3 drops saturated NaF (about 4%) added. The test is positive if the phenolphthalein turns pink in less than 60 seconds. In organic soils colour change is not visible. It suffices to observe the change in colour on the underside of the filter paper.

N.B. If the soil has been air dried, wettability is low and so the soil floats. In this case knead a pellet of moistened soil and do the test with that.

2.4.2 Total salinity

Principle
—measurement of conductivity at 1:10 or 1:2,
—measurement by refractometry (hand salinometer).

Procedure
1. —2 g soil + 20 mL water,
    —stir;
    —decant;
    —measure conductivity of the clear supernatant liquid.
2. Take a drop of the above extract and place in the salinometer.

2.4.3 pH

Principle
    —Measurement of pH (H_2O) and pH (KCl) on the same sample.
Procedure
    —10 g soil + 25 mL water;
    —stir;
    —let settle for 5 minutes, measure pH (H_2O);
    —using a scoop, add approximately 1 g powdered KCl;
    —stir, let settle for 5 minutes;
    —measure pH (KCl).

2.4.4 Remarks

Rapid determination of conductivity is only done if the tests for Cl^- and SO_4^{2-} and occasionally CO_3^{2-} (alkali soils) are positive and if analysis for soluble salts is not required; otherwise conductivity is determined at the time of regular analysis.

Determination of pH (H_2O) and pH (KCl) should be more than just a test. It could be the first link in the analytical chain with a rigid procedure. The pH in NaF is done if the 'allophane' test is positive.

3. Qualitative Tests and Selection of Methods for Soil Analysis

If the tests for CO_3^{2-} and Ca^{2+} are positive
    —measure the pH after 10 minutes of contact in absence of air to avoid carbonation;
    —determine total and, if required, active $CaCO_3$;
    —exchangeable Ca will not be significant by the standard method: use a differential double extraction;
    —select a suitable method for particle-size analysis;
    —select a suitable method for extractable P_2O_5 ;
    —watch carefully for a reaction from any/all methods of acid mineralization;
    —select a suitable method for cation exchange capacity (CEC);
    —select a suitable method for determination of carbon.

If tests for Cl^- and $SO_4{}^{2-}$ are positive
   —measure the electrical conductivity; (EC in dS m^{-1})
   —determine the soluble salts (total cations = total anions in miliequivalents per litre $\simeq$ 10 EC) and take them into account during determination of exchangeable bases;
   —determine pH in $CaCl_2$ medium;
   —determine gypsum if $SO_4{}^{2-}$ appears abundant.

If the tests for $CO_3{}^{2-}$ are positive (but Ca^{2+} negative or low)
   —check the pH: if pH > 9.0, alkalization (high Na^+) has occurred;
   —determine soluble salts.

According to the difference between pH (H_2O) and pH (KCl)
   —select the P_2O_5 method suitable for acid or alkaline medium as the case may be;
   —select method for extraction of exchangeable cations;
   —select method for particle-size analysis;
   —if pH (KCl) < 4.5, Al^{3+} and H^+ may be present;
   —if pH = 7.0, check for calcium carbonate;
   —if pH = 8-9, check for saline-alkali soil.

If NaF test is positive
   —preserve half the sample moist;
   —consider methods for selective dissolution;
   —determine pH of the soil in its original moisture condition;
   —determine variable charge and 'PZC';
   —CEC methods, unbuffered;
   —do not determine extractable P_2O_5 in presence of NH_4F;
   —determine reversion and fixation of P;
   —determine pH after 1 or 2 minutes in NaF.

According to value of pH (NaF)
   —if pH < 9.0 in one minute, scarcely any amorphous material;
   —if pH > 9.5-10.4 in one minute, allophane is present in large quantity.

If test for Mn is positive
   —determine easily reducible Mn;
   —the hydrogen peroxide used in particle-size analysis will be very rapidly destroyed;
   —allophanic soils unstable in particle-size analysis (eliminate the ferromanganic materials with a magnet);
   —select a method for organic substances and humic compounds.

If test for S is positive
   —forms of sulphur;
   —organic substances;
   —reducing environment if Fe^{2+} present—avoid oxidation.

Bibliography

Head K. H. 1996. *Manual of Laboratory Testing: Soil Classification and Compaction Tests.* John Wiley and Sons, New York (2nd ed.), 400 pp.

Head K. H. 1996. *Manual of Laboratory Testing: Permeability, Shear Strength and Compressibility Tests.* John Wiley and Sons, New York (2nd ed.), 454 pp.

Pentech Press. 1994. *Manual of Soil Laboratory Testing.* Pentech Press (2nd ed.).

6

Analytical Balances

1. Introduction

Although gravimetric analytical methods are less and less used, the weight of samples and reagents is the base for all quantitative analytical determinations, third in importance after collection and preparation of the sample (*see* Chapters 1 and 3). It is also directly involved as the measuring tool for certain gravimetric chemical and physical determinations on soils (density, particle-size distribution by the pipette method, porosity, specific surface area etc.).

Analytical balances have been greatly modified in the past two decades and double-pan beam balances have almost totally disappeared. In France, balances are subject to checking by the Department of Weights and Measures, especially when weighing pertains to commercial transactions. Effective from 1 January 1993 (notified 27 March 1991) balances are classified into two categories:

—balances of 'international' type that can be used only in certain domains, under the responsibility of the buyer;

—balances termed 'approved' are obligatory when the matter concerns determination of mass for application of a law, determination of mass for works impacting human and animal health, and also the weights in the framework of commercial transactions or those involved in pecuniary matters.

The GLP[1] balances should allow identification, by means of a printout, of all the true weight parameters to enable later verification. The small printer attached to this type of balance enables printing according to demand or automatically the following:

—date and time of weighing (internal clock run by a battery);

—type and number of the balance;

—identification of the sample with automatic numbering;

—calibration of the balance;

—results of the weighings;

—identity of the operator;

—transfer of data with test of plausibility through an RS232 interface.

The GLP standards require maintenance of a file of documents with statistical functions that can be done with the printer-calculator: number of determinations, mean, standard deviation, total, minimum, maximum etc.

2. Different Types of Analytical Balances

2.1 Mechanical Beam Balance with Two Equal Arms

Description

The concept of balances until around 1950 was based on the mechanical system of a beam with two equal arms (Fig. 6.1).

---

[1] GLP: Good Laboratory Practice.

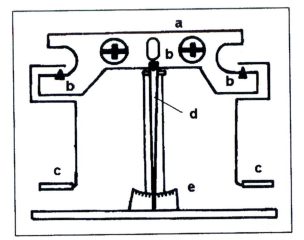

Fig. 6.1 Classic symmetrical mechanical balance with two pans and a short beam (before 1950).
a - beam; b - knife-edges; c - right and left pans; d - pointer joined to the beam; e - graduated scale.

The beam pivots on an axis of rotation constituted by a sapphire or steel knife-edge on a perfectly flat piece of agate.

Two knife-edges facing upwards, placed equidistant from the centre of the beam, support the two pans. The three parallel knife-edges are located on the same plane. Through appropriate design, the centre of gravity is pushed below this plane. The movements of the beam are amplified by a long pointer that moves against a micrometer scale. Reading is facilitated by a lens.

Errors of the lever are attenuated by adjusting screws on the beam and by double-weighing, placing an object on one pan, then on the other. A short, rigid beam is preferable to a long, more flexible one.

Sensitivity is affected by poor alignment of the blades of the three knife-edges with displacement of the centre of gravity. This error is a function of the load. It is therefore necessary to regulate the sensitivity for each load if simple weighing is done. Double-weighing with substitution or with constant load can then be followed.

These balances are low priced but tricky to operate. They are very fragile and their surroundings should be protected from corrosion, vibrations and air currents during manipulation. A constant temperature should be maintained to avoid changes in length of the beam.

It is thus necessary to establish a balance room equipped with a vibration-proof table. As the agate knife-edges are fragile, this type of balance should not be moved. Regular, frequent maintenance is necessary (cleaning of knife-edges, regulating sensitivity, levelling etc.).

Double-weighing with constant load

A known calibrated weight, greater than that of all the bodies to be weighed (sample + cupel, etc.), is placed on the right-hand pan. Equilibrium is established by an appropriate tare on the left-hand pan. The sample to be weighed is placed on the right-hand pan and equilibrium re-established with known weights. With double-weighing or constant load, the weight of the sample is the difference between the initial marked weight and the sum of fractional weights added to the pan with the sample.

The sensitivity remains constant since the load is constant.

2.2 Mechanical Single-pan Substitution Balance

This comprises a beam and two knife-edges (Fig. 6.2). Application of a force on the pan is balanced by a fixed counterweight within the balance. Any difference in weight between the forces causes angular displacement of the pointer and the micrometer scale. Equilibrium is re-established by adding or subtracting internal weights.

Reading is obtained by adding internal weights by means of a knob on the front of the balance and that indicated on the projected scale. The balance is always loaded with the same weight, the weight of the sample substituting for internal weights. Sensitivity is constant.

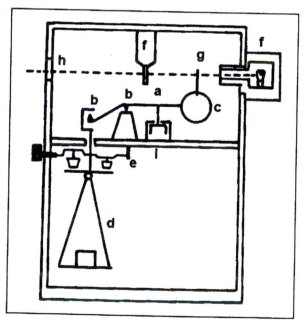

Fig. 6.2 Aperiodic single-pan mechanical (asymmetrical) balance.
a - beam; b - knife-edges; c - counterweight; d -pan; e - internal weights; f - optical lamp system; g - moving micrometer scale; h - projected scale; i - air damper.

2.3 Balances Based on Couple, Flexion and Torsion

Flexion balances with quartz fibre (Fig. 6.3), measuring changes in mass by deformation of the fibre, are mostly used for ultra-microweighing (elemental organic microanalysis for example).

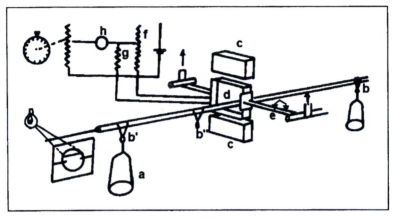

Fig. 6.3   Electromagnetic flexion balance (Cahn[1]).
a - sample; b - pan with counterweight; b' - lowest range; b'' - highest range; c - magnet; d - coil; e - strip suspension; f , g - potentiometers; h - galvanometer.

It may be recalled that low-priced torsion-wire balances were once available on the market but have now disappeared (Torsion Balance Co.).

2.4 Hybrid Mechanical-Electromagnetic Balances

These are identical in construction (Fig. 6.4) to the single-pan mechanical model but the beam of light does not show large angular displacements when weighing ('null-point' detector system).

The equilibrium of the beam is always re-established to the predetermined reference position by an electromagnetic force controlled by a servo-motor (magnet + coil + mirror galvanometer).

The slave system can be obtained by means of a direct current passing through the coil of the servo-motor, or by an alternating current that is easier to measure with digital displays. The unknown weight is then compared with an electromagnetic force (itself calibrated by comparison with standard weights).

Top-loading balances (Fig. 6.5) are based on the same principle. They have replaced the earlier assay balances. Rapidity of weighing, good precision and high range have made them very important laboratory instruments for routine weighings.

---

[1] See Appendix 6 for addresses.

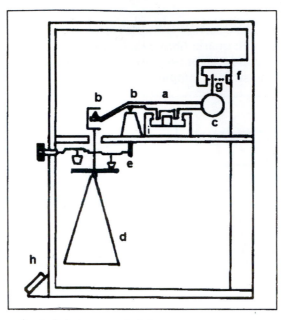

Fig. 6.4   Hybrid mechanical-electromagnetic balances.
a - beam; b - knife-edges; c - counterweight; d - pan; e - internal weight; f - 'null-point' detector, optical lamp system; g - micrometer scale; h - projected scale; i - electromagnetic servo-motor.

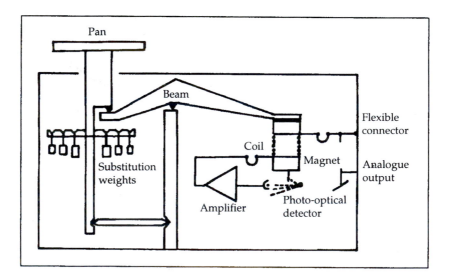

Fig. 6.5 Top-loading type balance.

2.5 Beamless Electronic Balances

The pan is placed directly on the moving magnet controlled by the coil (Fig. 6.6). Display of the coil current in the form of the corresponding weight is effected through capacitor-status detectors. This system has advanced spectacularly thanks to the introduction of the microprocessor and incorporation of weighing systems in the computer network handling laboratory equipment.

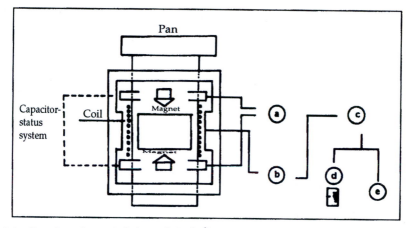

Fig. 6.6   Beamless electronic balance (Mettler[1])
a - system of capacitor-status detectors; b - amplifier; c - automatic taring; d - display; e - digital output.

These instruments are rather costly but the speed of weighing in series, low errors, ambiguity-free digital display, the possibility of attaching one-way or two-way series or parallel interfaces for recording on a microcomputer device (printer, recorder, calculator) make them incontrovertibly necessary for improvement in quality of data. Good Laboratory Practices (GLP) originated from this kind of instrument. These balances show little sensitivity to the environment. Mechanical vibrations have no effect but electrical variations necessitate efficient electronic regulation and filtration. The temperature should be corrected by thermostats; should overloading occur, the display shuts down. Telecommunication of weights is easy.

The ultra-micromodels allow measurements to 10^{-8} g. The tare can be 100% of the scale. These instruments should be protected against dust, which can enter betweeen the poles of the permanent magnet, slowly changing the precision and calibration.

Ferromagnetic materials can interfere with weighing by superposing their own magnetic field on the servo-motor. The zero reading of the balance gets displaced. Materials charged with static electricity can also cause the same type of interference. These phenomena can be moderated by conducting weighings below the measuring apparatus (types of balances open from

---
[1] See Appendix 6 for addresses.

below and provided with a hook for hanging). Radio-transmitters and equipment emitting electromagnetic radiation should be kept far away from electronic balances.

Development of these instruments is a continual process; for example, microprocessors have enabled improvement of the ratio between range and precision by maintaining the quality of data. The same balance can be used with the same precision for different ranges.

The balances have varied ranges and readabilities:

—semi-industrial balances for ranges 2 to 60 kg with readability of 1000 mg;

—laboratory balances with ranges 200 to 1000 g and 0.1 to 100 mg readability;

—medium-range balances, 10 to 200 g with 0.004 to 1 mg readability;

—lastly, microbalances and ultra-microbalances with ranges from 1 to 30 g and readability of 0.1 to 1 μg. Portable, battery-powered microbalances enable measurements in the field.

In a laboratory, the choice of balance should be based on usage, range, accuracy, precision, zero stability, speed of response, pan size, etc. Models with different specifications should be studied: flameproof, airtight, portable, with IR apparatus for drying, top-loading or not (some determinations require hanging of the material to be weighed, which is incompatible with top-loading balances), automatic with one- or two-way interface, GLP etc.

The balances can be connected to a furnace for DTA, TGA, EGA[1] controlled pyrolysis and so on.

3. Installation of Balances

Precision electronic balances can now be installed directly at the workplace if the environment is not too aggressive. Such is the case of flexible robotic installations wherein the balance is integrated with the system. Top-pan balances used every day are located near the place of their use. Airtight balances may be situated in an environment where dustproofing is not total, such as preparation rooms for soil and plant samples. However, mechanical balances with knife-edges and balances for high-precision weighing should be installed in constant-temperature balance rooms on heavy vibration-proof tables, with separate elbow and arm rests. When protected from corrosion, vibration and dust, changes in temperature and humidity, the instruments retain their quality for a long time.

Radiant and electromagnetic energy should be avoided and equipment that can transmit vibrations to the structures should be placed as far as possible from the balance rooms.

---

[1] See Appendix 2 for meaning of abreviations.

4. Fundamental Characteristics (Vocabulary)

Accuracy: ability to give values identical to the true value.

Capacity: maximum weight that can be placed on the pan without affecting weighing.

Detection limit: smallest load giving a significant reading. Electronic tare: maximum weight that can be subtracted electronically from the display.

Fidelity: reproducibility of the equilibrium position under a given load.

Mechanical tare: maximum weight that can be counterbalanced by substitution of weights.

Precision: indication of the dispersion of the measured value (standard deviation).

Resolution: number of counts (bits, points etc.) that can be determined on the display of the balance.

Sensitivity: the smallest mass causing the smallest displacement.

7

Separation by Filters and Membranes[1]

---

[1] Comprises certain microtechniques for purification and separation using phase chemistry. Does not cover techniques for separation by sedimentation and centrifugation.

1. Introduction

Separation techniques are very varied and a panoply of methods and procedures are available that can be adapted to the nature of the products to be separated.

Preparative separation carries out techniques in which the solid sample is separated on the basis of particle size, utilizing density and simple gravity or gravitational acceleration by centrifugation (mechanical analysis of soils, separation of clay, sands and silts, decantation etc.).

If a separatory medium (screen, filter, membrane etc.) is interposed, separation will be based solely on particle size (gravity, pressure or ultrasonic waves intervene only to accelerate passage of the phase to be separated).

Dialysis, based on the principle of diffusion across membranes, can be accelerated by the effect of an electrical field of appropriate intensity (electrodialysis, EUF[1]).

In the case of a liquid or gas, the constituents can be separated using the difference in speed of migration of a mobile phase in a static phase (liquid or gas chromatography, gel permeation chromatography, thin-layer chromatography etc.). If an electrical field is applied to a conducting liquid phase, separation of charged particles will occur (electrophoresis, isotachophoresis, capillary electrophoresis).

Lastly, certain phases can be physically freed or chemically transformed:
—ultrasonic waves for freeing elementary particles;
—mineralization for separating soluble minerals in a very aggressive medium;
—extraction with a solvent in a more or less aggressive medium (Soxhlet-Kumagava);
—distillation or entrainment in vapour for separating volatile products after chemical reaction.

Here we shall consider equipment for preparative separation using filters or membranes. This will be covered under two aspects:
—purification of a medium by eliminating particles that will vitiate the analysis;
—concentration of one phase to achieve better sensitivity and higher precision.

After passing through a filter, the filtrate is free of solid materials that could interfere with determinations:
—by chemical contamination (analysis by AA, ICP[1] etc.);
—by modification of the behaviour of a stationary phase (HPLC[1] etc.) through reduction or suppression of flow through capillaries;
—by abnormal propagation of radiation through diffusion (turbidimetry—the direction of radiation changes) or through slowing down (refractometry—same direction without phase change) or through absorption.

---

[1] See Appendix 2 for meaning of abbreviations.

For example, precision of the spectrocolorimetric method is greatly influenced by the phenomena of biochemical contamination, development in unfiltered medium of biologically active micro-organisms capable of modifying the medium or hindering conservation of the samples (natural waters, sugars etc.).

Quantitative recovery of solids, even in traces (atmospheric pollutants, suspended substances etc.), can be achieved on top of the filter. Recourse is had to filtering surfaces presenting a uniform pore density and through-flow to ensure a deposit as homogeneous as possible, with determinations done by weighing, counting particles on a grid sheet, observation by optical or electron microscope, microdiffraction, chemical analysis etc.

2. Preparative Filtration

2.1 Types of Separation

Sieving equipment (see Chapter 1.4) enable fractionation of particles about 50 μm in size through dry sieving. It is possible, however, to separate particles up to 5 μm in an ultrasonic field by using special sieves electroformed in nickel. In liquid medium, by using ultrasonic vibration, sieving filtration can be done on 1 μm Nylon sieves, or Teflon/polypropylene up to 70 μm, or polypropylene up to 5 μm. Stainless steel sieves are not recommended for chemical analysis of trace elements as they pose risk of contamination.

Filtration through filter papers or membranes enables separation of very fine particles that cannot be separated by sieving.

Many types of separation can be visualized:

—*Solid-gas separation:* this type of separation is done for determining particles in the air and the nature of aerosols, or for purifying air in clean rooms, laminar-flow hoods and glove boxes. The gas lines feeding analytical instruments should carry in-line filters with phase purification; where particles are separated for identification, screen filters, surface filters of in-depth filters soluble in certain solvents are used.

—*Solid-liquid separation:* this is applicable to all analyses of solid solutions in which the solid phase (clay, silts, sands etc.) or the physical filtrate (more or less selective extract from soils) is used for chemical or physical analysis. Fig. 7.1 shows the many parameters that control solid-liquid filtration systems.

—*Liquid separation:* this type of separation is used for soil compounds of various solubilities in the aqueous phase and the solvent-extractable phase. Hydrophobic filters (see Sec. 2.6 below) obviate the need for extraction glassware and reduce the volume of solvent (pesticide extraction etc.).

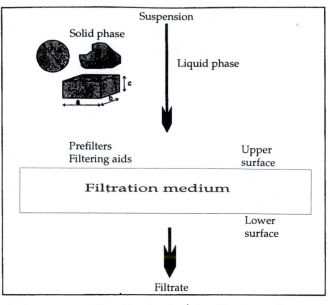

Fig. 7.1   Principal criteria controlling filtration through a filtration medium
—suspension: stability, solid-liquid interactions;
—solid phase: size and phase of particles, habit, rigidity, concentration, distribution, chemical properties, surface tension;
—liquid phase: density, viscosity, temperature, chemical properties, surface tension, dielectric constant;
—filtration aids, prefilters: porosity and depth of deposition of solid phase on top surface of filter paper;
—filtration medium: thickness, diameter of pores, retention, purity, ash content, mechanical, physical, biological and thermal resistance, surface tension;
—filtrate: quality of separation.

—*Liquid-gas or gas separation:* this type of separation is used for eliminating certain volatile compounds (gas or aerosol purification—desiccation). These separations use molecular sieves with a framework structure (3 Å, 4 Å, 5 Å,10 Å, type 13X—Fig. 7.2).

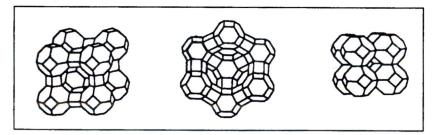

Fig. 7.2   Examples of molecular-sieve structures. From left to right: zeolite A, zeolite X and Y (faujasite), sodalite.

Filtration is accomplished on a separating medium with mesh designed for this purpose: mat of fibrous materials, polymer membrane, fritted-glass disc, etc.

If it is desired to filter large volumes of reagents or deionized water, in-line filters with filtration tubes (Fig. 7.3a) may be used or tangential filtration systems (Fig. 7.3b) if the liquids are heavily loaded (certain natural waters for example); the liquid to be treated passes with great speed parallel to the membrane (and not perpendicular to it as in most filters); deposition of sediment and clogging are reduced, thereby enabling maintenance of a high through-put.

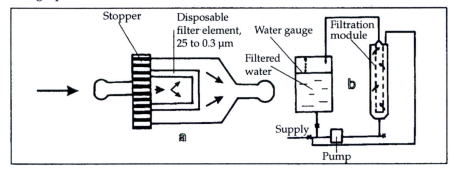

Fig. 7.3   In-line filtration systems. a - with filtration tube: gas— filtration-purification of gases for analysis (AA, ICP), combustion and IR analysis etc. liquid—large volumes; deionized water, reagents, b - tangential (loaded liquids, natural waters).

In chemical analysis, the composition of the filter can be limiting because oxidizing and strongly corrosive products can destroy it or because its purity is insufficient and leads to unacceptable contamination of the sample. The most common filtrations for soils are by gravity through fluted analytical filter paper with a large surface area or through flat filter paper folded in four if the residue is to be recovered (see Fig. 7.5). The filtration process can be speeded up by using a vacuum device on a flat filter paper placed on the perforated base of a Büchner funnel (see Fig. 7.6). Filtration occurs perpendicular to the filtering surface.

Ultrasensitive instrumental microanalysis requires that reagents and media be free of colloids that can contaminate the columns or hinder analytical procedures. As the quantities required are very small, syringe systems for filtration have appeared that permit filtration of liquids in volumes as small as 0.01 μL; they can be supplemented with chemical treatments in small purification columns particularly well suited to automated systems.

2.2 Selection of Filter Paper or Membrane

2.2.1 Aim of separation

—*Analysis of particles*: size limit, gravimetry, SEM/TEM[1] study, particle count, liquid scintillation (^{14}C, ^3H).

    —*Clarification*: elimination of particles that can interfere with instrumental determinations by absorption (UV, visible and IR spectrophotometry etc.) or by emission (liquid scintillation, fluorescence etc.);

    —*Sterilization and elimination of micro-organisms*

2.2.2 Liquid characteristics

The liquids should be compatible with the chemical resistance of the filter papers (acid, alkali, solvent etc.). Surface tension, temperature and viscosity modify the speed of filtration (increase of temperature reduces viscosity and improves the speed of filtration if the stability of the phases permits). The physical state of the precipitate (crystalline or amorphous, coarse or fine) affects the speed of filtration by clogging the pores and by creating a zone of variable permeability above the filter paper. The volume to be filtered and the presumed content of particles should also be considered.

2.2.3 Filtration media characteristics

A few parameters are to be taken into account (Fig. 7.4a, 4b):

    —equivalent pore size and efficiency of particle retention;

    —through-put, speed of filtration, loading capacity;

    —mechanical, chemical and thermal resistance, hydrophilic or hydrophobic surface;

    —permissible filtering pressure (normal pressure under partial Büchner vacuum or under pressure, especially when degassing of the liquid is apprehended);

    —chemical purity and ash content;

    —use of a prefilter or filtering aid;

    —behaviour when calcined at 1000°C.

2.2.4 GLP (Good Laboratory Practice) requirements

—For natural waters, the standard fixed for filtration is from 0.45 μm to 0.2 μm Van der Waals forces are very active at low limits).

    —For particles suspended in air, the standard is the 0.80 μm membrane filter (because of the electrostatic attraction between membrane and particles, some particles to 0.2 μm are retained by the membrane).

    Loose-textured rapid filter papers are used for precipitates too coarse to pass through them.

    Medium-porosity filter papers enable separation of bulky precipitates such as iron hydroxide or aluminium hydroxide or well-crystallized products (such as calcium oxalate).

---

[1] See Appendix 2 for abbreviations.

Particles				Separation medium					
				Fritted glass	Cellulose filter papers	Borosilicate glass	Membranes (µm)		
1	2	3	4				Hydrophilic	Hydrophobic	GLP standards

Clarification

1 (µm)	2	3	4	Fritted glass	Cellulose	Borosilicate	Hydrophilic	Hydrophobic	GLP standards
100	cosi / fs	Optical microscope	Pollen	1					
60				2					
40				3	Rapid				
20				4	Medium				
10							10		
6							8		
							5		

Microfiltration

4	fsi		Bacteria		Slow	2,7	3		
2				5					
1		TEM-SEM					1,2		
0,6	coc					0,7	0,8		Air–0,80
0,4							0,45	0,45	0,45
0,2							0,22	0,22	Water–0,22
0,1	fc						0,10		
0,06									

Ultrafiltration

0,04			Viruses				0,05		
0,02							0,025		
0,01							0,01 (PC)		
0,006									
0,004									
0,002									
0,001									

R.O.

Membrane materials	–PPFE	–Cellulose acetate/nitrate
	–Polyvinylidene fluoride	–Polysulphone
	–Polycarbonate	–Nylon
		–Silver

Fig. 7.4a Different types of filters and their use in relation to size of particles in soils (R.O. = reverse osmosis).
1 - Particle size, µm, 2 - Classification of soil mineral particles: coc - coarse clay; cosi - coarse silt; fc - fine clay; fs - fine sand; fsi - fine silt, 3 - Observation instrument: Optical microscope, TEM-SEM[1], 4 - Distribution of organic particles

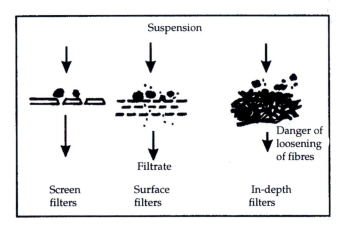

Suspension

Filtrate

Danger of loosening of fibres

Screen filters   Surface filters   In-depth filters

Fig. 7.4b Modes of particle recovery.

---

[1] See Appendix 2 for meaning of abbreviations.

Slow filter with dense structure can retain very fine precipitates such as barium sulphate (around 2 μm). In this size range, according to the quantities to be collected, one can use filters of borosilicate-glass microfibres of pore size 2.7 to 0.7 μm (or chemically less complex quartz fibres). Their speed of filtration is higher than that of α-cellulose filter papers and they can resist temperatures up to 500 to 550°C without significant weight loss.

Lastly, if the quantity of substances to be séparated is small and a very low cut-off threshold is required, one can use suitable membranes that will not release fibre and will allow retention of particles to 0.01 μm. As the membranes clog quickly, small prefiltered volumes are generally used.

2.3 Alpha-cellulose Filter Papers

2.3.1 Types and composition

These filter papers are fabricated from cotton wool with a high proportion of α-cellulose. Chemical treatments in 17.5% caustic soda have enabled taking the α-cellulose content of 95% by dissolving the lignin and part of the impurities. The cellulose paste is also subjected to acid washing, leading to a purity of 99.99% for analytical filter papers (ash content = 0.01%). A supplementary treatment with solvent is used to give lipid-free 'ashless' filter papers. The residual impurities are generally composed of complex silicates incorporated in the fibre during plant growth or minerals associated with the carboxyl groups present.

'Qualitative' filter papers, used only for certain non-critical analyses such as separation of soil particles for size analysis, do not have sufficient purity (98%) for analytical use. They may contain nitrogen and phosphorus in significant amounts. α-cellulose filter papers are characterized by two principal criteria:

—size of particles that can be retained on the filter paper;

—speed of filtration.

Filters that retain the maximum particles with the highest filtration speed are looked for. Retention of particles decreases with increase in porosity and speed of filtration. Conversely, the density and thickness of the paper reduce the porosity and speed of filtration and improve retention of particles; they also improve resistance and cohesion when wet (*see* Fig. 7.1).

The quality of a paper will thus be the result of fibre-compression treatments to increase density of structure and the thickness of lamination that leads to different porosities. These filter papers, composed of fibrous particles distributed in random fashion, are said to have 'in-depth action' (Fig. 7.4b). They are supplied in circles or sheets of varied qualities:

—'rapid'; filter papers retain particles with equivalent diameter of the order of 20 to 30 μm (Fig. 7.4a);

—'medium' filter papers—8 to 20 μm;

—'slow' filter papers enable separation of particles of the order of 2.5 μm.

A wide range of papers is available (Durieux, Fiorini, Macherey-Nagel, Schleicher and Schuell, Whatman[1] etc.). Some filter papers are subjected to treatment with concentrated HNO_3 to improve their toughness when wet and to reduce their ash content to less than 0.008% (hardened ashless filter papers). They are used for corrosive substances and when the deposited material is to be recovered for particle-size analysis by ignition or removal of water.

2.3.2 *Filtration through cellulose filter paper by simple gravity*

Fold the filter paper in two, then to 90° quadrants. Open the cone out and place in the funnel (Fig. 7.5). Moisten the filter paper with a jet of water from a wash-bottle and stick it into the funnel guarding against entry of air at the folds. Fill the funnel with distilled water and prime the filtration, driving out any air. The long stem should remain filled with liquid. The filter paper is now ready to receive the sample. Filtration can be speeded up with a Joulie funnel. The liquid should be poured non-stop into the filter paper. To avoid loss of liquid, the sample is poured against the stirring rod that touches the paper to at most 1 cm below the top edge of the filter paper (which should be 1 cm below the top of the funnel). In the final rinsing, the top edge of the filter paper should be thoroughly washed.

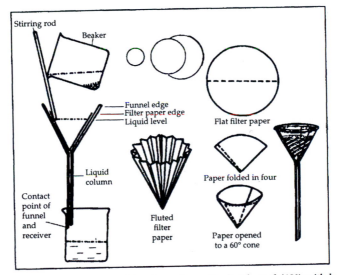

Fig. 7.5 α-cellulose filter papers, filtration using a 'Joulie' funnel (60°) with long stem.

2.4 Filter of Glass Fibre and Other Fibres

These have a high retentive capacity and can be used as final filters in applications which do not require a retention of 100% contaminations above

---

[1] See Appendix 6 for addresses.

a specific pore size. They resist temperatures up to 500°C at the most when made of 100% borosilicate-glass[1] without binder.

They are very stable in acid medium (except HF), but cannot be used with strong alkalis. They are preferably used for filtering gases (filtration of air pollutants and atmospheric dust) and liquids (natural waters, polluted waters, deionized water, reagents etc.).

These filters are supplied and used flat in Büchner (large-diameter papers) funnels supplied and used flat (large diameter) in Büchner funnels and semi-microcolumn filtering apparatus (Fig. 7.6). Demountable polypropylene Büchner funnels are easily cleaned.

Biologically inert and chemically resistant, glass-fibre filters perform better than those of α-cellulose and resist clogging better.

Some filters of borosilicate-glass fibres contain an organic acrylic binder that ensures better cohesion between the fibres, thereby precluding any separation later. Also, they have high retention capacity (less dense fibre matrix). In-depth filtration media are also available in quartz fibre, thermally stable up to about 550°C and chemically very pure ($SiO_2 > 99.0\%$), in siliconboride fibre, polyvinyl chloride (PVC), chemically very resistant

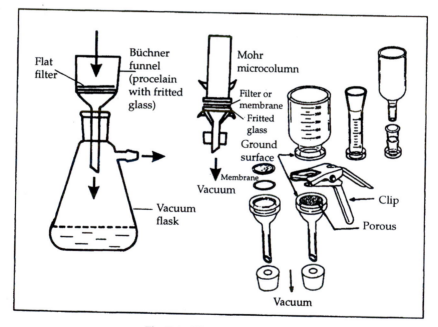

Fig. 7.6 Filtration under vacuum.

---

[1] Chemical composition of borosilicate-glass fibres (diameter 0.5-1.5 mm):

SiO_2 57.9%	Na_2O 10.1%	BaO 5.0%	B_2O_3 10.7%
K_2O 2.9%	ZnO 3.9%	Fe_2O_3 5.9%	CaO 2.6%
F 0.6%	MgO 0.4%		

polypropylene, hydrophobic polytetrafluoroethylene (PTFE) and polyvinyl difluoride (PVDF).

In-depth filters with various additives (activated charcoal, chemical sterilizants etc.) are used in-line in air-filtering systems for clean rooms and laminar-flow hoods.

Felts of soluble organic fibres with tetrachloronaphthalene base are used as in-depth filters to collect atmospheric dust, which is then recovered by dissolving the filter in a solvent.

2.5 Fritted Glass, Fritted Quartz or Porous Ceramic Filtration Media

Fritted glass (Fig. 7.7, Table 7.1) is chiefly used nowadays as the support for membranes or for in-line filtration of air or highly corrosive liquids such as sulphuric acid-chromic acid mixture, concentrated nitric acid etc.

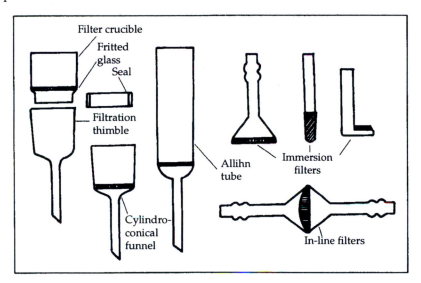

Fig. 7.7 Various apparatus of fritted glass.

Table 7.1 Porosity of fritted glass filters (dried at 150°C, ignited at 400°C)

Porosity	Pore diameter, μm
00	250-500
0	150-250
1	90-150
2	40-90
3	16-40
4	10-16
5	1-1.6

Types of crucibles and Allihn tubes that are very difficult to clean have been replaced by porous-ceramic crucibles with coarser particle size and greater thermal stability (Poretics[1]). These crucibles have a pore size of 12 to 0.7 μm and capacity from 20 to 60 mL.

Porous ceramics are also available as filter discs 10 to 50 mm in diameter and pore size 12 to 0.1 μm. They are particularly suitable for retaining expanding clay minerals for XRD[2] analyses and allow sequential chemical and thermal treatment of these samples. They are available in thicknesses of 3 and 6 μm.

2.6 Hydrophobic Filtration Surfaces

'Hydrophobic' cellulose or cellulose-nitrate filter papers are coated with water-repellent silicone compounds. They allow separation of water-solvent phases without using hard-to-clean separatory funnels and large quantities of solvent. The separation of water and solvent is complete and does not require special ability as required when working with separatory funnels. The mixture without preliminary separation of the phases is poured on the filter paper folded into quadrants in the funnel (Fig. 7.8). At the end of the

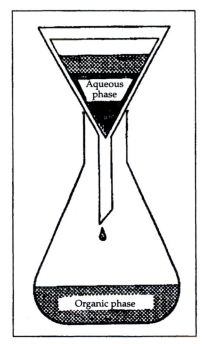

Fig. 7.8 Filtration through a hydrophobic medium.

---

[1] *See* Appendix 6 for addresses.

[2] *See* Appendix 2 for meaning of abbreviations.

separation the aqueous phase can be recovered without loss on the filter along with the residue. The separation is easily accomplished with emulsions as well.

The organic phase can be lighter (ether, petroleum ether, hexane etc.) or heavier (chloroform, carbon tetrachloride) than water. The particles present are retained on the filter along with the aqueous phase. These filter papers are resistant to mineral acids up to a strength of about 4 mol (H^+) L^{-1} and to strong alkalis up to about 0.5 mol (OH^-) L^{-1}. The presence of polar compounds and surfactants (Tween-polyoxyethylene or Span etc.) can lead to difficulties in separation. The presence of tin used for stabilization of the silicone may be a handicap in trace spectrometric analysis. A prewashing of the paper with the solvent used is done to limit contamination.

2.7 Complex-matrix Filters and Phase Purification

Filtration can be accompanied simultaneously by chemical separation or purification.

—*Ion-exchange filter paper*

Some manufacturers incorporate strongly acid (H^+-form) or strongly basic (OH^--form) ion exchange resins in their 'in-depth' filter paper. These papers can be used in the range of about pH 2 to 10, according to the nature of the supports and resins. Thus the cations or anions that could interfere with later analysis can be eliminated during filtration.

—*Filter paper with activated charcoal* (Macherey-Nagel)

These papers are suitable for filtration of samples containing organic materials in suspension. They are used in soil analysis for purifying extracts for determination of phosphorus and for decolorization of turbid liquids before spectrophotometric or polarimetric analysis.

Papers with Kieselguhr (diatomaceous earth) have high separatory ability but often dubious chemical purity.

2.8 Miscellaneous Filter Papers

—'*Chardin*' *papers*, thick and tough, and very spongy, can be used for filtration of large volumes of non-caustic liquids if their purity is not critical. 'Joseph' paper is used as fluff-free drying paper or for protecting work surfaces from dust during analysis and it should not be used as filter paper.

—*Filter papers with polyethylene support* of the Benchkote type are used for protecting work surfaces from liquids with biochemical or radioactive hazards. In case of accidental spills, the filter papers can be disposed of through selective safety systems provided for each technique. These filter papers also enable limiting the breakage of glassware by protecting large, heavy Pyrex receptacles from hard, rough surfaces. They are used in some continuous filtration systems, paper chromatography etc.

—*Filter papers for chromatography or electrophoresis* are specially manufactured to avoid the direct effect of the laminating process on fibre orientation.

—*Extraction thimbles* for Soxhlet-Kumagawa extraction, made of cellulose shaped into a tube closed at one end, are available in many qualities: ashless, decreased etc.

2.9 Membrane Ultrafiltration

2.9.1 Principle

The functioning of a membrane is very different from that of 'in-depth' filter papers. The test liquid moves in contact with the membrane whose pore size may be as small as 0.1 μm. The solutes pass through the membrane under the effect of pressure on the top surface or a vacuum on the bottom surface (Fig. 7.9), while colloids and occasionally macromolecules are retained on it. The particles thus retained accumulate on this surface and form an impermeable layer with a gel-like consistency: the formation of this layer is termed 'polarization of concentration'. It imposes a limit on filtration which can be pushed higher by increasing the pressure if the mechanical strength of the membrane is sufficient (limiting flow) except for certain applications (for example, some proteins may be degraded by excessive pressure).

This phenomenon of polarization of the surface could be caused by the osmotic pressure of solutions of macromolecules, which may increase exponentially with rise in concentration. The osmotic pressure opposes the pressure applied to the system and filtration comes to a stop when the pressures balance one another. In the mathematical models that have enabled

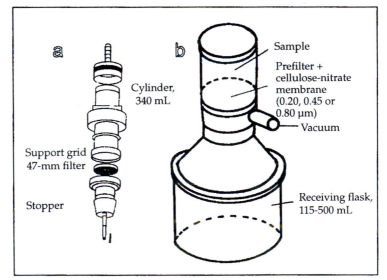

Fig. 7.9   Two commercial membrane filtration apparatus.
a - Millipore pressure-filtration system; b - Nalgene (Sartorius) vacuum-filtration apparatus for one-time use.

advances in the explanation of these phenomena, the layer that limits transport is combined with a static layer δ cm thick in which the concentration increases until 'gelling' of the compounds appears. In practice, it is possible to push this concentration threshold up by stirring the liquid or giving it an appropriate speed of circulation or by limiting the time of contact of the molecules with the membranes. The diffusion coefficient may thus be maintained at a normal value for a much longer period (increase in concentration leads to increase in viscosity). This has prompted manufacturers to focus their attention on tangential filtration apparatus, which can ensure rapid ultrafiltration of reagents or deionized water for example, by creating favourable hydrodynamic conditions (turbulent flow) near the membrane through preclusion of static laminar flow.

These systems can be likened to dialysis apparatus (for separation of macromolecules) especially when membranes of hollow fibres are used under pressure, but the principle differs as there is no diffusion involved. In general, the volumes filtered are kept small to avoid reaching the concentration threshold and, with highly loaded solutions, use of an in-depth prefilter could allow maintenance of 'instantaneous' filtration speed for a volume sufficient for instrumental analyses (HPLC[1] etc.).

2.9.2 Membrane characteristics

Filtration membranes are generally manufactured in the form of thin discs or hollow fibers. Most have a matrix with irregular, isotropic, continuous, spongy, inflexible structure with interconnected pores or showing an affinity for certain compounds (affinity membranes).

The structure may be regular in isoporous 'screen' membranes of polycarbonate. Some manufacturers supply 'layered' membranes composed of a filtering layer associated with a support of great mechanical strength.

The membranes trap particles on their surface like a sieve, unlike 'in-depth' filters. Several types are distinguished.

2.9.3 Membranes with random structure

These are used under pressure from 0.8 to 10 bars.

2.9.3.1 HYDROPHILIC MEMBRANES

—*Polysulphone membranes* are stable to 120°C, are hydrophilic and do not contain wetting agents. Their protein adsorption capacity is low. Their pore size can be as low as 0.1 μm. With good mechanical strength and chemical resistance, they can be used for weak acids and alkalis.

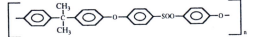

—*Acrylic-copolymer membranes* with Nylon support have high mechanical strength and chemical resistance. They can be sterilized by gamma rays or UV radiation. Hydrophilic, they have a pore size of 0.2 μm.

---

[1] *See* Appendix 2 for meaning of abbreviations.

—*Membranes of mixed cellulose esters* (cellulose acetate and nitrate), regenerated cellulose or cellulose triacetate are hydrophilic. Autoclave-sterilizable, they have a pore size of the order of < 0.45 μm and a void volume as high as 80%. They resist dilute acids and alkalis and certain solvents. A test is always necessary.
—*Nylon membranes*

$$[\text{—NH—CO—}(CH_2)_4\text{—CO—NH—}(CH_2)_6\text{—}]_n$$

Hydrophilic, they have chemical resistance suitable for solvents and chemicals; they can be used for filtering samples for HPLC analysis (Chapter 15). Concentrated acids and bases destroy these membranes as also certain halogenated hydrocarbons. They are not autoclavable.

—*Fluoropolymer (PVDF) membranes* are hydrophilic, resistant to solvents used in HPLC and autoclavable. *Vinyl-acrylic-copolymer membranes* are used when IR measurements are needed as this material allows high transmission. *Polyvinyl chloride* has low ash content. These membranes are often used for analysis of air particles and for scintillation counting.

2.9.3.2 HYDROPHOBIC MEMBRANES

—*PTFE membranes* (pure polytetrafluoroethylene or reinforced with a polypropylene weft)

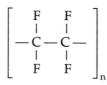

are resistant to all reagents used in analysis (acids including HF, alcohols, amines, alkalis, esters, ethers, ketones etc.) except concentrated HNO_3, tetrahydrofuran, certain halogenated hydrocarbons (such as carbon tetrachloride, chloroform and methylene chloride) and some hydrocarbons (benzene, toluene, xylene). They resist temperatures to 200°C and sometimes higher. Pore sizes to 0.2 μm are available. These membranes are used in HPLC, GC, PIXE, NAA, XRE[1] etc. analysis.
—*Hydrophobic polypropylene membranes*

$$[\text{—CH}(CH_3)\text{—}CH_2\text{—}]_n$$

are also used for concentrated acids and alkalis and oxidizing agents because of their high chemical and thermal stability. They are not recommended for nitric and sulphuric acids.

2.9.3.3 MISCELLANEOUS MEMBRANES

—*Binding-matrix membranes*, macromolecular, of nitrocellulose, are charged affinity membranes so modified that chemical structures with high density of positively or negatively charged bonds are obtained. The affinity for

---

[1] *See* Appendix 2 for meaning of abbreviations.

macromolecules (polyamines, proteins, nucleic acids etc.) is strong. Fixation is generally covalent:

$$-CHO \quad H_2N\diagdown \atop -CHO \quad + \quad H_2N\diagup protein\ NH_2 \longrightarrow \quad -CH=N\diagdown \atop -CH=N\diagup protein\ NH_2$$

Contrarily, for non-specific compounds with small molecules the binding properties are weak or non-existent, which is useful for separations and purification.

—*Ion-exchange membranes* with cage-type polysulphone matrix with cation (H^+ type) or anion (Cl^- type) exchange capacity are used for traces of cations or anions. Their mechanical strength is often improved by a polyester or Nylon support.

Membranes of Nylon 6.6 have amino and carboxyl groups that immobilize biomolecules.

Nylon 6.6 membranes with high density of quaternary ammonium groups are strongly cationic and the positive charge can be put to use between pH 3 and 10, whereby negatively charged compounds can be immobilized.

—*Metallic membranes* of silver, with pore size 5 to 0.2 μm, are used for retaining fine ash in the air (Poretics).

2.9.4 Polycarbonate 'screen' membranes with capillary pores

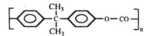

These membranes are isoporous, optically transparent and have a smooth surface. They have high mechanical and physical strength and good chemical resistance. Their chemical purity renders them suitable for trace-element analysis and other applications in optical and electron microscopy. They are biologically inert and non-hygroscopic, and their thermal behaviour allows use to 140°C. They are autoclavable but their resistance to organic solvents is low.

These membranes are made by subjecting a thin polycarbonate film of 10 μm thickness to a stream of charged particles in a high-energy nuclear reactor or cyclotron. The particles pass through the membrane leaving sensitized zones termed 'molecular-damage tracks'. The number of zones per cm^2 depends on the time of exposure in the reactor. Solvent attack will give uniform cylindrical tracks. By combining irradiation time of the film with the temperature and activity of the dissolving solution, pores of precise size between 14 and 0.01 μm can be produced.

This type of membrane with very low ash content and well-defined cylindrical pores is used for analysis using low-energy α-rays, β-rays, activation, X-ray fluorescence, HPLC, examination by SEM[1] etc.

—*Calibrated-porosity membranes* of the 'Anopore' type are obtained by anodic oxidation of aluminium. They have a regular 'beehive' structure and pore size that can be as low as 0.02 μm.

---

[1] *See* Appendix 2 for meaning of abbreviations.

2.9.5 Using membranes

Membranes are mounted flat on supports of fritted glass (Fig. 7.6), stainless steel, in-line microfiltration assemblies (Fig. 7.9) or single-use apparatus directly adaptable to syringes for example (Fig. 7.10). In principle, for small volumes of liquid, (1 mL to 50 mL syringes) membranes of 9 mm to 25 mm diameter are used with or without prefilter according to the load of the medium. For volumes of 300 mL a diameter of 47 to 90 mm is selected. Above two litres, diameters from 142 to 293 mm are available with special supports (expensive).

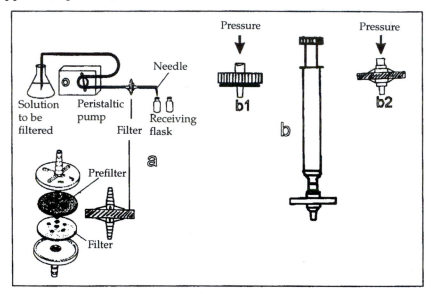

Fig. 7.10 Microapparatus for membrane filtration.
a - in-line microfiltration (one-time use), b - filters for syringes: b1 - on support b2 - single-use (Gelman, Millipore, Sartorius, Lida[1]).

When incinerating certain membranes such as those of cellulose nitrate one should guard against the thermal behaviour of these materials, which are often unstable above 200°C when dry and lead to loss of material, and against accidents by flaming, flash or explosion.

When filtering gases, the electrostatic resistance of the supports should be taken into account, as the filters could be charged and cause errors in weighing with electronic balances. Treatment with α-rays after deposition could regularize this stage of weighing.

Counting the particles retained on the membrane is facilitated by the colour of the support. White, black and green discs square-ruled at 3.1 mm are available.

---

[1] *See* Appendix 6 for addresses.

The optical properties of membranes are important for observing the particles with an optical microscope in direct or incident light. If the refractive index of the membrane is identical to that of the immersion reagent, the membrane becomes optically transparent. For example, cellulose-nitrate membranes (refractive index 1.50) are rendered transparent without change in pore structure by liquid having an appropriate refractive index such as sandalwood oil or immersion oil; cellulose-acetate filters with $n_D = 1.47$ are rendered transparent by paraffin oil, and those of regenerated cellulose or polyamide by immersion oil.

When the presence of pathogens necessitates sterilization, this is done in an autoclave at 120°C if thermal stability permits, otherwise by gamma rays or UV radiation. Examination by fluorescence enables differentiation of mineral substances and those of biological origin (biomolecules).

Substances labelled with ^{14}C or ^{3}H can be directly counted on the filters placed in contact with a liquid scintillation counter (aromatic hydrocarbons). For such use the support should be carefully tested

Some matrices, for which the cut-off threshold is very low, permit separation of colloids and even substances with high molecular weight, which may lead to inconvenience. But if this property is to be studied, some membranes can compete with the classic dialysis or gel methods, by enabling work on very small volumes and providing eight cut-off thresholds between 1000 and 100,000 daltons[1] . The technique of ultrafiltration has the advantages of not diluting the samples, enabling concentration of one of the phases and being much more rapid than dialysis.

2.10 Filtration Aids

These are less and less used in analysis on small volumes, for which less-contaminating prefilters are preferred. For large filtration volumes, however, they are of undisputed value.

—Very pure (99.99%) *α-cellulose pastes* may be added to media difficult to filter for holding coarse particles away from the filtering surface and restricting clogging. Sufficient speed of filtration and porosity of the filter are conserved. However, retention by the fibrous mass may be large and might require very large wash volumes (0.5 g dry powder, in tablet form, for 200 mL solution for example).

Glass or quartz wool, consisting of fibres 3 μm in diameter, are very useful for prefiltering gases or liquids. The risk of fibre loosening demands a final filtration through a filtration medium with well-defined cut-off diameter.

Polypropylene wool, highly resistant chemically, allows filtrates practically free of fibres to be obtained very rapidly.

---

[1] One dalton = unit of atomic mass = 1.66024×10^{-24} g.

Purified media such as *Celite, Celatom, Diatomite, Kieselguhr* and *Super-Cel* contain about 95% silica. They come from diatomaceous earth or infusorial earth treated with acid and separated according to particle size. *Perlites* are aluminium silicates; *Floridin* and *Fuller's earth* are very pure kaolins that retain their structure up to 130°C.

Phosphorus-free activated charcoals (the activation is often done by phosphoric acid vapour at high temperature) are used to purify and decolorize extracts containing organic substances. These are trapped in the pores of the charcoal. These charcoals are often loaded in columns for purifying deionized water. For example, the systems for production of ultrapure water use ultrafiltration columns and activated charcoal according to the constraints imposed by techniques such as HPLC or filtration of reagents.

Molecular sieves are often used to separate gas-gas systems in analysis (drying of gases). They are synthetic aluminates with a three-dimensional structure of interconnected cages (Fig. 7.2). The tunnels and pores with defined structure at Angstrom scale trap gaseous or liquid substances with molecules that can penetrate pores of size 3 Å, 4 Å, 5 Å and 10 Å and of 13-X type (Linde molecular sieves). Molecules of H_2O, CO_2, H_2S, mercaptans, NH_3 etc. can be retained in this manner. These synthetic zeolites can be regenerated by heating.

2.11 Systems for Extraction, Filtration and Purification of Liquids

Analysis of a large number of samples has prompted manufacturers to fabricate filtration ensembles adapted to the various uses and filtration media:

—controlled high-vacuum filtration systems are available for volumes as large as several litres and can be mounted in parallel (filtration ramps—multistation units);

—stainless-steel systems permit application of high pressures to increase the speed of filtration and preclude degassing of specific samples;

—syringe filter-holders enable filtration of small volumes under pressure by injection or under vacuum by aspiration.

Besides these filtration systems, the requirements of automation and fine analytical techniques (HPLC, chromatography, capillary electrophoresis etc.) have led to the development of a range of filtration and purification units for single-use to avoid all contamination. These units can be used under pressure or suction and even with centrifugation. They are almost universally used if one is able to adapt available phase-purification tubes and filtration systems.

Centrifugation is done directly with the tubes placed in rotors with swing-out shields or angular heads. The centrifugal force field is of the order of 3000 to 5000 G and enables sedimentation of colloids and monodispersed macromolecules.

2.11.1 Equipment for microfiltration by centrifugation

Prep-set equipment (Hamilton[1])

Samples are prepared by dissolution, ultrasonic dispersion, sedimentation, filtration, centrifugation, solid-liquid extraction or liquid-liquid extraction. The technique is manual. The equipment consists of polypropylene tubes (50 mL) with screw-cap, a sample carrier, tungsten-carbide or stainless-steel balls and filtration cylinders.

The sample is placed with a ball and the extractant in the polypropylene tube. After sealing hermetically, it is shaken by rotating the carriers (10 samples). After dispersion, the filtration tube is introduced into the grinding tube. The liquid passes through the filter, which is selected on the basis of its porosity (Fig. 7.11a).

Microfilter (Schleicher and Schüll[1])

This apparatus can be reused after changing the 9 mm diameter filter (Fig. 7.11b). It is used for ultrafiltration of samples for HPLC, TLC and GC (Chapters 15 to 17). Centrifugation at 3500 rpm can be substituted by a *Luer*-type syringe. Hydrophilic (cellulose esters, nitrate or acetate, regenerated cellulose etc.) or hydrophobic (PTFE) filters can be used. Pore sizes available: 12, 10, 5, 1, 0.45 and 0.20 μm.

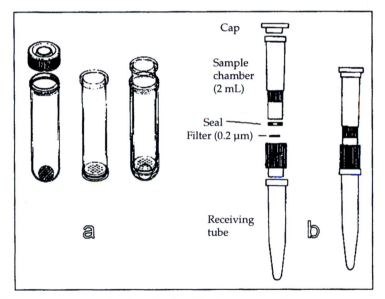

Fig. 7.11 Preparative apparatus for microfiltration by centrifugation.
a - Prep-set (Hamilton); b - Microfilter (Schleicher and Schüll)

---

[1] *See* Appendix 6 for addresses.

Centriprep system (Amicon[1]) for separation by centrifugation
This apparatus (Fig. 7.12) enables purification, concentration and desalination
with two cut-off thresholds of 10 kD and 30 kD. Centrifugation is done at
3000 G. Two or three successive centrifugations may be done. The final
volume of the concentrated phase is of the order of 0.5 mL. Other systems
working in the same manner are also available, such as the *Ultrafree MC* of
Millipore suitable for small volumes with cut-offs between 10 kD and 300 kD.

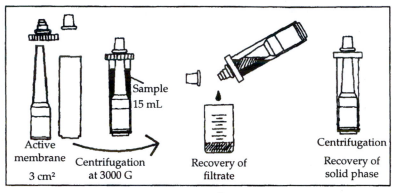

Fig. 7.12   Centriprep (Amicon[1]) preparation system.

Microsep system (Filtron[1]) with centrifugation
This apparatus (Fig. 7.13) consists of elements for eight cut-offs (molecular
weight 1 kD, 3 kD, 10 kD, 50 kD, 100 kD, 300 kD, 1000 kD). A centrifuge
with angle-head rotor (45° or 34°) is used at 5000 G.

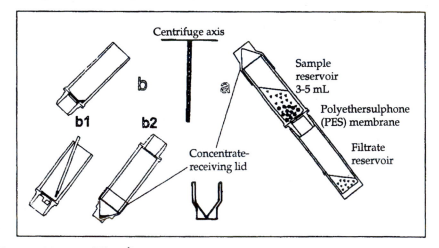

Fig. 7.13   Microsep (Filtron[1]) system.
a - filtration b - recovery of concentrate: b1 - with a micropipette b2 - in the receiving
lid

---

[1] *See* Appendix 6 for addresses.

Apparatus for separation by centrifugation with Millipore-Waters[1] polysulphone membrane with cut-offs from 10 kD to 30 kD
The sample (up to 20 mL) is subjected to horizontal centrifugation at 2500 G. The membrane is oriented parallel to the direction of the centrifugal force to prevent its clogging. Molecules smaller than the cut-off dimension of the membrane pass through and are collected in the filtrate tube. Macromolecules and/or particles are collected at the bottom of the membrane-support tube.

These can be used between pH 1 and 14, up to 50°C; their resistance to organic solvents is low.

Millipore[1] ultrafiltration unit for microsamples
A centrifugal force of 2000 G is applied to six 1.5 mL microtubes (Fig. 7.14a) carrying an inner tube with 0.22 μm or 0.45 μm Durapore membranes or polysulphone membranes with three cut-off values: 10 kD, 300 kD and 100 kD.

Hettich[1] single-use cell for study of solids
The suspended sample is placed in the sample cell (Fig. 7.14b) and centrifuged at 6000 rpm. Four 1 mL to 8 mL cells allow collection of the sediment in a thin uniform layer on a square-ruled membrane or thin object-slide. The excess liquid is drawn off with a pipette or recovered by decantation if the sediment is very sticky.

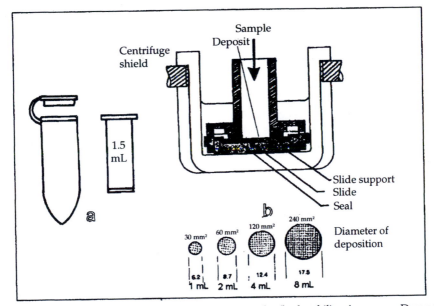

Fig. 7.14 a - Millipore ultrafiltration unit for microsamples (hydrophilic microporous Durapore or regenerated-cellulose membranes); b - Hettich cell for mineralogical preparation.

[1] *See* Appendix 6 for addresses.

2.11.2 Apparatus for filtration through a hollow-fibre pencil

The Mediakap system is used for in-line filtration of reagents down to 0.2 μm through a pencil of hollow, hydrophilic fibres made of cellulose esters. A hydrophobic 0.2 μm fibre pencil allows passage of bubbles of sterile air. The ensemble works under a pressure of 2 to 3 bars. The range of apparatus (Fig. 7.15a) allows filtration of volumes of 50-200 mL.

For small volumes the *Dynagard* system based on a hollow porous fibre mounted in a Luer-type syringe (Fig. 7.15b) is preferred. Two qualities of fibre are available:

—hydrophilic cellulose-ester fibre (for filtration of aqueous solutions);
—hydrophobic polypropylene fibre (for filtration of organic solvents).

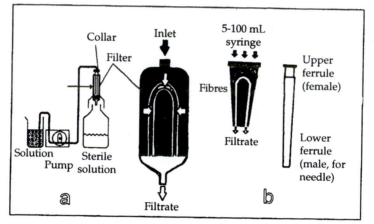

Fig. 7.15   Filtration systems with hollow-fibre pencil.
a - *Mediakap* (Merck); b - *Dynagard* syringe system.

Pore sizes of 0.20 and 0.45 μm as well as filter sizes of 0.8 cm^2 (for 5 mL volume) and 3.9 cm^2 (for volumes from 25 to 100 mL) are available for both fibre types. The system can work by aspiration or by injection. It is particularly well suited to determinations by HPLC etc.

2.11.3 Filtration using phase chemistry

Other systems designed for the simultaneous preparation of microsamples use filtration under pressure or suction and techniques of phase purification of extracts, mineralized solutions or distillates (syringes, vacuum and multisample apparatus).

Liquid-solid and liquid-liquid extractions are the techniques used in the determination of pesticide residues[1] in soils.

---

[1]   Waters-Millipore Sep-pak
β-alumina for pesticides, herbicides; α-alumina for herbicides; silica for pesticides; Florisil for pesticides, herbicides; Cyanopropyl for CN pesticides; C18 for organic traces in water; Accell Plus CN for cationic compounds; Accell Plus QMA for anionic compounds; Diol for trace elements in water.

Retention mechanisms are influenced by the combined action of the molecules to be extracted, the matrix of the sample, the extraction phase (solvent) and the stationary phase (retention-adsorption and selective elution).

The stationary selective phases are non-polar or semipolar supports or strongly polar ion exchangers. They are loaded in tubes permitting direct coupling without adapter.

After dissolution, the substance to be determined can be adsorbed on the stationary phase which enables separation from the matrix, or the matrix itself adsorbed on the stationary phase.

The Sep-pak[1] (Waters-Millipore), Extra-sep (Lida[1]), Adsorbex (Merck[1]) etc. cartridges are disposable minicolumns functioning on the principle of liquid chromatography. They contain fillings analogous to those used in HPLC (see Chapter 15) to retain the compounds of interest or, on the contrary, trap the interfering substances before injection of the purified sample. Extraction on the stationary solid phase is rapid and efficient and allows integration with automated systems (see Chapter 17) without difficulty, along with the possibility of limited manual handling for small series (Figs. 7.16 and 7.17).

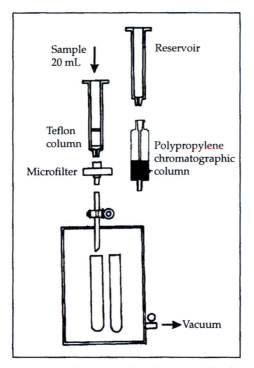

Fig. 7.16 Microfiltration and purification with columns.

---

[1] *See* Appendix 6 for addresses.

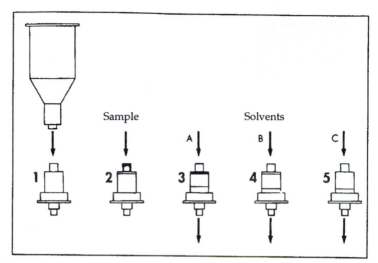

Fig. 7.17 Example of microseparation of compounds by elution (Waters - Millipore system). 1 - treatment with water or a buffer solution, 2 - solute adsorption, 3 - washing to eliminate interferences, 4 - elution of compound 1, 5 - elution of compound 2
These columns can serve as storage to avoid transportation of voluminous solutions, for example concentration of water for sending it to laboratories. The microcolumns after elution are used for determinations by HPLC, GC, MS, RIA, IR, UV, AA[1] etc.

3. Permeable Membranes: Dialysis, Electrodialysis, EUF, Osmosis

Ultrafiltration systems based on classic 'thick' filter papers or membranes depend on the simple force of gravity (or centrifugal forces) and on pressure gradients to force the particles through the separatory medium. Every suspended particle or macromolecule smaller than the cut-off value passes through the filter along with the suspending liquid.

In dialysis systems, the substances (colloids, micelles, molecules or ions) diffuse the semipermeable membrane. The diffusion coefficient varies with the size, nature and shape of the particles, their molecular mass and, of course, the pore size and thickness of the membrane. Thus one can, in theory, effect summary separations based on molecular weight. Nevertheless, it should be understood that the molecular mass is significant for a substance only if there is isomolecularity, which is not always the case in natural systems in which the macromolecules can be polydispersed. The molecular mass has no meaning in such cases except for differentiating molecular entities composed of mixtures of monomers and polymers of one or more substances and for evaluating their evolution under given conditions.

Dialysis enables detection of macromolecular compounds and colloidal substances, defined as complex disperse systems of submicroscopic dimensions (about 100 μm or less).

---

[1] *See* appendix 2 for meaning of abbreviations.

3.1 Dialysis

3.1.1 Principle

When in contact with the solution to be analysed (Fig. 7.18), the solvent, usually water, is enriched in the substance with the highest diffusion coefficient allowed by the pore size of the membrane. A substance of high molecular weight may be purified in this manner by eliminating small molecules or ions that will pass through the membrane diaphragm. Macromolecules and micro-organisms (proteins, bacteria etc.) remain trapped. The process comes to a halt when the concentrations of the solutions on either side of the membrane become equal. The method is not suitable for very dilute solutions. Very large particles with diameter larger than that of the pores of the membrane diaphragms separating the phases cannot be separated unless the pore size is enlarged. Dialysis enables concentration of a solution of macromolecules, desalinating a solution, determination of the components in turbid extracts, modification of a buffer by diffusion etc.

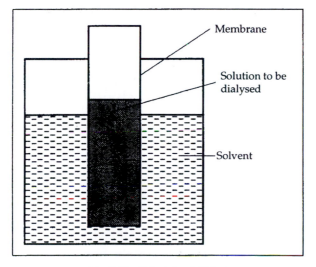

Fig. 7.18 Principle of dialysis.

The thinner the membrane and the greater the difference in concentration, the more rapid and more efficient the operation. Such a membrane cannot ensure a very precise cut-off value partly because of its heterogeneity—the more or less irregular pores permit passage of molecules of slightly different size—and partly because of the phenomenon of selectivity due to ionic charge (positive or negative membranes) or to solubility differences within the membrane. Electrolytes with low molecular weight, organic acids and bases, sugars and free amino acids in solution diffuse rapidly across the membrane. But substances giving scarcely diffusible colloidal solutions such as tannins, dextrins, polyholosides, proteins, enzymes etc., can be separated only by adjusting the pore size of the membranes.

Certain dissolved mineral compounds in the medium to be dialysed can be modified during the operation and precipitated giving colloidal solutions: iron and aluminium hydroxides, silica etc.

Since dialysis is slow, the components can also be modified under the influence of hydrolysis or biochemical reactions. To restrain these phenomena, buffers and protective antiseptic substances should be used.

To reduce dialysis time, the area of the membrane should be as large relative to the sample volume as possible, and the temperature as high as is compatible with the stability of the compounds since molecular motion increases with temperature.

3.1.2 Dialysis membranes

These membranes are very thin to allow relatively rapid diffusion and to present pores of such diameter as will enable cut-off in molecular mass.

Cellophane is a derivative of viscose, a by-product of treatment of cellulose with caustic soda. Its thickness is of the order of 0.02 mm and the pore diameter around 25 nm. It has low mechanical strength and chemical and thermal resistance despite additives that enhance these properties.

Hollow fibres of regenerated cellulose are preserved moist in aqueous medium in the presence of glycerol ($CH_2OH—CHOH—CH_2OH$) or polyethylene glycol to retain their flexibility and are sometimes filled with isopropyl myristate [$CH_3(CH_2)_{12}CO_2—CH(CH_3)_2$] to prevent deformation by flattening. To avoid biochemical degradation (attack by cellulolytic micro-organisms) an inhibitor can be added (sulphidic or sulphitic reducing compound, 0.05% sodium azide (NaN_3), 1% sodium benzoate ($C_6H_5CO_2Na$) or dilute formaldehyde). These membranes can also retain trace elements. Before being put to use they should be treated with EDTA and rinsed with water. Resistance of regenerated celluloses to organic solvents (hydrocarbons, halogenated hydrocarbons, alcohols, ketones, esters etc.) is satisfactory. Certain dilute (25%) organic acids can be used, but inorganic acids (HCl, HNO_3, H_2SO_4) and strong alkalis are prohibited.

Cellulose esters without traces of metals, free of sulphite and glycerol, provide a more limited choice of solvents and are attacked by inorganic acids and strong alkalis even when diluted to 1 mol (H^+/OH^-) L^{-1}. They are stored in 0.05% EDTA at 4°C and can be used up to 37°C. The molecular weight cut-off (MWCO) of these cellulosic membranes varies from 1000 to 50,000 D: 1, 2, 3, 5, 6-8, 10, 12-15, 25 and 50 kD. Cut-off values of 0.5 kD and 100 kD can be found, but rarely. The precision is of the same order as for separations through gels.

Membranes for ultrafiltration (see Sec. 2.9 above), mechanically much stronger but generally thicker, can be used if their pore size is suitable for the types of separation planned. Cellulose-derived membranes (regenerated cellulose, cellulose esters, acetate and nitrate) are commonly used, as also membranes of chemically very resistant polymers. Semipermeable fibres with a very large diffusion surface area are also used. They should be preserved wet in the presence of bactericides to avoid biochemical degradation.

Hollow porous-glass fibres are hardly used nowadays for soils; they are reserved for medical uses (artificial kidneys for example). The cut-off values are less variable than with other types of membranes. Flat diaphragms of porous glass allowing 'free diffusion' of liquid through pores 1-2 μm in size are also used in analytical equipment.

3.1.3 Dialysis Apparatus

The dialysis apparatus for analytical use should enable maintenance of the concentration of the diffusing substance in the solvent at a low level. This is achieved by constant renewal of the solvent through cycles of dialysis and counter-dialysis in the same direction (Technicon system, see Chapter 17), or by a countercurrent (Fig. 7.19), or by introducing the sample in a large volume of solvent in order to minimize the relative concentration at the diffusion surfaces, if need be with a multisample circulatory system (Fig. 7.20). Ion-exchange resins can also be added to the solvent for fixing the ions passing into the compartment.

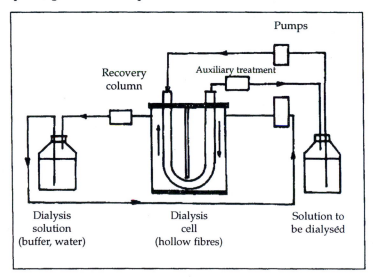

Fig. 7.19 Rapid countercurrent dialysis. Recycling of the liquid and control of the retained substance is effected by a peristaltic pump.

3.2 Electrodialysis

3.2.1 Principle

This is a rapid technique for separation by dialysis that enables extraction of ions from a liquid medium without modification of the phases. The ions of one solution are transported under the effect of a direct electric current into

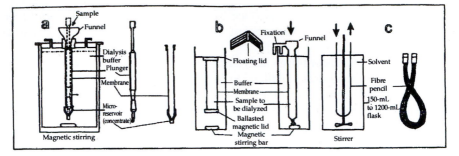

Fig. 7.20 Miscellaneous dialysis equipment.

    a: Micro-Prodicon system

        Vertical concentration with accumulation of concentrate by gravity. The thickness
        of the dialysis film is around 2 mm because of the Teflon plunger. The vacuum in
        the solvent compartment prevents any osmotic diffusion.

        Molecular weight cut-off: 10, 15 or 30 kD

        Capacity: 0.05 to 40 mL

        Maximum volume exchange: 30 to 120 mL

        Buffer volume: 500 to 2000 mL

    b: Membrane tube for dialysis (Spectra-por) made of regenerated cellulose or cellulose
       ester (cut-off 0.5, 1, 3.5, 8, 10, 15, 25 or 50 kD).

    c: Hollow-fibre separator (Bio-rad, Spectra-por etc.) for dialysis media
       (0.5 mL to 5 L). Hollow fibres of regenerated cellulose or cellulose acetate
       (A or C) or silicone polycarbonate.

       Diameter 150 to 200 μm. Thickness 10 to 40 μm. Cut-off 1, 6 or 9 kD.

another solution separated from the first by a membrane. The positive ions
move to the cathode and the negative ions to the anode.

The presence of the semipermeable membrane slows down diffusion.
This is related to the thickness of the membrane, tortuosity of the pores and
their interconnections, and chemical inertness of the support. This technique
is more rapid than simple dialysis and enables purification of a soil sample
and recovery of the extracted ions that pass through the membrane so that
they can be determined.

This fraction, under defined analytical conditions, represents the easily
extractable elements present in the soil solution and is related to the supply
of nutrients or less readily available elements that are more slowly released
from the exchangeable phase. Electrodialysis is also used for desalinating
natural waters and sea-water.

To avoid modification it is necessary to keep the temperature from rising
excessively, control fluctuations in voltage during dialysis (the composition
of the medium can vary with the relative diffusion rates of the ions, thus
setting up gradients of electrolytic concentration and modifying the pH),
continuously agitate the central compartment to minimize the effect of
colloids, constantly renew the water at the level of the membranes in the
two side compartments, correct changes in pH caused by transport of H^+
and OH^- ions and avoid flocculation of components.

3.2.2 Electrodialysis apparatus

The simplest apparatus are the static systems with one membrane (Fig. 7.21a). These apparatus are seldom used in analysis today.

Analytical apparatus are most often composed of electrodialysis modules with two membranes. They comprise three compartments separated from one another by two membranes, M_1 and M_2 (Fig. 7.21b). The central compartment contains the sample to be dialyzed, for example a soil suspension. It is provided with a stirrer to continuously homogenize the sample and a heating coil with thermostat to maintain constant temperature throughout the operation.

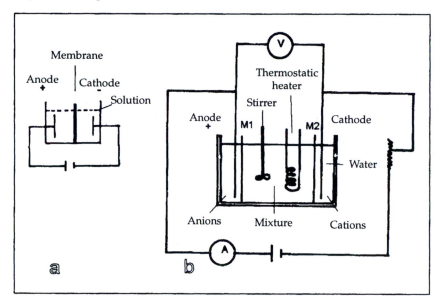

Fig. 7.21 Electrodialysers.
a - with one membrane; b - static electrodialyser with two vertical membranes (Pauli type).

The two side compartments contain the electrodes fixed close to the 'downstream' surfaces of the membranes. These electrodes are gratings made of platinum, stainless steel, carbon or titanium. The two compartments are continuously fed with deionized water, and the separated fractions (anions and cations) are collected below for later analysis.

A direct-current generator enables supply to the electrodes of defined voltage and current as a function of concentration, diffusion effects and convection transport. Tension and current are modulated by a rheostat and measured by a voltmeter (30 to 5000 V) and a microammeter (9 mA to 5 A).

When the suspension to be dialysed is highly conducting, the current is quite large, then decreases to reach a fairly constant value, indicating that equilibrium has been attained. This means that the content of ions in the central compartment has been gradually reduced to the benefit of the two side compartments according to a law of variation of concentration as a function of time:

$$C = C_0 - k\,(\eta)it$$

where C_0 is the initial concentration; i the current; k the rate constant; η the efficiency; and t the time.

These operations can be done at constant voltage (the current decreases until the extractable ions disappear). The phenomenon of retrodiffusion may occur at the end of the cycle if the side chambers are not kept under reduced pressure. Dialysis can also be done at constant current, the rate of extraction of ions from the central compartment being maintained constant. In this case the voltage should not exceed the maximum authorised by the generator. The maximum is rapidly attained by eliminating parasitic diffusion.

A large potential difference applied between the walls of the electrodialysis cell can lead to rise in temperature of the solution, necessitating temperature regulation of the sample chamber as a function of the resistivity of the solution. The phenomenon of polarization can occur near the membranes, ensuring the spatial regulation of the ions displaced by the electrical field.

Large ions can accumulate on the membrane and reach saturation (limiting concentration). Therefore agitation should be continuous and water in the side compartments constantly renewed. Membranes with selective permeability allow concentrating certain ions in the side chambers. A problem with this type of electrodialyzer, e.g. the EUF (see Sec. 3.2.4 below), is the change in pH in the vicinity of the electrodes, which sets certain troublesome precipitations in motion. This is why a more complex system was proposed for working with soil solutions (Doulbeau, 1991).

3.2.3 Electrodialysis membranes

The membranes used in electrodialysis are identical to those used for dialysis (see Sec. 3.1.2 above).

Neutral membranes enable separation of small ions from large molecules. They are made of cellulose acetate, cellulose nitrate, regenerated cellulose, cellophane, glass fibre etc. These membranes can be degraded by 'poisoning' consequent to reactions with molecules that can pass through the pores, or by clogging after variation in pH etc.

Membranes with selective permeability can be cationic or anionic, hydrophilic or hydrophobic (by grafting a film with charged groups). The counter-ions (ions of sign opposite to those of the site) can be freely displaced through the membrane while co-ions (ions with the same sign as the site) are electrostatically repelled. Their thickness is 100 μm to 500 μm and their strength is improved by adding a supporting grill.

Certain membranes are amphoteric or composed of two joined or mixed layers (piezodialysis).

3.2.4 Electro-ultrafiltration (EUF)

This technique, developed for soil analysis chiefly by Nemeth (1979, 1985), is directly derived from electrodialysis and the apparatus, with some slight difference, similar to a two-membrane electrodialyzer. Automatic regulation devices and the possibility of collecting separate fractions with two turntables able to carry many tubes (for anions and cations) enable easy execution of the extraction cycles recommended for this technique since 1980 (at the first International Symposium on EUF), that is:

—EUF I: 0 to 10 or 30 minutes at 200 V ($\leq$ 15 mA) at 20°C;

—EUF II: 30 to 35 minutes at 400 V ($\leq$ 150 mA) at 80°C.

These different temperatures and voltages are used to define the EUF fractions related to the concentration of nutrients in the soil solution.

a) The fraction obtained at 20°C under 200 V for 10 minutes can be correlated for example with the actual contents of N, P and K in the soil solution, but does not guarantee maintenance of these values during the plant growing season.

b) By keeping the sample for 30 minutes under these conditions, values close to those of the so-called 'available', i.e., extractable elements are obtained for a given plant species.

c) EUF at 80°C and 400 V for 30-35 minutes (2 extractions) enables definition of the less reactive elements termed nutrient reserve.

The combination of parameters (current, voltage and temperature) enables establishment of pertinent correlations [for example, EUF 20°C and (EUF 80°C)/(EUF 20°C)] and study of the significance of the fractions extracted (nutrition and plant growth, mobility in soils). For K, for example, the nutrient dynamics can be established with 50 V at 20°C, 200 V at 20°C and 200 V at 80°C, the last extraction giving results close to those obtained by Mehlich's (1980) method-2. The study can be complemented by a chemical method (boiling 1.0 M HNO_3) for long-term K-supplying power. Another example is the study of sodium in saline soils (Kolbe, 1989).

Thus, by applying potential differences at a given temperature, the gradual release of quantities of nutrients identical to those potentially allowed by the matrix in a period of time related to the plant growing cycle can be simulated. The amount of migration of ions in EUF is proportional to the electrical field and temperature and inversely proportional to the bonding energies of the ions in the soil.

3.3 Osmosis

This method puts to work the colligative properties of liquid media but is not used in analysis except for pre-purification by reverse osmosis of water for laboratories.

Osmosis is a phenomenon of transport between two liquids with different concentrations, separated by a semipermeable membrane. A flow of solvent is set up from compartment A to compartment B with higher concentration (Fig. 7.22). To prevent the flow, pressure can be applied on the more concentrated solution up to the equilibrium value, termed osmotic pressure (OP).

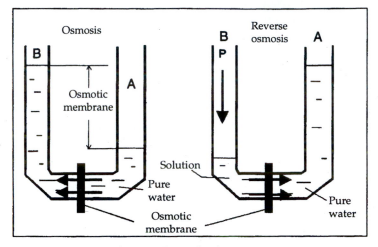

Fig. 7.22   Principle of osmosis.

In reverse osmosis, a pressure P greater than OP is applied. The solvent crosses the membrane and the process is reversed since the medium B, in which the unwanted particles are found, gets concentrated.

Tangential ultra-microfiltration systems require good understanding of the phenomena of osmotic pressure and the values of P required to optimize the process (area of membrane, pressure needed, mechanical strength of the membranes, polarization of concentration, clogging etc.).

References

Doulbeau S. 1991. Séparation par électrodialyse des éléments retenus par un sol. *Cah. ORSTOM Ser, Pédol.*, XXXVI (3): 213-225.

Mehlich A. 1980. In *Handbook of Reference Methods for Soil Testing*. The Council on Soil Testing and Plant Analysis, pp. 112-121.

Nemeth K. 1979. The availability of nutrients in the soil as determined by electroultrafiltration (EUF). *Adv. Agron.*, **31**: 155 - 187.

Nemeth K. 1985. Recent advances in EUF research. *Plant and Soil*, **83**: 1-19.

Kolbe D.E. 1989. *Essai de séparation des formes solubles et échangeables du sodium par E.U.F.* Mémoire de maitrise de biologie des organismes et populations, Orstom-Université (USTL) Montpellier.

Bibliography

American Water Works Association. 1994. *Evaluation of Ultrafiltration Membrane Pretreatment and Nanofiltration of Surface Waters.*
Cheryan M. 1986. *Ultrafiltration Handbook.* Technomic Publ. Co.
Cheryan M. 1998. *Ultrafiltration and Microfiltration Handbook.* Technomic Publ. Co.
McGregor W. C. (ed.). 1986. *Membrane Separations in Biotechnology.* Marcel Dekker.
Nemeth K. 1982. *Application of Electro-ultrafiltration in Agricultural Production.* Martinus Nijhoff.
Zeeman L. J. and Sydney A. L. 1996. *Microfiltration and Ultrafiltration.* Marcel Dekker.

PART II

Instrumentation and Quality Control

8

Introduction to Analytical Techniques

1. Introduction

1.1 Generalities

The soil is a complex dynamic medium, subject to various climatic influences and to chemical, physical and biological processes. To study the phenomena observed in nature, analysis has at its disposal a wide range of chemical, physicochemical and physical methods that can be used to characterize and evaluate fertility and to identify and quantify pollutants and, equally, to investigate fundamental theories of development.

In the soil, atoms combine to form simple substances or compounds and more or less complex molecular structures such as the clay minerals or humic substances. According to the law of thermodynamics, every system spontaneously evolves towards its most stable state, corresponding to its lowest enthalpy relative to the temperature, pressure, etc.

In the present century, the selectivity, sensitivity and precision of determinations have enabled finer and yet finer approaches, chemical methods being progressively replaced by reliable physical or physicochemical methods that are more sensitive, faster and more selective. These methods allow moving the scale of determination towards detection thresholds of the order of 10^{-15} g for the currently best-performing methods.

The dendrogram in Appendix 1 highlights the gap between chemical methods and physical methods, rapidly widening because of computers and development in analytical equipment.

1.2 Constitution of Matter: Brief Summary

Matter is composed of atoms dynamically arranged in space and in time. Mendeleïev organized the elements in the Periodic Table (Appendix 7), which explained the complexity of atomic systems and similarities in physicochemical properties. A central nucleus, surrounded by electron orbitals corresponding to several energy levels, characterizes each element (Fig. 8.1).

The atomic number corresponds to the number of protons (+) in the nucleus and the number of electrons (−) balancing the charge. The atomic mass takes into account the neutrons and the existence of isotopes. The state of the electrons is characterized by the quantum number n, which determines their distribution in successive shells with varying energy—K, L, M, N, O, P, Q—each shell being filled with $2n^2$ electrons, thus:

$n = 1$	2 electrons	(K shell)
$n = 2$	8 electrons	(L shell)
$n = 3$	18 electrons	(M shell), and so on.

The filling of shells with electrons takes place in the order of increasing energy of the shells. When one shell is filled (saturated), its electrons form a compact assembly that does not participate in chemical bonds. Only the

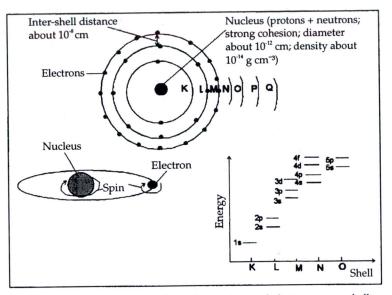

Fig. 8.1 Schematic summary of atomic structure and electron energy shells.

electrons of the outermost shell represent the chemically active level (valence shell).

Each shell is composed of *s, p, d* and *f* subshells (defined by the quantum number *l*), the orbitals of which have increasing energy and may, sometimes, overlap the next shell, starting from the *M* shell (Fig. 8.1). This modifies the order of energies, the *4s* orbital having an energy slightly lower than the *3d* orbital, implying therefore that *4s* would be filled before *3d*. These anomalies are put to use in analysis, the cohesion of electronic structure being related to the stability of the element and its behaviour under the influence of radiation. The electrons (and the nucleus) are also dependent on the quantum number *s*, which corresponds to the true angular moment of the electron, or its spin. The energy is determined by these three quantum numbers *n, l* and *s*, and the magnetic moment by the quantum number *m*. Thorough understanding of the structure of atoms enables us interpret emission spectra and to predict the chemical behaviour of a substance.

The methods listed in the dendrograms in Appendix 1 are based on the wave nature of propagation of light (refraction, diffusion, diffraction, interference, etc.) and on various radiations; precise information can be obtained on the structural chemistry of atoms and molecular structures from the energies. Chemical and physico chemical methods have a limited range of application as they only use the outermost shell of the atoms for identification and quantification.

2. General Principles of Instrumental Techniques

2. 1 Methods of Separation

Despite the development of microscopic techniques and direct analysis of uncoated thin sections or undisturbed samples, separation techniques are very widely used because of the complexity of the soil material. They are based on very varied physical or chemical principles: mechanical sieving, densimetry, filtration, various extractions, selective chemical attack, distillation, chromatography, electrophoresis, etc. Some techniques are more specifically used for preparing the sample prior the actual analysis: the most popular are sieving (*see* Chapter 4), sedimentation and centrifugation, and filters and membranes (*see* Chapter 7). Other techniques are concomitantly separatory and analytical: the most common are the various chromatographic techniques (*see* Chapters 13 and 15) and the C, H, N, O and S analyses (*see* Chapter 16).

2.2 Chemical Methods

At the level of atomic structure, these methods put to use reactions that essentially pertain to the outermost electron shell (valence shell).

These methods have been found to be inadequate by themselves and are now associated with more sensitive reaction detectors. They are more and more used for digestion, extraction, separation, purification and concentration (preparative analysis) prior to instrumental analysis or for immediate repetitive analysis in agronomy, for example, where they remain indispensable. Chemical methods have the advantage of requiring only relatively inexpensive equipment and of allowing selection of samples.

In chemical reactions, we try to achieve rapid equilibrium when possible, in order to apply the Law of Mass Action and to establish constants that can be used in practice. Activity and concentration should be related by a known constant depending on the nature and charge of the ions:

$$A = \gamma C,$$

where A = activity, C = concentration, γ = activity coefficient.

The activity coefficient γ will be close to unity in very dilute solutions, in which the ions are dissociated, allowing the activity to be nearly equal to the concentration. In very concentrated solutions the activity coefficients are quite different. The reactions used are (a) electron transfer (oxidation-reduction), (b) proton transfer (acidimetry) and (c) ion transfer (complexometry).

(a) *The oxidation status* is a fundamental property of soils. Oxidants are substances capable of accepting electrons while reductants are able to surrender electrons. For example,

$$Fe^{3+} + e^- \underset{\text{oxidation}}{\overset{\text{reduction}}{\rightleftarrows}} Fe^{2+}$$

The oxidation status cannot exceed the number of electrons in the outermost shell or the number of electrons necessary to saturate this shell. For an oxidant to be able to accept electrons, it should be placed in the presence of a reductant able to surrender electrons to it. The field of oxidation-reduction comprises methods known earlier by the names manganimetry, cerimetry, iodimetry, etc., which require only basic volumetric apparatus.

It should be noted that electrometric methods (electrolysis) also allow exchange of electrons: oxidation takes place at the anode and reduction at the cathode. The redox potential of a system can be determined, the equivalence point being the equilibrium point at which neither reductant nor oxidant is formed.

(b) *Acids and bases* react by proton exchange. Acids surrender protons (H^+) and bases accept them. For an acid to surrender protons, it should be placed in contact with a base able to accept protons.

Water acts as an acid: $H_2O \rightleftarrows OH^- + H^+$

or as a base: $H_2O + H^+ \rightleftarrows H_3O^+$.

The equilibrium of an acid in water is written

$$acid + H_2O \rightleftarrows base + H_3O^+.$$

The acidity constant is given by

$$K_a = \frac{[base][H_3O^+]}{[acid]}$$

Quantitative relations permit determination of acids and bases by volumetry (titrimetry) using an indicator. The precision is improved by using a potentiograph that draws a curve of changes in pH until the equivalence point is reached.

Several reactions are conducted in media with constant pH by using buffer solutions. These media are indispensable for quantitative spectrocolorimetric determinations, for extraction of exchangeable cations at defined pH, determination of cation and anion exchange capacities, of extractable P_2O_5, etc.

Strong acids and bases are used for wet digestion (single acids or mixed acids, oxidizing or otherwise, etc.) or dry fusions (alkali fusions). Various digestions and fusions enable bringing most of the soil elements into solution and their quantitative determination by spectrometric methods. Chemical and physical methods are closely associated and complementary here.

(c) *Transfer of ions and polar molecules* is effected by acceptor-donor pairs, the donor being a complex with individuality different from that of the constituents of the molecule and the acceptor being a metallic ion [the ethylene diamine-tetraacetate (EDTA) ion is one of the most commonly used in the laboratory].

Coloured complexes, the most easily quantifiable, can be formed (*see* Chapter 9) and so can colourless complexes that are made visible at the turning point by an indicator.

Complexes can become more stable or less under the influence of factors such as temperature, pH, etc., which cause partial or complete dissociation. Perfect complexes enable masking of metallic ions, a property widely used for determinations by absorption spectrometry or emission spectrometry.

(d) *The combination of proton transfer and electron transfer* leads to variation in apparent potential with pH of a system with oxidation-reduction properties. These reactions are extensively used in physical chemistry. Similarly, electron transfer and ion transfer enable study of the function of complexes in reducing or oxidizing media. The properties will be used, for example, in methods for extraction of humic compounds (pyrophosphate method, etc.). The combination of ion transfer and proton transfer enables demonstration of the effect of pH on the equilibria of complexes.

All the chemical reactions can be applied in the following systems:

—solid-liquid: extraction and digestion, selective precipitation, ion exchange reactions on resins, solvent extraction (Soxhlet, Kumagawa);

—liquid-gas (or liquid-gas-solid): fractional distillation, entrainment in vapour, collection and determination of gases released in reactions;

—liquid-liquid (immiscible): extraction and selective purification, organic or inorganic, in separatory funnels (partition coefficients), establishment of relations between structure and physical properties (boiling point, etc.).

2.3 Electrochemical Methods

Quantitative determinations are based on measurement of small quantities of electricity associated with electron transfers in reactions. These reactions too pertain to the outermost valence shell. Variables such as current and potential are standardized in conjunction if possible with the time factor in transitory regime. Chemical and physical relations are close and inseparable. The most common and simplest method used is ionometry using ion-specific electrodes (*see* Chapter 12). The following techniques are also distinguished.

2.3.1 Voltamperometry — polarography — amperometry

In voltamperometry, the substances to be quantified should be electroactive or compatible with amperometric sensors. The intensity of the electrolysis current, a function of the concentration and potential, is measured. Determination of the equivalence point is the end of the analysis.

The term polarography is applicable when a dropping-mercury electrode is used. This method is used in organic or inorganic chemistry for measurements on a small number of samples. Polarographic apparatus requires constant maintenance for retaining optimum performance. The determination limit is of the order of 10^{-5} g and the precision is 1 to 5% according to the concentration and complexity of the solutions.

The term amperometric titration is applicable when volumetry is used, in which known quantities of reagents are added. Availability of an amperometric sensor allows many determinations to be done. The instrument is simpler than that for polarography and calibration is simple. The determination limit is of the order of 10^{-5} g.

2.3.2 Coulometry

When electrolysis is used, the term is 'coulometry'. Nowadays the instruments are automated. Their selectivity depends on half-wave potentials and the possibility of chemical displacement of the curves. A constant voltage is applied between electrodes (generally one or two indicator electrodes and one reference electrode). The quantity of electricity consumed during the electrochemical reaction is measured.

Electrolysis can be done at constant current or at constant potential as a function of time. The end of the reaction is indicated by an exponential decrease in the current related to the reduction of the concentration of the substance to be determined. The precision is directly related to the quantity of electricity necessary for complete transformation of the analyte and to the quantity of electricity consumed in the blank solution.

2.3.3 Potentiometry

This method is based on the determination of equilibrium potentials: potentiometry at zero current or potentiometry at constant non-zero current. The electrode potential under a current I is a function of the concentration and enables quantitative analysis. Titration is done by volumetric addition of the reagent; the end point is calculated from the values of the potential.

The simplest ionometric techniques use the determination of potentials of membrane electrodes (such as glass electrodes for hydrogen-ion potential) with moderately priced instruments (*see* Chapter 12).

2.4 Methods Using Radiations

2.4.1 Principle

Any physical measurement done by a reaction between a radiation and a target is based on a signal originating from an electronic or nuclear transition. It is possible to quantify the reaction by means of a suitable sensor and by taking into account the ranges of distance and time, the magnitude of the responses, the sensitivity of the instruments and techniques, the stability of the target to the radiation, space-time correlations, etc.

The use of radiations (Fig. 8.2) enables determination of the physical properties, nature and crystallochemical constitution of minerals, behaviour of clay minerals and weathering products (level of organization, paragenesis, chronology and dating, etc.).

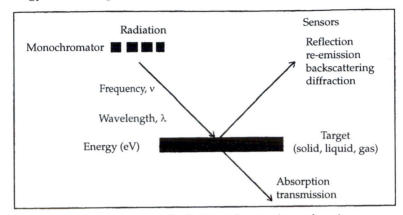

Fig. 8.2   Interaction of radiation and matter (general case).

It is possible to adapt the scales of study of minerals (macro, micro and submicro) and to combine the imagery (hand lens, optical microscope, SEM, TEM[1], etc.) with elemental analysis *in situ* (EDXRA, WDXRA, EELS[1], etc.) and microdiffraction (X-ray, electron) and even to IR[1]-vibration determinations (IR and Raman spectra). Thus the complex assemblages of mineral phases (crystalline and amorphous phases, substitution in phyllosilicates, various oxides, etc.) can be evaluated. These determinations are possible on particles smaller than 50 μm for 10^{-5} to 10^{-14} g according to the method and sensor used.

Unlike these very fine determinations carried out without much disturbance of the samples, most analyses are destructive. By reacting the sample with a liquid phase of known chemical properties, one can determine the liquid-solid interactions and the charges acting, conduct total chemical analysis and determine balances, selectively dissolve certain phases, concentrate and separate; instrumental physicochemical analysis permits determination of various elements even in infinitesimal traces in these extracts, with allowance made for matrix effects and contamination caused by various treatments.

The physical methods based on radiation-matter interactions for the non-destructive study of soil minerals should be adapted at the appropriate scale to characterize details:

—for studies of structure, diffraction methods use X-rays, neutron beams or electron beams that permit investigation of the spatial characteristics of mineral systems without change in frequency (elastic or coherent diffraction);

---

[1] *See* Appendix 2 for meaning of abbreviations.

—for characterizing mineral components and organic compounds, optical, IR, UV and X-ray absorption spectroscopy enable excitation of an internal process (electron transitions, molecular vibrations, nuclear transitions, Mössbauer, etc.). This response can be quantified;

—emission spectroscopy (X-ray, vibrational Raman, far-UV, etc.) makes use of the radiation re-emitted following an internal process. This re-emission can be of the same nature as the incident radiation (coherent scattering) or different from it (inelastic scattering-loss of energy, luminescence, photoemission, EXAFS[1], etc.).

Nuclear reactions modify the nature of the atom and can cause a change in the atomic number or atomic mass and emission of radiation; these phenomena are widely applied in analysis, especially in the areas pertaining to history of the earth (trace elements by neutron activation, natural isotopes, isotopic tracers). The disintegrations can be natural or artificially induced. In the latter case, a source of nuclear radiation must be used, which generally is heavy equipment: nuclear reactors, accelerators, etc., to which access is reserved for very specialized laboratories. Sealed sources with limited power are more commonly used, but with strict safety precautions.

X-rays of atomic origin are generated by the slowing down of a fast electron in the nuclear field (continuous radiation) or by the rearrangement of the inner electron shells when a gap is produced under the action of α-, β- or γ-radiations. The X-ray emission or Auger electrons can be measured.

2.4.2 The case of incident electronic radiation

The electron comprises one negative charge and has a spin of $1/2$. Electronic spectra are varied and are differentiated by the nature of the electrons analysed (Fig. 8.3): Auger electrons, photoelectrons, secondary inner or valence-shell electrons, etc.

The incident beam of electrons with controlled energy (electron gun) causes backscattering, and emission of secondary electrons and various radiations that can be exploited for analysis. Secondary Auger-transition electrons are generated in AES[1]. Electrons scattered without loss of energy (elastic scattering) are accompanied by phonons measurable by photo-acoustic spectrometry. Electrons scattered with loss of energy are the basis for EELS or CELS[1]. Backscattered secondary electrons are used in electron microscopy. These methods are non-destructive.

2.4.3 The case of incident ionic radiation

The ion beam excites high-energy Auger transitions (Fig. 8.4) but causes surface pulverization, which enables attainment of concentration profiles with progressive destruction of the object (destructive method).

---

[1] *See* Appendix 2 for meaning of abbreviations.

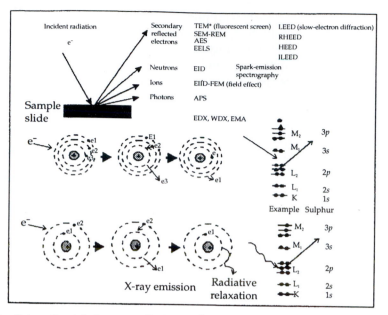

Fig. 8.3 Interaction of electron radiation with matter (*see* Appendix 2 for meanings of abbreviations).

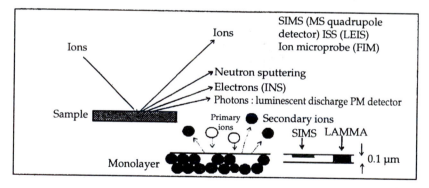

Fig. 8.4 Interaction of ionic radiation and matter (*see* Appendix 2 for meaning of abbreviations).

2.4.4 The case of photon radiation (Fig. 8.5)

The photon source of the incident radiation can have varying wavelength, from IR[1] to X-ray, and the energy of these radiations can be very high (LASER[1], with anticathode X-ray tube, etc.).

---

[1] *See* Appendix 2 for meaning of abbreviations.

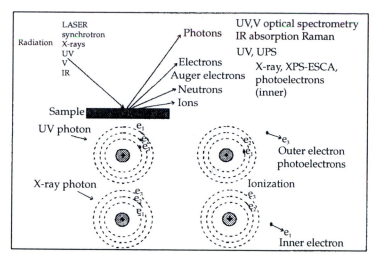

Fig. 8.5 Interaction of photonic radiation with matter (*see* Appendix 2 for meaning of abbreviations.).

2.5 Scales of Determination

Each determination should be selected based on the size of the analytical sample available and the concentration of the element or constituent to be determined (Table 8.1).

Today's most sensitive instrumental techniques permit determination of quantities down to a picogram and even to a femtogram under the best conditions (Table 8.1). The range over which each method functions well is variable; it depends on the element or compound to be determined, the matrix of the sample and performance of the instrument: excitation level, intensity of incident flux, quality of vacuum, focussing, detector sensitivity, etc. For example, the limit for determination in a PIXE[1] method could be 10^{-7} to 10^{-8} g in the micromethod, but could be as low as 10^{-12} g using a macrobeam.

Table 8.1 'Sample weight' and 'accessible mass' scales of the element of study; the sample weight analytical grouping depends on the technique used.

Sample weight scale		Element amount scale		
Analytical grouping	Weight	Unit	Gram	Molecules
Macroanalysis	>10 to 0.2 g	Milligram (mg)	10^{-3}	10^{18}
Microanalysis	1 mg	Microgram (μg)	10^{-6}	10^{15}
Ultramicroanalysis	1 μg	Nanogram (ng)	10^{-9}	10^{12}
Submicroanalysis	10^{-5} g	Picogram (pg)	10^{-12}	10^{9}
For notes		Femtogram (fg)	10^{-15}	10^{6}
		Attogram (ag)	10^{-18}	10^{3}
		Mologram (?)	10^{-21}	10
		?		1

[1] *See* Appendix 2 for meaning of abbreviations.

Bibliography

Chemical and physical techniques
Cairns T. and Sherman J. (eds.). 1992. Emerging strategies for pesticide analysis. In *Modern Methods for Pesticide Analysis*, CRC Press, Boca Raton FL, USA, 352 pp.
Eberhart J.P. 1976. *Méthode physique d'étude des minéraux et des matériaux solides.* Doin, 507 pp.
Eberhart J.P. 1989. *Analyse structurale et chimique des matériaux.* Dunod, 614 pp.
Fripiat J.P. 1982. *Advanced Techniques for Clay Mineral Analysis.* Elsevier, 235 pp.
Khan S.U. 1980. *Pesticides in the Soil Environment.* Elsevier, 240 pp.
Rouessac A. and Rouessac F. 1998. *Analyse chimique.* Dunod, 4th edition, 424 pp.
Stucki J.W. and Banwart W.L. 1980. *Advanced Chemical Methods for Soil and Clay Minerals Research.* Reidel, 477 pp.
Tarbuck E.J., Lutgens F.K. and Pinske, K.G. 1996. *Applications and Investigations in Earth Sciences.* Prentice-Hall, 313 pp.
Téchniques de l'Ingénieur. 1992. Analyse chimique et caractérisation, Séries P1 - P2 - P3 - P4, 1300 pp.

Electrochemical methods
Brett C. and Brett M.O. 1998. *Electroanalysis* (Oxford Chemistry Primers No. 63). Oxford University Press.
Galus Z., Chalmers R.A. and Bryce W.A.J. 1994. *Fundamentals of Electrochemical Analysis,* Ellis Horwood Ltd.
Ivaska A., Lewenstam A. and Sara R. (eds.). 1991. *Contemporary Electroanalytical Chemistry.* Plenum Publ. Corp.
Kalvoda R. 1987. *Electroanalytical Methods in Chemical and Environmental Analysis.* Plenum Publ. Corp.
Vanysek P. (ed.). 1996. *Modern Techniques in Electroanalysis* (Chemical Analysis, Vol. 139). John Wiley & Sons, 369 pp.
Wang J. 1994. *Analytical Electrochemistry.* VCH Publishers.
Yu T.R. and Ji G.L. 1993. *Electrochemical Methods in Soil and Water Research.* Pergamon Press.

Spectroscopy, surface analysis
Briggs D. and Seah M.P. (eds.). 1990. *Practical Surface Analysis: Auger and X-Ray Photoelectron Spectroscopy.* John Wiley & Sons, 2nd ed., 674 pp.
Calas G. 1986. *Méthodes spectroscopiques appliquées aux minéraux,* Société française de minéralogie et cristallographie, 1-2, 680 pp.
Creaser C.S. and Davies A.M.C. (eds.). 1988. *Analytical Applications of Spectroscopy.* Scholium International.
Ebdon L. and Evans H.E. (ed.). 1998. *An Introduction to Analytical Atomic Spectrometry.* John Wiley & Sons.
Engelhart G. and Michel D. 1987. *High Resolution Solid-state NMR of Silicates and Zeolites.* John Wiley & Sons, 485 pp.
Grimblot J. 1995. *L'analyse de surface des solides.* Masson, 232 pp.
Günther H. 1996. *La spectroscopie RMN: Principes de base, concepts et applications de la spectroscopie de résonance magnétique nucléaire du proton et du carbone-13 en chimie.* Masson, 576 pp.
Hollas J.M. 1998. *Spectroscopie, cours et exercices.* Dunod, 416 pp.
Howard M. 1996. *Principles and Practice of Spectroscopic Calibration* (Chemical Analysis, Vol. 118). John Wiley & Sons.
Jones C., Mulloy B. and Thomas, A.H. (eds.). 1993. *Spectroscopic Methods and Analyses: NMR, Mass Spectrometry and Related Techniques* (Methods in Molecular Biology, Vol. 17). Humana Press.
Klockenkamper R. 1996. *Total-reflection X-ray Fluorescence Analysis* (Chemical Analysis, Vol 140). John Wiley & Sons, 245 pp.

Montana A. and Burragato F. 1990. *Absorption Spectroscopy in Mineralogy*. Elsevier, 294 pp.

Saisho M. and Gohshi Y. (eds.). 1996. *Applications of Synchrotron Radiation in Materials Analysis* (Analytical Spectroscopy Library, Vol. 7). Elsevier Science Ltd.

Schlag E.W. 1998. *Zeke Spectroscopy: the Linnett Lectures*. Cambridge University Press.

Shagidullan R.R. 1991. *Atlas of IR Spectra of Organophosphorus Compounds (Interpreted Spectrograms)*. Kluwer Academic Publishers.

Silverstein R.M. and Webster F.X. 1997. *Spectrometric Identification of Organic Compounds*. John Wiley & Sons, 6th ed., 496 pp.

Watts J.F. 1990, *An Introduction to Surface Analysis by Electron Spectroscopy* (Royal Microscopical Society, Microscopy Handbook No. 22). Oxford University Press.

White R. 1989. *Chromatography/Fourier Transform Infrared Spectroscopy and Its Applications*. Marcel Dekker.

Wilkins C.L. (ed.). 1994. *Computer-enhanced Analytical Spectroscopy* (Modern Analytical Chemistry). Plenum Publ. Corp.

Molecular Spectrometry

1. Principle

1.1 Definitions

This technique makes use of the emission or, more often, the absorption of electromagnetic radiation resulting from the excitation of electrons in molecular structures. It is thereby distinguished from techniques using atomic electron transitions (*see* Chapters 10 and 11). Molecular spectrometry, more commonly called spectrophotometry, often has for its objective the determination of the concentration in solution of a substance that is usually coloured or capable of forming a soluble coloured compound, whence the other term 'colorimetry' or spectrocolorimetry. It is often also applied to gas analysis.

The term colorimetry is more commonly used for determinations with white or filtered radiation, and spectrophotometry for measurements with monochromatic radiation. Spectrophotometry is not limited to the visible spectrum and is applied to the determination of colourless compounds capable of absorbing a radiation outside the visible spectrum. The widest range of applications pertain to the portion of the spectrum from 200 to 1000 nm, that is, from the near ultraviolet to the beginning of the infrared, the range concerning most of the reactions of inorganic compounds.

Spectrophotometry is widely used and sensitive enough for trace analysis; it is rapid, thus allowing monitoring of the progress of reactions. The effect of interfering ions can be reduced without having to eliminate them beforehand by appropriate choice of reagents, complex formation, pH, Eh or wavelength of radiation used or more simply, by use of a 'blank'.

A colour reaction cannot serve as a basis for quantitative determination unless the following conditions are met:

—the reaction is stable enough to provide sufficient time for a measurement;

—it should be true, that is, reproducible;

—the intensity of the colour should increase regularly with increase of solution concentration.

1.2 The Fundamental Law

When a beam of light of intensity I_0 passes through a coloured layer (Fig. 9.1) of thickness e, the rays of wavelength λ are absorbed, resulting in a diminution of light energy and a different transmitted intensity denoted by I.

Lambert in 1760 established the relation expressing the law of absorption:

$$\log (I_0/I) = Ke$$

where K is a constant characteristic of the substance, and e is the thickness (in cm) traversed.

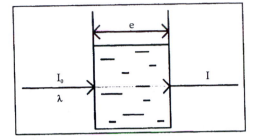

Fig. 9.1 Colorimetric cuvette.

Beer in 1852 found an extension to this law by noting that this absorption is proportional not only to the thickness of the layer traversed, but also to the concentration C of the dissolved substance. This new relation is the Beer-Lambert law:

$$\log (I_0/I) = keC \qquad (1)$$

The ratio I/I_0 represents the transmission T, often expressed in per cent. Log (I_0/I) is termed the optical density D, the extinction E or, more generally, the absorbance A.

The coefficient of proportionality k is the extinction coefficient; when the concentration C is expressed in g L^{-1}, k is the *specific extinction coefficient*. If C is expressed in g ions or g mol L^{-1}, k is the *molecular absorption coefficient*. The thickness e is always expressed in cm.

1.3 Validity of the Law; Interferences

Beer's Law is a limiting law requiring ideal conditions, particularly with respect to the light beam, which should be monochromatic, and the solutions, which should be of low concentration and in which there should be no foreign effect on the absorption of the substance to be determined. The coefficient k is independent of the concentration; it depends on the dissolved substance, the wavelength of the radiation and the temperature. Theoretically, the solvent has no effect.

When the compound does not obey the Beer law, the calculation of concentration cannot be done from equation (1). It then becomes necessary to experimentally obtain a curve from a range of standards by marking the absorbance A on the ordinate and the corresponding concentration on the abscissa. However, a range of concentrations should be chosen for which the curve does not deviate too much from a straight line. The principal interferences can be grouped as given below.

1.3.1 Chemical interferences

Law (1) is not valid when the solutions are too concentrated, there being a validity range of concentration. Reactions can also be produced by dilution, which causes dissociation. The absorbance then ceases to be proportional to the concentration and it is necessary to take into account the dissociation constant α of the analyte; the new relation becomes

$$A = keC(1 - \alpha) \qquad (2)$$

The presence of non-absorbing compounds capable of affecting the electronic state of the molecule or differences in solvent can cause variation in the coefficient k that can be experimentally determined.

It should also be noted that the Beer-Lambert relation is strictly obeyed only when the solution studied is free of fluorescence or colloids. This point can be critical in the case of soil extracts.

1.3.2 Physical interferences

It is difficult to obtain ideal conditions of monochromaticity of the radiation, which has a certain band width; conditions are best when the wavelength range overlaps the maximum or the minimum of the curve $A = f(\lambda)$.

The coefficient k depends on the refractive index n; also, it is not k but the function $kn/(n^2 + 2)^2$ that is independent of the concentration. In usual cases, if the concentration does not exceed 10^{-2} mol L^{-1}, the error is less than 0.1%.

The Beer law is obeyed if the solution traversed retains the same thickness over the entire width of the beam; change in e affects the reading in the same way δ as change in k; it is therefore necessary to use only very uniform cuvettes.

2. Instruments

2.1 Colorimetry by Visual Comparison

These rapid and inexpensive methods can be used in small laboratories or in the field; they can also be used to confirm certain results obtained by more sophisticated methods; their principle is based on visual comparison between a solution of known concentration and that of unknown concentration to be analysed; sensitivity to differences in intensity depends on the wavelength. The most favourable for the human eye is found around 550 nm and, in the best of conditions, a relative error of 1% may be expected.

2.1.1 Examination of coloured solutions of the same thickness

The method of standard scales consists of preparing a series of test tubes (well calibrated, of the same thickness) and filling them with coloured standard solutions of increasing concentration. The difference in concentration between two tubes should be as regular as possible. For example, a geometric progression of common ratio of 1/2 gives for a stock solution of 1 g L^{-1} the range 1 -- 0.5 – 0.25 – 0.125 – 0.0625 and so on. Observations are made under

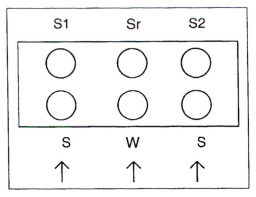

Fig. 9.2 Walpole comparator, viewed from above, showing the arrangement of tubes; the arrows indicate the direction of visual observation: S - unknown solution alone; Sr- unknown solution + reagent; W - water; S1 and S2 - standard solutions of concentrations 1 and 2.

the best conditions with a Walpole comparator (Fig. 9.2). The mean value of two concentrations on either side of the colour is taken.

In the *dilution method* the colour of the standard solution is compared to that of the solution to be analysed in tubes of the same diameter. After examination, the strongest-coloured solution is diluted until the colours appear equal. If W is the weight of the substance contained in the standard of volume v, and x the concentration in the final volume v_1 at equal colour,

$$x/v_1 = W/v,$$
$$\text{whence } x = v_1 W/v \tag{3}$$

2.1.2 Examination of coloured solutions at different thicknesses

This method is much more precise than the ones described above. Examination at different thicknesses is done with the Duboscq-type immersion colorimeter (Fig. 9.3). Equality in colour is achieved by changing the thickness traversed by one of the light beams. If l_1 is the height read by vernier for the standard solution of concentration c, and l_2 the height for the solution of unknown concentration x,

$$x = \frac{cl_1}{l_2} \tag{4}$$

2.2 Photoelectric Methods

2.2.1 Detectors

Use of colorimetric methods has been broadened and considerably improved by replacing the eye with a photoelectric detector. This detector enables comparison of unequal light intensities without being restricted to visible

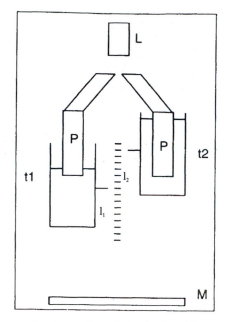

Fig. 9.3   Schematic diagram of an immersion colorimeter: t_1 and t_2 - comparison and measuring tubes; l_1 and l_2 - distances (read on a vernier) between the base of each tube and the fixed transparent immersion cylinder transmitting each beam of light (P); M = mirror.

radiations (Table 9.1) and gives a precision of 0.2% in determination of concentrations.

Table 9.1   Details of the principal photoelectric detectors.

Characteristic	Photoemissive cells	Selenium photodiodes	Silicon photodiodes
Upper limit of λ	1.2 μm (caesium oxide)	0.8 μm	1.1 μm
Lower limit of λ	180 nm with quartz window	350 nm	200 nm sensitized
Highest sensitivity	UV and IR	Middle of visible	580 nm
Additional voltage	Always necessary	Superfluous	Superfluous
Amplification	Easily adaptable, transistorized	Difficult, galvanometer	System optimized with operational amplifier
Sensitivity in current for $T = 2600$ K	20-100 μA lm^{-1}	400 - 500 μA lm^{-1}	120 μA lm^{-1}
Sensitivity in voltage	5 V mlm^{-1}	0.01 V lm^{-1}	
Inertia	Zero	Larger	Usable to 1000 Hz

Photovoltaic cells (called 'barrier-layer' cells; Fig. 9.4a) are the simplest and the most popular, and have the advantage of not requiring an external

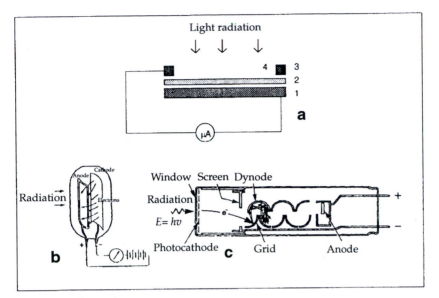

Fig. 9.4 Photoelectric detectors:
    a - photoelectric ('barrier-layer') cell
        1. metallic base plate
        2. semiconductor
        3. current collecting ring of the front electrode
        4. transparent front electrode
    b - photoemissive cell: the cathode releases electrons when the light radiation strikes
        it; they are attracted to the anode and give rise to a photoelectric current.
    c - photomultiplier tube: similar to the photoemissive cell in principle but the electronic
        current is amplified by emission of new electrons when the electron beam strikes
        each dynode.

voltage source, the cell itself functioning as a generator. The oldest of these cells are those of selenium; for a very long time they equipped most colorimeters. The more recent silicon diodes are also designed for photovoltaic-type functioning without external polarization, and have a wider range of response in the red and near infrared. Certain photodiodes are also sensitized for blue and ultraviolet.

Photoemissive cells (Fig. 9.4b) make use of the fact that certain metals when taken to a negative potential release electrons under the action of light. Sensitivity to wavelength varies with the nature of the metal used for the cathode. It has been proved that the current generated is proportional to the intensity of light. Deviations in determinations are lower than with certain barrier-layer cells. Nonetheless these models are little used. Photomultiplier tubes (Fig. 9.4c), which enable multiplication of the number of electrons emitted by a factor of 10^8, are preferred.

2.2.2 Source of radiation

The source differs according to the desired wavelength region:

—from 350 to 1000 nm, incandescent tungsten lamps with continuous spectrum at 3200-3400 K are used;

—below 350 nm, absorption by glass reduces the intensity of light and annihilates it at 300 nm; the light source, for example a hydrogen lamp with quartz walls will be richer in UV radiation; it is also necessary to use tubes and optics of quartz or at least of special glass transparent to near-UV.

Voltage fluctuation in the current supplied to the lamps is the most important cause of errors in determination of optical density and hence the supply should be strictly stabilized. Differential equipment that enable elimination of variations in source are preferred.

2.2.3 Filters and monochromators

A simple solution for isolating the spectral region from a spectrum is to use coloured filters, which are generally composed of a thin layer of coloured gelatine sandwiched between two glass sheets (Kodak), or of totally coloured glass. The band-pass width is often larger than 20 nm and may reach 40 nm in the most popular types supplied with simple colorimeters.

Interference filters allow a band pass of 10 nm. For better resolution, a dispersion system, with a prism or diffraction grating and an adjustable slit is used. The greater the dispersion of the system and the narrower the slit, the closer the approach to monochromaticity. Grating instruments have a band pass of 2 nm with spectral precision of ±0.1 nm.

The instruments are generally calibrated for wavelength, but it is often necessary to check calibration. For this check-filters are available that can be used in the tube compartment. An unusual set-up that deserves attention is the Christian Sen-Waigert apparatus composed of a cylindrical glass tube of 94 mm diameter, filled with glass powder and benzyl benzoate. The refractive indices of these substances vary with temperature. At 15°C they are identical for red light, at 40°C for blue. Thus, by raising the temperature from 15°C to 40°C all spectral colours from red to blue can be obtained with the same spectral purity as obtained with a good filter of band pass 10-20 nm.

2.2.4 Principal types of instruments

The problem faced is always that of determining the ratio of the two light intensities I_0 and I. The instruments can be grouped into apparatus with one or two detectors and with or without optical/electrical compensation device (Fig. 9.5).

The apparatus with a single photoelectric cell (Fig. 9.5a, 9.5b) is the most popular for routine work and often has the classic galvanometer replaced by a digital display. It is used with simple filters from 400 to 750 nm with a band pass of 40 nm.

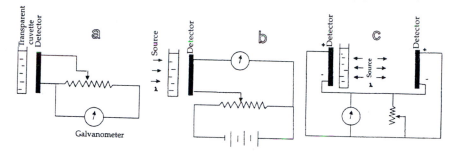

Fig. 9.5 Different photoelectric colorimeter set-ups:
a - direct-reading colorimeter; b-compensation-type colorimeter; c-two-detector system, differential reading.

The instrument with two photoelectric detectors (Fig. 9.5c) is much more precise and allows elimination of variations in intensity of the light source. One of the detectors and its light beam serves as reference, the other receives the beam that has passed through the cuvette containing the coloured solution. The same filter is placed in the light path leading to both cells. In many instruments a second cuvette can also be placed between the source and the reference detector, this system allowing elimination of certain absorptions caused by the solvent or the matrix of the test solution.

2.3 Instrument Selection

The precision sought will determine the selection; in the simplest case, recourse is had to a system of visual comparison with a range of standards. For some determinations instruments for visual comparison are provided with coloured discs, more stable than standards. Reproducibility with a relative standard deviation (*see* Chapter 18) less than 5% can be attained.

For more precise measurements with repeatability of the order of 1%, there is a choice of a range of photoelectric colorimeters that can be used for routine work. Apart from the price the major criteria for choice of instrument are stability and reproducibility of the instrument, type of display (digital or analogue), recording outlet or RS232 interface (for use with a printer or for connecting to a microcomputer), zero adjustment for absorbance and transmission, quality of coloured or interference filters, cuvette holders for 10-20-40 mm cuvettes and provision for using test tubes of 22 mm diameter. A colour filter with the complementary colour of the test colour (Table 9.2) is chosen, which ensures the maximum light absorption with the narrowest band pass. In this category of instruments we should mention the photometers with light cables of optical fibre, which enable continuous measurement in the reaction vessel (Fig. 9.6). A flexible lead about a metre long ends in a system acting as a cuvette with optical path of 10 or 20 mm.

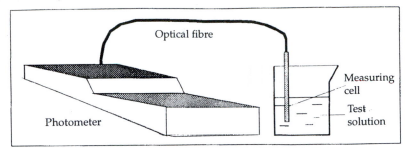

Fig. 9.6 Photometer with optical-fibre light cable (Type 662 - Metrohm[1]).

Table 9.2 Selection of filter

Colour	λ (nm)	Complementary colour	λ (nm)
Violet	433	Yellowish green	564
Ultramarine	482	Greenish yellow	574
Blue	485	Yellow	585
Greenish blue 1	490	Orange red	608
Greenish blue 2	492	Red	656
Green	495-560	Purple	

Single-beam or double-beam spectrophotometers are the instruments of choice for analytical work or research requiring high spectral purity with a band pass of 5 nm. These features are obtained with an instrument with dispersion by prism or more often by grating. All these instruments work with a wavelength from 325 to 1000 nm.

For analytical work requiring very high photometric stability double-beam instruments are available, in which the band pass does not exceed 1 or 2 nm with spectral accuracy of ~1 nm. Such instruments enable measurement of ultraviolet and infrared absorption in the wavelength range 190 to 9500 nm, with many optional accessories.

2.4 Estimation of Precision of an Instrument

The extinction coefficient k is determined at the same wavelength for different concentrations of a substance that obeys Beer's law. The deviation between the values of ke and the average curve $ke = f(C)$ indicates the precision. In the ideal case, $ke = C^{te}$ (Fig. 9.7) with slight dispersion around this value; if the deviation is large, greater than the dispersion, the difference defines a systematic error that be determined.

For this operation, a potassium permanganate solution, which obeys Beer's law even for a wide band of radiation around the absorption maximum 520-550 nm, is used; with a band pass of 50 nm the law is still obeyed within 0.2%.

---

[1] *See* Appendix 6 for addresses.

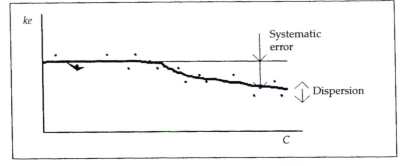

Fig. 9.7 Determination of systematic error and dispersion.

The permanganate solution is about 15 mg Mn L^{-1} in 0.3 mol $(1/2\ H_2SO_4)^{-1}$ containing 20 mg potassium periodate per 100 mL. Necessary dilutions are made with 0.3 mol $(1/2\ H_2SO_4)\ L^{-1}$. The same determinations can be repeated with different wavelengths. The results allow refining of operational protocols and monitoring of causes of error.

3. Methodology

3.1 Precision, Sensitivity and Detection Limit

The precision desired will depend on the purpose in view. Where the presence of an element is to be detected, the deviation of the determination is often of the order of the content of the element. A very simple instrument suffices, without amplification or monochromator. Contrarily, for precise quantitative determinations, the selection of instrument can become important and the operations should be carefully done.

The measurement should be done at a wavelength at which $dk/d\lambda$ is small; with this condition a wide band and a greater initial intensity I_0 can be used.

The effect of temperature will be negligible if two solutions of very similar composition are compared. The cuvettes should be perfectly identical, and this should be well tested.

During measurement the absorbance should be periodically checked with a very stable, known solution (chromates, cobalt salts, copper salts, etc.). The relative error is generally limited by calibration of volume measurements, weighings, etc. It is not expected to be greater than 0.1%.

Sensitivity (see Chapter 18) is given by $\Delta A/\Delta C$, with $A = keC$, showing that the sensitivity can be improved by adjusting k (selection of reaction and wavelength) and e (thickness traversed by the light beam).

From equation (1) the relative error in $\Delta C/C$ is equal to that in absorbance $\Delta A/A$. It thus increases very rapidly towards low absorbances. To compensate for this it is necessary to choose an instrument that can measure very small values of $\Delta(I/I_0)$. The *determination limit* (*see* Chapter 18) can be defined as the point at which the error becomes equal to the measured value, that is, $\Delta C/C = 1$ or $\Delta A = A$.

3.2 Colour Reaction

Elements or compounds transparent to radiation should be taken to a molecular state at which they absorb radiation, mostly in the visible range, whence the popular but restrictive term 'colour reaction'. Very often a single element or compound may undergo several 'colour reactions'. It is necessary to conduct a comparative study in which the following factors should be considered:

—*reproducibility*: equal intensities of colour should be obtained under identical operational conditions;

—*specificity*: this presumes a study of the behaviour of substances with similar properties with respect to the colour reaction; the most specific reaction should be preferred;

—*stability*: the reaction should be stable enough to give time for doing the measurement;

—*sensitivity*: the greater the sensitivity the lower the minimum concentration that can be determined;

—*simplicity*: it is advantageous to select the technique that avoids too many manipulations for doing the measurement.

All these factors are also influenced by the type of instrument available:

—for measurements made by visual comparison, reactions having absorption maximum in green or, if not possible, in blue should be selected;

—for colorimeters with barrier-layer cells or voltaic photodiodes, colour reactions should, as far as possible, have absorption maximum in yellow, as the sensitivity of these cells increases from green towards red;

—lastly, among the colour reactions, preference should be given to those that follow the Beer-Lambert law.

After the colour reaction has been selected, the operating conditions should be fixed after study. The concentration should correspond to a coloration easily measurable by the instrument used; the reading should be in the middle of the scale. Optical densities around 0.1 to 0.5 should be used.

Development of colour should be done with the greatest care; the following are recommended.

—*Sequence and rate of addition of reagents*: the sequence of addition of reagents is critical; also some reagents should be added slowly, others rapidly.

—*Effect of temperature*: temperature affects colour development; in some cases heating endangers stability; this factor plays a very important role.

—*Effect of time*: the optimum period of contact between the substance to be determined and the reagents should be adhered to; this requires a prior study of the kinetics of the reaction to find out exactly the time taken for colour development and stability or the linearity of the phenomenon as a function of time (non-equilibrium studies, FIA method, *see* Chapter 17).

—*Drawing of calibration curve*: for a given method, it is examined whether the Beer law is obeyed. If not obeyed, it is necessary to draw a graph relating absorbances (optical densities) to the corresponding concentrations. A logarithmic scale is used, using a linearization programme.

3.3 Determinative Increments

3.3.1 Direct method

The absorbance $\log (I_0/I)$ of the dissolved substance is determined and the unknown concentration is deduced from it. In practice, the transmitted intensity is what is left of the I_0 after absorption by not only the coloured solution but also by the solvent and the faces of the absorption cuvette (Fig. 9.8). For correction, a blank determination is done, which is composed of the cuvette containing just the solvent. The absorption given in this test is I_0. It is automatically deducted in double-cell differential spectrophotometers or in single-cell systems provided with adapted electronics.

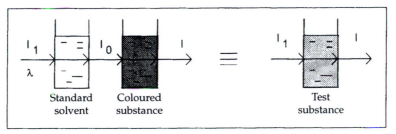

Fig. 9.8   Direct increments.

It is difficult, if not impossible, to use values of k or ke from the literature for a known substance because the analysis is not always done with strictly monochromatic radiation and the coefficient obtained varies with the instrument used; it is therefore necessary to determine k or ke for each experiment with the instrument used.

3.3.2 Differential method

In this method, I_0 is the light emerging from the cuvette when filled with a coloured solution of known concentration, slightly lower in intensity of colour (therefore of concentration) than the unknown solution. The new absorbance $\log (I_0/I)$ due to the difference in concentration ΔC between the two solutions is measured. The value of C_x can be deduced where the Beer law is obeyed. If not obeyed, a calibration curve of $I = f(\Delta C)$ is constructed. This method is more precise than the direct method; it also permits analysis of more concentrated solutions.

3.3.3 Balancing the optical absorbance of two solutions

This is similar to the dilution method (*see* §2.1.1). The instrument is set with the test solution; in another cuvette containing the same volume of solvent the same intensity *I* is attained by addition of a standard solution of the analyte.

This method is well suited to automated determinations.

3.3.4 Method of increments

This method is applicable to determinations when the optical density is proportional to the analyte concentration and other reactions do not occur. Since variation in optical density is most often limited to a small increase, the law of proportionality may be presumed to be obeyed.

The optical density *D* is determined before and after addition of a known quantity of the substance to be determined (Fig. 9.9); if *a* is the increase in concentration,

$$D = klC \quad \text{and} \quad D' = kl(C + a),$$

whence $D/D' = C/(C + a)$, from which *C* can be obtained

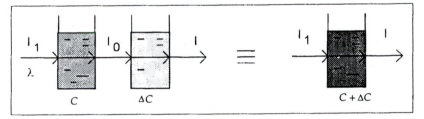

Fig. 9.9   Method of increments.

This presumes that the dilution error either has been eliminated by so managing the addition of a sufficiently concentrated known solution, or is very small in comparison to the total volume of the unknown solution.

This method does not preclude the use of a standard, above all if there are other coloured substances in the mixture; its advantage is that it takes into account the effect of the spectral background on the wavelength used.

4. Fluorimetry

4.1 Principle

Fluorescent substances are able to absorb radiation of a given wavelength and then emit other radiation of less energy, that is, of longer wavelength. After selection of the wavelength of maximum emission, variation of fluorescence intensity *F* with concentration *C* (Fig. 9.10a) is studied. This intensity generally decreases at high concentrations; at very low

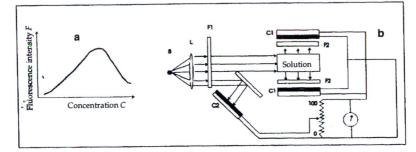

Fig. 9.10 a. Variation in intensity of fluorescent radiation with concentration.
b. Schematic diagram of a filter fluorimeter: S - UV radiation source; L - collimator; F1 - exciting-radiation filter permitting UV radiation to pass; P2 - fluorescent-radiation filter; C1 - measuring photoelectric cell; C2 - comparison photoelectric cell.

concentrations and at small thicknesses the law obeyed is linear, $F = kC$; for many substances the determination range is from 10^{-4} to 10 mg L^{-1}. Foreign substances in the solution can cause variation in emitted intensity. Furthermore, pH and temperature are factors to be controlled. These different interferences necessitate drawing of a calibration curve for each system studied.

The precision is of the same order as for absorbance measurements and the detection limit is usually lower. Fluorimetry is thus a technique well suited to trace determinations, which have become important in biochemical analyses.

4.2 Instrumentation

The instruments are not much different from other spectrophotometers. The radiation source should provide a beam of monochromatic radiation and the emission is measured at right angles to the incident beam (Fig. 9.10b). Some spectrophotometers can function in both fluorescence and absorption modes.

The illumination source is mostly a UV lamp. Some precision instruments use two monochromators, one for selecting the irradiating wavelength and the other for isolating the fluorescent wavelength.

5. Turbidimetry and Nephelometry

If the analyte can be precipitated in the form of a very fine and very stable suspension (stability is often increased with certain materials, the most common being gelatine), a method similar to absorptiometry can be used. The light beam of intensity I_0 traverses the solution (suspension). When I is measured in the same direction, it is *turbidimetry*; when it is measured in another direction it becomes *nephelometry*.

As in colorimetry the ratio I_0/I and the intensity I_D increase with concentration of the precipitated substance.

Colorimeters and spectrophotometers can be used for turbidimetry without modification. For nephelometry in its simplest form, a fluorimeter enabling measurement of laterally transmitted light is used.

Bibliography

Bartos J. and Pesez M. 1980. Colorimétrie et fluorimétrie en analyse functionelle. *Techniques de l'Ingénieur, Paris,* **P 3255.**

Burgess, C. and Jones D.G. (eds.), 1995. *Spectrophotometry, Luminescence and Colour: Science and Compliance.* Papers presented at the Second Joint Meeting of the UV Spectrometry Group of the UK and the Council for Optical Radiation Measurements of the USA. Elsevier Science Ltd.

Burgess C. and Mielenz K.D. (eds.). 1987. *Advances in Standards and Methodology in Spectrophotometry* (Analytical Spectroscopy Library, Vol. 2). Elsevier Science, Ltd.

Burns D.A. and Ciurczak E.W. (eds.). 1992. *Handbook of Near-infrared Analysis.* Marcel Dekker.

Charlot, G. 1952. *Dosages colorimétriques.* Masson, Paris.

Connes J. and Bouchareine P. 1989. Applications de la transformation de Fourier. *Techniques de l'Ingénieur, Paris,* **P 210.**

Disant C. 1981. Spectrométrie d'absorption dans le visible. *Techniques de l'Ingénieur, Paris,* **P 2805.**

Duval C. 1975. Colorimétrie, fluorimétrie, turbidimétrie, néphélométrie, photométrie. *Techniques de l'Ingénieur, Paris,* **P 2255.**

James, G.E., 1987. Les progrés en spectroscopie UV-visible. *Analusis,* **15:** 78-84.

Levillain R. and Fompeydie D. 1986. Spectrophotométrie dérivée: intérét, limites et applications. *Analusis,* **14:** 1-20.

Malingrey B. 1987. Spectrométrie d'absorption dans l'UV et le visible. *Techniques de l'Ingénieur, Paris* **P 2795.**

McHale J.L. 1998. *Molecular Spectroscopy.* Prentice-Hall Press, 463 pp.

Mottana A. and Burragato F. (eds.). 1990. *Absorption Spectroscopy in Mineralogy.* Elsevier Science Ltd.

Oehmichen J.P. 1977. *Applications des dispositifs photosensibles.* E.T.S.F., Paris.

Pelikan P., Ceppan M. and Liska M. 1994. *Applications of Numerical Methods in Molecular Spectroscopy (Chemometrics).* CRC Press, Boca Raton FL, USA.

Ramond D. 1988. La spectrophotométrie UV/visible. *Analusis,* **16:** 25-49.

Sandell E.B. 1978. *Photometric Determination of Traces of Metals.* John Wiley and Sons, New York.

Sommer L. 1990. *Analytical Absorption Spectrophotometry in the Visible and Ultraviolet: the Principles* (Studies in Analytical Chemistry, Vol. 8). Elsevier Science, Ltd.

Susini J. 1979. Réalisation d'un colorimétre photoélectrique. *Cah. ORSTOM Sér. Pédol.* **XVII** (1): 65-76.

Talsky G. 1994. *Derivative Spectrophotometry of First and Higher Orders.* John Wiley and Sons, 229 pp.

Urban M.W. 1993. *Vibrational Spectroscopy of Molecules and Macromolecules on Surfaces.* John Wiley and Sons, 384 pp.

10

Atomic Absorption Spectrometry

1. Definition and Principle

1.1 Principle

The principle of atomic absorption spectrometry (AAS) is based on application to emission spectrometry of the Kirchoff law. The technique was proposed simultaneously in 1955 by Walsh in Australia, and by Alkemade (Netherlands) and Milatz, the last two presented in the *Journal of the Optical Society of America*, the first double-beam spectrophotometer. After several improvements (Pinta, 1985), AAS 'became the principal elemental-analysis technique for agricultural and environmental science' (Ure, 1991).

In fact AAS is quite complementary to molecular absorption spectrometry (*see* Chapter 9) and also to emission techniques such as flame atomic emission spectrometry (FAES: *see* Chapter 11). Atomic absorption instruments are also simultaneously flame spectrometers, but the reverse is not always true. It is necessary to briefly review the basic theories of FAES and AAS to understand the difference between them.

In atomic emission, the atoms are excited by an energy source (flame, arc, spark or plasma). The excited atoms return to their ground state and, in the process, emit a radiation of frequency v, related to the energy level E_m of the excited state m compared to the ground-state energy E_0:

$$E_m - E_0 = hv \tag{1}$$

where h is Planck's constant. The corresponding intensity I_m of the emitted radiation is given by

$$I_m = BN_0\, e^{\left(-\frac{E_m}{kT}\right)} \tag{2}$$

where B and k are constants, T the absolute temperature and N_0 the number of atoms at the ground state, proportional to the concentration. For a given element, the high-energy electronic transitions are those corresponding to the highest frequencies (eqn 1) and therefore to the shortest wavelengths. Rise in the energy level E_m causes an exponential decrease in emission intensity (eqn 2), which should be compensated for by increasing the temperature T in the same proportion. In practice, air-acetylene or nitrous oxide-acetylene flames are not hot enough to excite elements giving emissions at wavelength shorter than 250 nm. In such cases it is necessary to rely on emission sources with very high temperature, such as plasma (ICP[1]: *see* Chapter 11) or on atomic-absorption or atomic-fluorescence techniques.

Atomic absorption spectrometry consists of passing monochromatic radiation of variable wavelength through the source containing atoms of the element to be determined. When the energy of incident photons becomes equal to the atomic-transition energy (eqn 1), the photons are absorbed by the atoms, which are excited in turn. Two measurement techniques are available:

---

[1] *See* Appendix 2 for meanings of abbreviations.

—atomic fluorescence spectrometry (AFS): measurement of radiation emitted when the excited atoms return to the ground state; this is done perpendicular to the direction of the incident radiation of intensity I_s; the intensity of the fluorescent radiation is given by

$$I_f = I_s k C \qquad (3)$$

where C is the concentration of the element and k, a coefficient related to a given transition of that element;

—atomic absorption spectrometry: measurement of the absorption of the incident beam during transition of the analyte atoms to an excited state; this is done in the direction of the exciting rays.

In practice, AAS is used much more than AFS. The law of absorbance (A) is comparable to the Beer-Lambert law of molecular absorption (*see* Chapter 9):

$$A = \log (I_0/I) = keC \qquad (4)$$

where I_0 and I are respectively the intensity of incident radiation before and after absorption, k the atomic absorption coefficient, e the thickness of the population of absorbing atoms and C the concentration of the element. This equation (4) shows that the absorption changes linearly with concentration of the element and does not depend on the atomization temperature, unlike the emission law (eqn 2). In general, high values of k correspond to the atomic excitation level closest to the ground state (resonance radiation). Voïnovitch (1988) recalled an interesting observation of the pioneers of the technique: the value of k is generally between 10^7 and 10^9 when C is expressed in gram-atoms per litre. In molecular absorption (visible, IR and UV spectrometry) the coefficient k is replaced by the molar extinction coefficient with a maximum value around 10^5; the sensitivity of AAS is therefore 100 to 10,000 times better, but this comparison is dangerous, considering the definition of sensitivity of a method.

When the technique was proposed, Walsh suggested the following conditions (Fig. 10.1) for useful measurement of atomic absorption.

$-\lambda A_{max} = \lambda E_{max}$: the absorption wavelength λA_{max} of the vapourized atoms should be equal to the wavelength λE_{max} corresponding to the maximum emission intensity of the source; this is a direct corollary of the Kirchoff law enunciated in 1860;

$-\delta E \leq \delta A$: the band width of the emitted rays should be at least four times smaller than that of the absorption rays; if this condition is not obeyed, the atoms will absorb only a small quantity of the source radiation because its contour is wider than that of the absorption rays.

The classic methods of filtration of radiation (prisms, gratings, interference filters) do not give a beam of sufficiently narrow band pass and it is necessary to use special light-emitting sources with a band pass narrower than 0.005 nm (discharge lamps, hollow-cathode lamps, etc.). This feature makes the

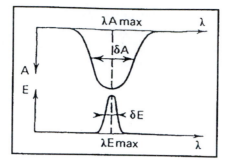

Fig. 10.1 The two conditions for atomic absorption.
λA_{max} and λE_{max} are the respective wavelengths of the absorption maximum of the atomic vapour (A) and the emission maximum of the source (E). δA and δE are the widths at half-height of these radiations.

technique very selective and the risk of superposition of rays, which is quite substantial in emission spectrometry, is negligible. The specificity is further heightened by the monochromaticity of the incident radiation, which can be modulated for being electronically recognized where it exits, thus eliminating continuous radiation capable of affecting the signal.

Sensitivity of the method depends on obtaining *free atoms in the ground state* (N_0, eqn 2) which, in turn, depends greatly on the temperature: room temperature for mercury, about 900°C for the hydride method and temperatures ranging from 2000 to 3000°C for most other elements that can be determined by AAS. Other analytical parameters also have influence (solution matrix, instrument, etc.).

1.2 General Scheme of an Instrument

There are several more or less sophisticated instruments, single-beam or double-beam, but all of them have nearly all the components shown in Fig. 10.2.

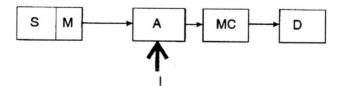

Fig. 10.2 Diagram of an atomic absorption instrument
S - radiation source (*see* §2.1); M - light modulator (or alternating or pulsed current for the source (*see* §2.1); A - atomization system (*see* §2.2 and §2.3); I - injection of sample into the atomizer; MC - monochromator (*see* §2.4); D - detection, amplification and data acquisition system (*see* §2.4).

2. Elements of an AAS Instrument

2.1 Radiation Source

The role of the source is to produce a narrow beam of radiation of the element to be analysed in order to optimize the absorption conditions (*see* § 1.1, Fig. 10.1). For this, it is necessary that the source functions at
—as low a temperature as possible to avoid broadening of the beam caused by random movement of the emitting and absorbing atoms relative to the observer (Doppler effect),
—under low pressure to reduce broadening due to collisions between atoms and molecules (pressure effect or Lorentz effect).

2.1.1 Hollow-cathode lamps

These are the sources most used in AAS. They are composed of
—a cylindrical lamp with glass body and quartz window filled with a rare gas-argon or neon-under a pressure of 1-2 mm mercury;
—a cylindrical hollow cathode positioned to optimize emission of the light towards the atomizer; the cathode should be made of the same metal as the one to be determined, but multielement lamps are also available;
—an anode, usually consisting of a simple tungsten wire arranged so that it does not interfere with emission from the cathode.

A continuous direct current (200-300 volts) is applied between the cathode and the anode to cause ionization of the rare gas. The voltage is then moderated and stabilized at 100-200 V. Positively charged ions bombard the cathode at high velocity. Atoms are freed and excited to the resonance level closest to the ground state. Rapid return of the excited atoms to the ground state gives rise to a radiation composed of narrow rays, the most intense of which is used to measure atomic absorption. As the cathode temperature is lower than 200°C and the pressure within the lamp very low (1-2 mm of mercury), broadening of the emitted ray (Doppler effect + Lorentz effect) is of the order of, or lower than, 10^{-3} nm, while the width of the absorption radiation is about 10^{-2} nm. Thus the second condition laid down by Walsh (*see* § 1.1, Fig. 10.1) is obeyed.

The lamp current ranges from 3 to 10 mA for a primary voltage of 200 V. Stability of current should be ensured to within ±0.1% and it is advisable not to exceed 40-50% of the maximum permissible current (indicated on the lamp). The life of hollow-cathode lamps ranges from 500 to 5000 hours of operation.

The emitted light should be modulated for being electronically recognized when it leaves the monochromator and to eliminate the continuous emission from the flame. Modulators were earlier always composed of an electromechanical device placed where the radiation leaves the lamp (a half-opaque, half-transparent disc rotating at constant speed). Such systems have mostly been replaced by pulsed-current supply for the hollow-cathode lamps and, more recently, by square-wave alternating current.

At present hollow-cathode lamps are available for more than 70 elements. There are also multielement lamps whose use is controversial. Instruments are provided with turrets for holding several preheated lamps, thereby enabling multielement analysis by programming the rotation of the turret simultaneously with that of the monochromator and measuring system at the exit.

2.1.2 Other sources
Tentative attempts at use of sources that have had no more than ephemeral success will not be mentioned here. *Electrode-less lamps* are composed of a sealed quartz tube of a few mm internal diameter filled with a rare gas under a pressure of 2-3 mm mercury and containing about 10 mg of the element to be determined. This tube is placed at the centre of a coil fed by a high-frequency generator. The induced electrical field ionizes the rare gas and the ions produced atomize and excite the atoms of the element present in the lamp. These lamps are difficult to use because they require a high-frequency generator. However, they are complementary to the sources described earlier to the extent that elements not well suited to hollow-cathode lamps, in particular the volatile elements (Hg, As, Se, Te, Bi, etc.) are well analysed with such lamps. Nonetheless, these lamps are being abandoned because of the progress achieved in the manufacture of hollow-cathode lamps, especially for the volatile elements.

Use has been proposed of *coherent lasers* that have the advantage of being usable for very many elements by providing very narrow rays of high intensity in conformity with the requirements of atomic absorption (Fig. 10.1). However, these sources have not been retained by manufacturers because of their cost.

2.2 Atomization in Flames

2.2.1 Flames
In atomic absorption, since the inception of the technique, atomization has been most commonly accomplished in a flame, the sample (in liquid form) being introduced into the flame through a nebulizer. The role of the flame is to yield the maximum number of atoms of the element to be analysed in the form of a vapour of atoms in the ground state (*see* § 1.1 above).

The energy of the flame, and hence its temperature, should be considered first: if too low, atomization will not take place; if too high, ionization of the analyte may result and thus reduction of the yield of neutral atoms and therefore of sensitivity.

The nature of the fuels and oxidants to be used should be considered next. In general, the flame should have a high transmission coefficient and low self-emission in the spectral domain of the radiation traversing it.

Lastly, it is necessary to consider the chemical reactivity of the flame components *vis-à-vis* the element to be determined, some reactions being favourable for formation of neutral atoms, others unfavourable. This reactivity depends not only on the nature of the gases, but also on their proportion. Thus for certain elements reducing flames are preferred and for others, oxidizing.

The flame with lowest temperature used in AAS is the *air-propane flame*. Its temperature is around 1900°C; it is most useful for atomization of alkali and alkaline-earth metals, and also of Cu, Au, Ag, Zn, Cd, Mn and Fe. It has a high transmission coefficient around 220 nm, low self-emission and also low power to ionize atoms; caesium is partly ionized in this flame.

The commonly used flame is the *air-acetylene flame*, with temperature around 2300°C. It is very stable; the fuel : oxidant ratio can be controlled over a wide range, from highly oxidizing to highly reducing, without greatly changing the temperature. It has high transmission around 200 nm and low self-emission. It gives good atomization of more than 20 elements: alkalis, alkaline earths, Cu, Pb, Mn, Fe, Ag, Ni, Co, Zn, etc., albeit with partial ionization of the alkali metals.

The *nitrous oxide-acetylene* (N_2O-C_2H_2) *flame* enables extension of the range of AAS to many elements. Its temperature, nearly 2950°C, places it among the hottest of all the known laminar-flow flames and, furthermore, it has high atomization power. However, it has the disadvantages of high self-emission and high ionization energy for use with elements with ionization potential less than 5 eV. The air-acetylene and nitrous oxide-acetylene flames complement each other very well and enable determination of more than 70 elements. The latter is not mentioned without the former because it is very dangerous to light a nitrous oxide-acetylene flame directly. The air-acetylene should be lit first, as indicated in § 2.2.2 below. In analysis the N_2O-C_2H_2 flame is always used as a reducing flame. It is also possible to use a propane-nitrous oxide flame with a temperature of 2600°C.

2.2.2 Burners

The function of the burner is to produce a stable laminar-flow flame from premixed fuel, oxidant and an aerosol of the element to be determined, and then to take the element to the atomic vapour state. The material of the burner and its geometry are thus very important. Burners are often provided with long (10 cm), narrow (1 mm) slits in order to:

—attain the maximum sensitivity by increasing the length of the optical path across the population of neutral atoms (eqn 4);

—reduce the interfering emission by narrowing the width of the flame.

As the temperature at the level of the burner slit can reach 300-400°C, burners should be made of a chemically inert material; they are currently made most often of titanium.

Since burners almost always function in a laminar-flow regime, it is important that the rate of flow of the gas mixture across the slit be two or three times the speed of propagation of the flame. If it is higher than this, the flame becomes unstable and may be extinguished. If it is lower, the flame may strike back into the burner and cause an explosion. An appropriate burner is recommended for each gas mixture. Although the speed of propagation of the flame is independent of the fuel : oxidant ratio if the oxidant is air, the same is not true for nitrous oxide flames; speed of flame propagation then depends on the composition of the gas mixture and is least with reducing flames. That is why, to avoid a violent explosion it is imperative that the following protocol be strictly adhered to:
—light the air-acetylene flame;
—increase the flow of acetylene as instructed by the manufacturer;
—substitute the air with nitrous oxide;
—extinguish the flame in the reverse sequence.
In modern instruments these operations are automated.

Another problem regarding nitrous oxide should also be highlighted; the expansion of this gas, highly endothermic, causes intense cooling at the outlet of the regulator; the outlet should be warmed slightly to avoid condensation, which interferes with its functioning.

2.2.3 Nebulizers
As the sample is generally in solution, the role of the nebulizer is to introduce it uniformly throughout the flame. The sample is transformed into an aerosol by pneumatic pulverization by means of a nebulizer (Fig. 11.7 in Chapter 11). The solution is nebulized by the oxidant gas in the form of droplets that get entrained in the flow of expanding gases, and finally mixed in the burner with the fuel gas (acetylene).

The nebulizer chamber allows sorting of the mixture by separating the large droplets from the fine; a maximum diameter of 10 μm is recommended for proper operation, leading to atomization in the flame. A popular nebulization chamber has two small fans the blades of which change the direction of flow of the gas and homogenize it, while concomitantly eliminating large drops that adhere to the walls.

Although these nebulization systems are the most commonly used, they have the disadvantage of low yield as only a small fraction of the injected sample is atomized in the flame. Other techniques have been devised to send the sample in gaseous form to the fuel-oxidant mixture. In particular, ultrasonic nebulizers give increased sensitivity and abatement of certain matrix effects by virtue of better nebulization of the injected liquid (smaller droplet).

2.2.4 Hydride and cold-vapour techniques
Another well-known technique is that of formation of hydrides by the action of very strong reductants such as sodium borohydride in acid medium.

Under these conditions many elements (As, Sb, Se, Te, Bi, Sn, Ge, etc.) form volatile hydrides that are entrained in an inert gas in a cell heated to 900-1000°C, the temperature at which the hydride decomposes into neutral atoms:

$$As_2O_5 \rightarrow As_2O_3 \rightarrow AsH_3$$

and near 900°C

$$AsH_3 \rightarrow As^0 + 3/2\ H_2.$$

The sensitivity can be increased by a factor of 10 to 50 compared to the classic method described earlier (§ 2.2.3 above).

Lastly, it is useful to separately treat the case of mercury, which is a very good illustration of the theory of atomic absorption (*see* § 1.1 above). Actually, atomization of the element is accomplished in 'cold vapour' at room temperature. The mercury atom in ionic state is displaced from the medium by reduction to oxidation state zero:

$$HgCl_2 + SnCl_2 \Leftrightarrow Hg^0 + SnCl_4.$$

A certain vapour pressure of the element is created in the gas in equilibrium with the cell in a quartz cell traversed by the rays from a mercury lamp.

2.3 Mechanisms of Atomization in Flames

When a solution is introduced in the flame in the form, for example, of a mist, a sequence of physical and chemical reactions takes place leading to atomization-in particular, fusion, volatilization, dissociation and decomposition reactions leading to formation of free atoms and recombinations (especially with the combustion products) giving oxides and hydroxides, reduction reactions in presence of reducing compounds and radicals favouring formation of free atoms and, lastly, reactions that lead to ionization or deionization of the atoms.

In defined flame conditions there is always a tendency towards thermodynamic equilibria regulated by the Law of Mass Action. Knowledge of these equilibria enables understanding and correction of interactions that interfere with measurement of absorbance of an element in a complex medium (matrix effect).

2.3.1 Principal reactions (Pinta, 1978)

For a salt of formula MA in solution, the first step leads to volatilization of the compound MA; the reactions are the following:

$MA_{solution} \rightarrow MA_{solid}$: desolvation in the flame, formation of a solid-gas aerosol;

$MA_{solid} \rightarrow MA_{liquid}$: fusion (liquid-gas aerosol);

$MA_{liquid} \rightarrow MA_{vapour}$: volatilization.

Also taking place are:

$MA_{solid} \rightarrow MA_{vapour}$ (sublimation);

$MA_{vapour} \rightarrow M^0 + A^0$ (dissociation);

$MA_{vapour} \rightarrow M^* + A^*$ (excitation);

$MA_{vapour} \rightarrow M^+ + A^-$ (ionization);

$MA_{vapour} \rightarrow MA^*$ (molecular excitation).

All these reactions presume direct decomposition of the original salt into atoms. These are, in fact, simple theoretical reactions; the process is not always simple and intermediate compounds do form.

2.3.2 Reactions with flame constituents

These encompass all possible reactions with the compounds and free radicals that can result from combustion:

$M + O \rightarrow MO$ (oxide formation)

$M + OH \rightarrow MO + H$ (oxide formation)

$MO + C \rightarrow M + CO$ (reduction)

$M + C \rightarrow MC$ (carbide formation).

All these reactions are actually equilibria resulting from the nature of the flame and combustion conditions (ratio of gas flows).

2.3.3 Reactions with elements of the complex medium

An element is generally determined starting from a complex medium and results in atomization of a large quantity of foreign elements.

The reactions observed are:

$M + M' + O \Leftrightarrow MOM'$ (double oxide);

$MA + B \Leftrightarrow MB + A$ (anion exchange);

$MB \Leftrightarrow M + B$.

These reactions may be summarized by the scheme:

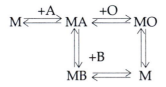

2.3.4 Practical examples

Alkali metals

Taking the example of sodium chloride, atomization passes through the following stages: desolvation-fusion-volatilization; it is the NaCl vapour that will be dissociated in the flame. The melting and boiling points of sodium chloride are respectively 1100 K and 1750 K, lower than the temperature of traditional flames.

What generally happens with alkali halides is dissociation of the original salt in the gaseous phase.

Alkaline-earth metals

The following reactions are observed with calcium chloride in the condensed phase and then in the vapour phase:

$$CaCl_2.H_2O_{liquid} \Leftrightarrow CaCl_{2(vapour)} + H_2O \tag{1}$$

$$CaCl_2.H_2O \Leftrightarrow CaO + 2HCl \tag{2}$$

But as the boiling point of $CaCl_2$ is lower than that of CaO, it is always reaction (1) that favours formation of atoms:

$$CaCl_2 \Leftrightarrow Ca^0 + Cl_2 \quad \text{and}$$

$$CaO \Leftrightarrow Ca^0 + O$$

Thus atomization can take place in two different ways. The most favourable pathway is volatilization of $CaCl_2$ followed by decomposition, while oxide formation yields a product that is not volatile in flames.

With calcium nitrate, decomposition to CaO is almost total[1]:

$$Ca(NO_3)_2 \rightarrow CaO + N_2O_5, \text{ then } CaO \Leftrightarrow Ca + O$$

Decomposition of CaO takes place before volatilization (melting point of CaO is 2850 K). Therefore, atomization of calcium in a given flame is better from the chloride than from the nitrate.

Let us now consider the atomization of calcium salts in the presence of an aluminium salt. In hydrochloric acid medium:

$CaCl_2.H_2O \Leftrightarrow Ca + Cl_2 + H_2O$
$CaCl_2.H_2O \Leftrightarrow CaO + 2HCl$
$CaO \Leftrightarrow Ca + O$
$2AlCl_3.6H_2O \Leftrightarrow Al_2O_3 + 6HCl + 3H_2O$
finally, $CaO + Al_2O_3 \Leftrightarrow CaAl_2O_4$

At the temperature of the flame, the double oxide formed will be CaAlO, which is more stable than $CaAl_2O_4$.

In nitric acid medium:

$Ca(NO_3)_2.3H_2O \Leftrightarrow CaO + N_2O_5 + 3H_2O$
$2Al(NO_3)_3.9H_2O_{(vapour)} \Leftrightarrow Al_2O_3 + 3N_2O_5 + 9H_2O$

In nitric acid medium, calcium atoms result from decomposition of CaO, but the number of atoms is smaller than with calcium chloride. Furthermore, the presence of an aluminium salt interferes with atomization of calcium by formation of a thermally stable double oxide of calcium and aluminium, which complexes part of the calcium atoms. Interference with atomization

---

[1] As reaction mechanisms at high temperature are sometimes poorly understood, some of the reactions given are only indicative.

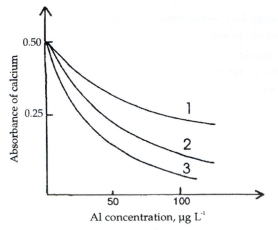

Fig. 10.3   Interaction of aluminium (1-100 μg L^{-1}) on absorbance of calcium (1 μg L^{-1}): 1 - hydrochloric acid medium; 2 - nitric acid medium; 3 - sulphuric acid medium.

of calcium by aluminium is greater in nitric acid than in hydrochloric acid medium (Fig. 10.3). In other words, for the same concentration of calcium, when large quantities of aluminium are present, the number of neutral calcium atoms formed in the flame is greater in hydrochloric acid than in nitric acid medium.

2.3.5 Chemical interactions

Molecular combinations can be formed from matrix compounds:

$$M_1 + M_2 + O \Leftrightarrow M_1M_2O$$
$$M_1 + M_2 \Leftrightarrow M_1M_2$$

Equilibrium reactions: the degree of dissociation increases with rise in temperature; it is also a function of the dissociation constant of the complex formed. Some indicative examples are

$$2CaCl_2 + 2H_3PO_4 \Leftrightarrow Ca_2P_2O_7 + 4HCl + H_2O \text{ (stable pyrophosphate)};$$
$$CaCl_2 + Na_2SiO_3 \Leftrightarrow CaSiO_3 + 2NaCl$$

Hydrofluoric acid also has an effect on aluminium and titanium but, as Al and Ti fluorides are more volatile than the oxides, absorption is exaggerated.

Other examples in the air-acetylene flame can be cited:

$$CaX + 2MoX + 7O \Leftrightarrow CaMo_2O_7 + 3X \text{ (or } CaMoO + X)$$
$$CaX + 2AlX + 4O \Leftrightarrow CaAl_2O_4 + 2X \text{ (or } CaAlO + X)$$

These compounds are thermally stable in the flame. In nitrous oxide-acetylene we have:

$$AlX + 2VX + 4O \Leftrightarrow AlV_2O_4 + 3X \text{ } (\rightarrow AlVO)$$
$$AlV_2O_4 \Leftrightarrow Al^0 + 2VO + 2O$$

But in this case the absorbance due to Al^0 is higher than that resulting from dissociation of AlX:

$2AlX + 3O \Leftrightarrow Al_2O_3 + 2X$

$Al_2O_3 \Leftrightarrow 2Al^0 + 3O$

The compound AlV_2O_4 (or AlVO) is more volatile than the oxide Al_2O_3 (or AlO).

Again:

$CaX + TiX + 3O \Leftrightarrow CaTiO_3 + 2X$

The compound $CaTiO_3$ (perovskite) is more stable than CaO.

$2CaX + 2FeX + 5O \Leftrightarrow Ca_2Fe_2O_5 + 4X$ (or CaFeO + 2X)

$2CrX + FeX + 4O \Leftrightarrow FeCr_2O_4 + 3X$ (or FeCrO + 2X)

$FeX + TiX + 3O \Leftrightarrow FeTiO_3 + 2X$

The compounds formed are thermally stable and thus responsible for interactions.

2.3.6 Elimination of chemical interactions

It must be realized that the presence of a foreign element is likely to modify the volatilization of the compound studied and also the dissociation of the latter. This principle will be put to use for improving absorbance or for eliminating certain interactions. Thus, for example, addition to the test solution of a lanthanum salt eliminates the effect of aluminium on absorbance of alkaline-earth elements. The lanthanum salt should be in an amount 10 to 20 times higher than the interfering element.

2.3.7 Ionization reactions

The energy of flames is often sufficient to displace and remove one of the outer electrons of neutral atoms; this causes ionization. Actually ionization is caused when the energy of the flame is greater than the ionization potential of the element. Some values of ionization potential are given in Table 10.1 with the percentage of ionized atoms (N^+) relative to the number of neutral atoms (N^0) in flames of different temperatures.

Table 10.1 Percentage of ionized atoms in flames.

	Ionization potential, eV	Air-propane flame, 2200 K	Oxyhydrogen flame, 2900 K	N_2O-C_2H_2 flame 3200 K
Lithium	5.37	0.01	1.0	16
Sodium	5.12	0.30	5.0	26
Potassium	4.32	2.50	31.0	82
Rubidium	4.16	13.50	44.0	89
Caesium	3.87	28.30	69.0	96
Calcium	6.11	—	1.0	7
Strontium	5.69	—	2.7	17
Barium	5.21	—	8.6	42

But the degree of ionization of an element is modified by the presence of another ionizable metal; the following reactions occur:

$M \Leftrightarrow M^+ + e^-$ and

$M' \Leftrightarrow M'^+ + e^-$.

Electroneutrality is written as

$[M^+] + [M'^+] = [e^-]$.

If M' is present in large quantity in the matrix, release of a large number of electrons $[e^-]$ displaces the ionization equilibrium $M^0 \Leftrightarrow M^+ + e^-$ in the direction of increased formation of neutral atoms. The element M is then said to suffer ionization interference from the metal M'. Ionization reactions exaggerate absorbance of the analyte element.

The elements most sensitive to ionization interactions are the alkali and alkaline-earth metals with low ionization potentials. Ionization interactions are particularly sensitive in the nitrous oxide-acetylene flame.

Elements such as the rare earths, aluminium, titanium, etc., may give a high percentage of ionized atoms in the nitrous oxide-acetylene flame.

Ionization thus depends not only on flame temperature but also on chemical composition of the matrix, which may contain ionizable elements as well, that is, those capable of liberating electrons and ionizable molecules. Thus there are physicochemical equilibria in addition to dissociation equilibria.

2.4 Electrothermal Atomization

2.4.1 Atomizers

Electrothermal atomization was proposed by L'vov of the University of Leningrad in 1959, about five years after the birth of AAS. This system, which effects great improvement in sensitivity of determination of many trace elements, has come to be known as electrothermal atomic absorption spectrometry (EAAS) and is recognized as a very important development (Hoenig and Kersabiec, 1990). The L'vov chamber is a small dish-shaped graphite cylinder on which the sample is placed and volatilized rapidly at high temperature by an electric arc.

The process was perfected by Massmann, and Massmann furnaces are today the most popular for EAAS instruments. In this system, the sample is introduced manually or automatically in a graphite furnace heated by the Joule effect.

The furnace is protected from burning by a continuous stream of inert gas. The ideal inert gas is helium because of its high thermal conductivity, but argon is often preferred because of lower cost. The current may range from 0 to 400 A under a voltage of 0 to 10 V, allowing working at up to about 3000°C, with external cooling provided by circulation of water in a sleeve surrounding the furnace.

In soil analysis, flame AAS is the tool recommended for many major and exchangeable elements with contents of a few mg kg^{-1}. For elements with lower contents (mg kg^{-1} to µg kg^{-1} or less), flame AAS is no longer used, and EAAS is the method of choice.

2.4.2 Programming the electrothermal cycle

Although some techniques are available for the introduction of solid or gaseous samples, the samples are generally in solution, just as in classic AAS; a volume of 5 to 20 µL, accurately measured, is introduced directly into the cold furnace and subjected to a programmed heating cycle leading to atomization of the test element (Pinta, 1979):

—drying phase: first step to around 100°C for evaporating the solvent giving non-specific absorption (Fig. 10.4);

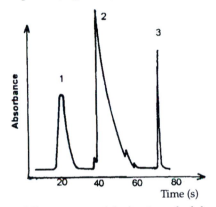

Fig. 10.4   Absorbance during different stages of the heating schedule
1 - drying or desolvation; 2 - decomposition; 3 - atomization.

—thermal pretreatment phase (decomposition, pyrolysis, thermal cracking): very variable according to nature of the sample (analytical matrix) between 200 and 1000°C; this is for destroying the organic salts or compounds by eliminating the fumes prior to atomization; this results in strong non-specific absorption;

—atomization phase: ranges from 1500 to 3000°C according to the element to be analysed; absorbance is measured during this phase.

For each determination a heating cycle optimizing the sensitivity and reproducibility should therefore be programmed. The settings are delicate enough for most of the temperatures indicated above; other settings can be selected for temperature programming. In practice, for a sufficiently rapid rise in temperature and sufficient drying and decomposition time to evacuate the fumes, it is above all the two variables 'decomposition temperature' (θ_1) and 'atomization temperature' (θ_2) that are the most important.

The process can be illustrated by considering these two variables θ_1 and θ_2, assuming there is no interaction between them. For a constant drying

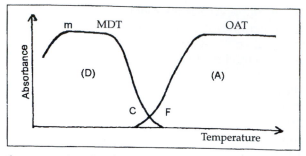

Fig. 10.5 Thermal programming: D - decomposition curve: variable decomposition temperature, and fixed atomization temperature; A - atomization curve: variable atomization temperature, and fixed decomposition temperature.

temperature (100°C for 15-20 seconds for 10 mL), the influence of decomposition temperature on absorbance is studied, with the atomization temperature fixed at an arbitrary value. Then the absorbance is studied at variable atomization temperature, the decomposition temperature being kept constant. Variations in absorbance as a function of these temperatures are shown in Figure 10.5.

Decomposition (see Fig. 10.5)
The maximum decomposition temperature (MDT) marks the start of volatilization of the molecule MX or MO containing the test element. Beyond this temperature some of the molecules are volatilized during the decomposition step and are lost at the time of atomization. The temperature at the point F marks the end of the decomposition curve. Beyond this temperature all the molecules are found volatilized during the decomposition step. The starting point (m) of decomposition seen here is observed mostly with complex media. Below this value (m) heating is insufficient to destroy all the substances. To summarize, interpretation of the curve of decomposition *vs* temperature θ_1 is as follows:

θ_1 < m: incomplete volatilization, insignificant absorbance;

m < θ_1 < MDT: total volatilization and decomposition, maximum absorbance;

MDT < θ_1 < F: volatilization losses in the decomposition cycle; insignificant absorbance;

θ_1 > F: total loss by volatilization during the decomposition cycle.

Atomization
A decomposition temperature slightly lower than the MDT is selected. The optimum atomization temperature (OAT) corresponds to total decomposition of the MX molecules into free atoms; below this, heating is insufficient to destroy all the gaseous molecules. The temperature of the point C marks the start of the atomization curve, that is, the beginning of degradation of gaseous molecules. The temperature of the points MDT and C should coincide and so should those of the points F and OAT if the gaseous molecules have

decomposed immediately after their volatilization. Interpretation of the atomization curve enables selection of the atomization temperature θ_2 relative to the temperatures OAT, C and F:

θ_2 < C: incomplete volatilization, no dissociation; zero absorbance;

C < θ_2 < OAT: incomplete dissociation; insignificant absorbance;

θ_2 > OAT: total volatilization and dissociation; maximum absorbance.

Thus the optimum analytical conditions are:

decomposition $\theta_1 \leq$ MDT; atomization $\theta_2 \geq$ OAT.

The time taken by each of these three stages is a function of the volume, medium and analyte; the decomposition time should then be studied carefully. The rate of rise in temperature of the tube is the maximum speed compatible with the instrument : $2000°C\ s^{-1}$.

2.4.3 Optimization of the electrothermal cycle

Although data are available in specialized literature on the recommended settings for each element, it is preferable to optimize the analytical conditions for the available instrument from time to time. The programming technique described in § 2.4.2 is illustrative of the phenomena. It lacks rigour, however, when the variables to be optimized (in this case, decomposition and atomization temperatures) interact between themselves. Furthermore, the experiment is not very economical.

Optimization techniques should thus be employed. Among them, experimental design and the methodology of response surfaces are the tools of choice.

Teaching the use of these techniques in AAS for several years has proved their never-failing efficiency, as also the reproducibility of the results obtained with simultaneous optimization (Fig. 10.6) of decomposition and atomization temperatures in the determination of lead (CNRS-Formation, France, 1986-1997).

2.5 Optical and Detection Systems

The *monochromator* serves to select the wavelength corresponding to the analytical radiation from among the various wavelengths emitted by the source. Although these are few in number, the quality of the monochromator is not as essential as in emission spectrometry (*see* Chapter 11). The monochromator can be a dispersive prism, but is most often a grating. The quality of a grating depends on the number of lines per mm, which ranges from 600 to nearly 3000 in various instruments. The dispersion of the rays depends on this number of lines per mm and the focal length of the objective lens.

Many types of more or less sophisticated instruments are available: single-beam or double-beam, single-channel or double-channel, etc. The reader is referred to specialized publications such as that of Pinta and Laporte (1979) or to manufacturers' catalogues.

Factor Midpoint Step	DT, °C 450 200	AT, °C 1850 250	Peak height
Experiment			
1	250	1600	707.5
2	650	1600	665
3	250	2100	942.5
4	650	2100	970
5	450	1850	907.5
6	450	1850	907.5
7	733	1850	892.5
8	167	1850	902.5
9	450	2204	980
10	450	1496	542.5
11	450	1850	940
12	450	1850	920

(table heading: Experimental Design)

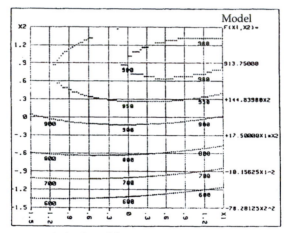

Fig. 10.6 Simultaneous optimization of decomposition temperature (DT→X1) and atomization temperature (AT→X2) in the case of determination of lead by EAAS. The experimental design (entire central composite) and the determination of the optima by the method of response surfaces: DT = 450 + 0.9 the 200 = 630°C, AT = 1850 + 1 × 250 = 2100°C (Pansu, 1985).

The *detection systems* almost always use an electron photomultiplier (PM: Fig. 9.4, Chapter 9). The incident radiation strikes an electrode (cathode k1) coated with a thin layer of an alloy containing easily ionizable elements (Cs, Rb, Li, etc.). The incident photons cause ejection of electrons that then strike another electrode (k2) at great speed, each electron in turn expelling 2 to 5 electrons towards the next electrode and so forth, to finally reach the anode. Photomultipliers with multielement (alkali metal) cathode ensure high sensitivity in the wide spectral range used in AAS.

However, for some specific determinations (As, Se, Li, Rb, Cs) using the lower limit (190-200 nm) or the upper limit (750-850 nm) of the spectrum, photomultipliers specially designed for these ranges are recommended.

2.6 Correction for Non-specific Absorption

Some compounds or particles in the atomization source may cause non-specific absorption and interfere with determination of test elements. There are two principal types of optional apparatus for correction of the spectral background: a deuterium lamp and application of the Zeeman effect.

The deuterium lamp (Fig. 10.7) emits a continuous spectrum and allows measurement of non-specific absorbance on both sides of the wavelength of the test element. This absorbance can then be deducted from the total absorbance to give the absorbance of the test element alone.

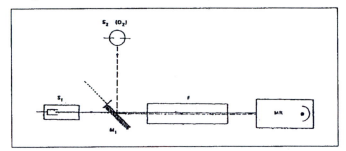

Fig. 10.7 Optical scheme of a single-beam spectrometer with background correction (Pinta and Laporte, 1979)
S_1 - hollow-cathode lamp; S_2 - deuterium lamp; M_1 - pivoted half-silvered mirror; F - atomization source; MR - monochromator and acquisition system.

The Zeeman effect enables correction for non-specific absorption at wavelengths closer to that of the analyte. The correction is then more exact especially with a non-uniform background. Application of the Zeeman effect in atomic absorption has been well described (Pinta, 1985; Hoenig and Kersabiec, 1990). Briefly, the principle of decomposition of spectral lines in a magnetic field (Fig. 10.8) is used. Along the magnetic field axis the original

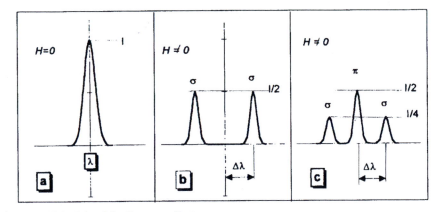

Fig. 10.8 Principle of the Zeeman effect
a - spectral range in absence of magnetic field; b - longitudinal Zeeman effect; c - transverse Zeeman effect.

line λ disappears, to give rise to a doublet in which the components are polarized in opposite directions. In the direction perpendicular to the field axis a triplet with three plane-polarized components is observed. In AAS the transverse Zeeman effect is most commonly used. The magnetic field can be applied directly on the emission source (direct Zeeman effect) or on the atomization source (reverse Zeeman effect). The total absorbance is measured without the magnetic field (A_λ), then the non-specific absorbance is measured on the radiations σ when the field is applied (A_σ). The specific absorbance is then $A_\lambda - A_\sigma$. By this procedure almost total correction for non-specific absorption is obtained.

3. Practical AAS

3.1 General Scheme of an Analysis

Although it is possible by specific techniques to conduct analyses in a solid or gaseous medium, the range of AAS chiefly concerns analysis of solutions. Therefore the first step in analysis is always taking the sample in an appropriate solution.

In the case of soils and rocks, it is necessary to proceed with extractions or total or partial dissolution with various reagents; Pinta (1980) has listed 11 dissolution methods. The preparation of samples of water and other extracts of soil could be simpler, but always require thoughtful planning (filtration of microparticles or otherwise, acidification, change of medium, etc.). Once the sample is in solution, the next step is correction of the medium: dilution, addition of a matrix corrector, etc.

Along with sample preparation, standard solutions will also have to be made either for external calibration or by division of the samples by the method of standard additions (*see* Chapter 18). Standard solutions should contain a medium as close as possible to that of the samples, the medium being exactly the same in the method of standard additions. In external calibration care should be taken to add the same reagents to the standard solutions as to the samples. Also, analysis should be done as far as possible in the ranges where law 4 (§ 1.1) is followed, that is, the optical density varies linearly with concentration. Figure 10.9 shows that the concentration ranges can be very variable according to the element determined; it also shows the great difference in sensitivity between FAAS and EAAS. Calibration curves of polynomial type enable noticeable extension of the analytical range of each element, but it must be kept in mind that precision is always the best at concentrations near the middle of the range of linearity (*see* Chapter 18).

The settings of the spectrometer should be made considering the actual analytical conditions of the element to be determined: selection of flame type and composition (*see* § 2.2 above), selection of the heating schedule in

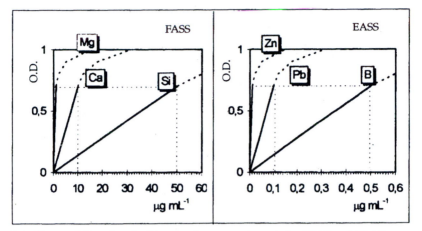

Fig. 10.9 Range of linearity of calibration for some elements by FAAS and EAAS; examples of curves (C. Riandey, IRD, Bondy, France, pers. comm.).

EAAS (§ 2.3), selection of emission lamp and settings appropriate for the element to be determined (*see* § 2.1), setting of slit width (band pass), attenuation, etc. Thereafter the actual analysis of standard solutions and samples is done.

The settings of a spectrometer can be obtained from the literature or by laboratory experimentation. But these conditions can vary from one instrument to another, and with age of an instrument. So it is advisable to periodically re-optimize certain conditions (Figs. 10.5 and 10.6).

3.2 Interferences in AAS

3.2.1 Generalities

Although sensitive and precise, atomic absorption spectrometry is, like many other analytical techniques, subject to various interferences, the correction for which should be known. These problems were reviewed by Riandey (1979). Interferences can be of various physical and chemical types. Only the principal interferences that affect atomization and thus the sensitivity of determination are touched upon here. They are
—dissociation interferences,
—ionization interferences,
—redox interferences,
—chemical interferences.

3.2.2 Dissociation interferences
To achieve atomization, the molecules should be dissociated at high temperature in accordance with reversible reactions of the type $NaCl \Leftrightarrow Na^0 + Cl$.

Excess chlorine, such as in hydrochloric medium, displaces the equilibrium to the left and sodium in the atomic state is decreased. These interactions reduce the absorbance.

However, these effects are slight and do not lead to errors in analysis if standard solutions with the same concentration of hydrochloric acid as the sample solutions are used. For improved sensitivity, it is even advised that analysis should be done in a hydrochloric acid medium as dilute as possible (1%).

3.2.3 Ionization interferences

It was shown earlier that ionization is a troublesome phenomenon because it diminishes the atomic absorption by neutral atoms. If the atom of the test element only is present, it is easy to predict ionization as a function of temperature, and tables give the ionization potentials of elements (Table 10.1).

The ionization equilibrium of the analyte M can be expressed by

$$M^0 \Leftrightarrow M^+ + e^-$$

When another ionizable element is present along with the test element, the concentration of free electrons rises in the flame and, in accordance with the Law of Mass Action, the equilibrium is shifted to the left; the concentrations of M^+ ions in the flame decreases and, simultaneously, the concentration of neutral atoms M^0 increases and so does the absorption due to these atoms. The risk of error is thus now of excess.

Other ionization interactions can also take place. For example, the electron affinity of halogens is high and can lead to reactions of the type

$$X^0 + e^- \Leftrightarrow X^-$$

or, if hydrochloric acid is present,

$$HCl + e^- \Leftrightarrow Cl^- + H$$

Electrons are thus captured instead of being released, and the equilibrium of the test element is shifted in the direction of its ionization, which leads to a negative error. Other interactions of this type can also take place but, in sum, ionization interferences are frequent.

Such interferences are readily corrected, however. First of all, excessive atomization temperatures should be avoided, especially for the alkali metals and other easily ionizable elements. Addition is often recommended of an *ionization buffer* in the form of an easily ionizable salt such as those of the alkali metals: CsCl, 0.5 mg mL^{-1}; KCl, 1 mg mL^{-1}. Lanthanum is also an ionization buffer, especially in nitrous oxide-acetylene flames. Ionization buffers should be added in the same concentration to all solutions (standards and samples). It is also useful to apply the same rule to prevent other effects; thus the significant effect of acids on precision will be eliminated if the acidity of standard and sample solutions is kept at the same level.

3.2.4 Redox interferences

These are caused by lowering of the oxidation state of a metal in the presence of another. This interference occurs chiefly in reducing flames in which the partial pressure of oxygen is as sensitive to influence as it is low. For example, consider the equilibrium

$$FeO \Leftrightarrow Fe^0 + O$$

In a reducing air-acetylene flame, the equilibrium is shifted to the right and absorption is augmented. When aluminium is present, this effect is further amplified because of fixation of oxygen atoms existing or formed:

$$FeO + Al \Leftrightarrow Fe^0 + AlO$$

The great affinity of aluminium atoms for oxygen also explains why this element is not atomized in the air-acetylene flame.

In the N_2O-C_2H_2 flame, oxidation is equally diminished by the reducing character of the flame and the presence of free radicals such as CN and NH. The interference can be as high as 30% of the absorption signal. It affects the elements that form oxides stable at high temperature: Mg, Ca, Fe, Al, Ti, V, etc.

It is possible to partially correct this interference by adjusting the reducing character of the flame, and also by adding the interfering element to the standard solutions. But the most classic correction is obtained by adding an excess of lanthanum to all the solutions (standards and samples). The stability of lanthanum monoxide is such that it fixes the oxygen of the flame according to the reaction:

$$2LaCl_3 + O_2 \Leftrightarrow 2LaO + 3Cl_2$$

This element thus behaves as a veritable redox buffer. It is generally used at a concentration of 5-10 mg mL^{-1} in the solution.

3.2.5 Chemical interferences

These interferences are caused by formation of stable compounds of the test element with another. According to whether the compound formed has greater or less tendency than the original salt to release neutral atoms, the effect heightens or lowers the signal (Table 10.2).

Most compounds formed in the flame and identified at the ambient temperature are double oxides of the general formula $A_xB_yO_z$, among which can be counted some major crystalline groups. One group has the structure of spinel $MgAl_2O_4$, another that of perovskite $CaTiO_3$ or sometimes of α-Al_2O_3, etc. In the flame, the compound formed would be simpler (CaAlO instead of $CaAl_2O_4$, MgAlO instead of $MgAl_2O_4$).

Organic substances may form condensation compounds that interfere with atomization of the test element.

Chemical interactions are the most common and the most troublesome. According to the volatility of the compound formed, errors of underestimation or overestimation in the range of 30-100% can be caused. These interactions vary with temperature, magnitude of the measurement in the flame, and its reducing character (decrease in formation of oxides); these parameters can therefore be adjusted for certain corrections.

Table 10.2 Chemical interferences (Riandey, 1979).

Interaction	Compound	Effect (– depression, + increase, c correction)
Al/Ca	$CaAl_2O_4$	–
Al/Mg	$MgAl_2O_4$	–
Al/V	AlV_2O_4; AlV_2O_6	+
Ca/Mo	$CaMo_2O_7$	–
F/Al	AlF_3	+
F/Mg	MgF_2	–
F/Ti	TiF_4	+
F/Zr	$ZrOF_2$	+
Fe/Ca	$CaFe_2O_4$	–
	$Ca_2Fe_2O_5$	–
Fe/Cr	$FeCr_2O_4$	–
Fe/Li	$Li_2Fe_2O_4$; $LiFe_5O_8$	–
Fe/Ti	$FeTiO_3$	–
Hf/Sr	$SrHfO_3$	–
La/Al	$LaAlO_3$	c
La/PO$_4$	$LaPO_4$	c
	$LaPO_4$; $3La_2O_3.2P_2O_5$	–
Mo/Ca	$CaMoO_4$	–
PO$_4$/Ca	Apatites	–
	$CaCl_2$	c
	$CaHPO_4$; $Ca_2P_2O_7$	–
	$Ca_3(PO_4)_2$; $Ca_2P_2O_7$	–
PO$_4$/Er, Eu, Y, Yb	MPO_4	–
PO$_4$/Sn	$Sn_3(PO_4)_2$	–
SO$_4$/Er, Y	$M_2(SO_4)_3$	–
SO$_4$/Eu, Yb	MSO_4	–
Ti/Ba	$BaTiO_3$	–
Ti/Ca	$CaTiO_3$	–
V/Ca	$Ca_2V_2O_7$	–
Zr/Ca, Sr	$MZrO_3$	–

Another highly recommended remedy is the addition of 'chemical-interaction buffers'. Many elements can be used for this—two are most efficient: strontium and, more so, lanthanum. The additive has the effect of combining with the interfering element in preference to the test element, which is then released. For example, phosphorus is fixed by lanthanum in the form of $LaPO_4$, aluminium as $LaAlO_3$. Table 10.3 shows the very high stability of this compound at high temperature compared to other possible combinations.

Preferential fixation of the interfering element over the test element by the buffer is stronger still when the buffer is greatly in excess. A concentration at least 10 to 20 times that of the interfering element is recommended. Thus, excess of strontium leads to preferential formation of $SrAl_2O_4$ compared to $CaAl_2O_4$ despite the slightly lower stability of the former (Table 10.3), but in conformity with the Law of Mass Action. In the case of lanthanum, additions giving a final concentration of 5-20 mg mL^{-1} are recommended. Figure 10.10 reproduces an example of corrections carried out by addition of different amounts of lanthanum for the determination of calcium.

Table 10.3   Thermodynamic stability of various compounds of some elements (Riandey, 1979).

Compound	Energy of formation, kcal mol^{-1}; 2000 K	Energy of formation, kcal mol^{-1}; 2400 K
BeO	−96	—
$BeAl_2O_4$	−345	—
$MgCl_2$	—	−73
MgCl	—	−29
MgO	−77	−57
$MgAl_2O_4$	−333	−284
CaO	−96	—
$CaAl_2O_4$	−348	—
SrO	−89	—
$SrAl_2O_4$	−339	—
BaO	−91	—
$BaAl_2O_4$	−340	—
$AlCl_3$	—	−104
Al_2O_3	−248	−216
AlO	—	−18
Al_2O	—	−67
Al_2O_2	—	−78
La_2O_3	−293	−266
4/3 $LaAlO_3$	−360	—

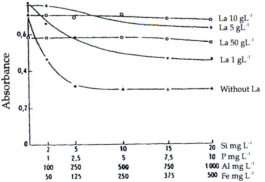

Fig. 10.10   Simultaneous correction of the interference of silicon (Na_2SiO_3), phosphorus (H_3PO_4), aluminium ($AlCl_3$) and iron ($FeCl_3$) in the determination of calcium (15 mg mL^{-1} $CaCl_2$, 422.6 nm) by increasing concentrations of lanthanum in the form of $LaCl_3$ (Riandey, 1979).

3.2.6 Interferences in EAAS

Mechanisms of the same type as described above for flames operate in furnace atomizers too. They are quantitatively very different: clearer separation between decomposition and atomization processes, formation of carbides with the walls of the graphite furnace, etc. Programming of the electrothermal cycle described in § 2.3.2 enables minimization of the interferences. We shall only mention here that the two major types of interference most often observed are:

—spectral interferences that can be manifested by non-specific absorbances of value often much greater than in flames; they can be corrected by using the deuterium lamp or by application of the Zeeman effect; the non-specific absorption is measured by these methods and then subtracted from the total absorption (*see* § 2.5 above);

—chemical interferences related to the atomization process and shifting of dissociation equilibria; their correction also requires use of matrix and analyte modifiers.

3.3 Some Practical Hints

The list of interferences presented in § 3.2 could give the impression that AAS is very complicated and subject to a large number of errors. In fact, the interferences in this technique are very often lower than in most other analytical methods when they are also subjected to detailed study.

In FAAS, it is important to simplify the anionic content of the medium by having just one predominant anion (1% HCl or HNO_3).

It is useful to keep in mind the principal interferences before perfecting an analytical method. Then the interaction caused by certain additions can be checked; for example, addition of a potassium salt will enable the analyst to decide between two principal kinds of interactions—ionization and chemical. Also, judicious addition of lanthanum is an effective way of correcting three of the four kinds of interactions, especially in the nitrous oxide-acetylene flame.

The buffers should be added to samples and standards in minimal and uniform quantity, and blanks containing all the reagents should be run (values to be subtracted from the results). The selection of technique for dissolution of the sample prior to analysis can also be defined as a function of possible interferences: for example, a flux used for the dissolution such as strontium metaborate may also serve as an interference buffer.

Generally dilution is always advisable; it reduces interferences and allows the ratio of buffer to interfering element to be increased. A general method for minimizing error consists of working with a standard solution in a complex medium, containing the same elements as the samples. In practice, it suffices to add to the standards the most abundant major elements; an alternative is to replace external calibration by the method of 'standard additions' (*see* Chapter 18).

In EAAS, especially with 'volatile' elements, selection of the matrix modifier is important.

Also available in the literature are specific methods for correction of the absorption signals. For example, the important work accomplished by Bois *et al.* (1984) evaluates the magnitude of a large number of binary interferences by attempting to relate them to various atomic parameters.

Another technique used to correct for non-specific absorption consists of using the Zeeman effect of decomposition of spectral lines in a magnetic field (*see* §2.6; also Pinta *et al.*, 1982).

A highly recommended procedure for ensuring precision of results consists of including in each set of analyses one or two reference samples of the same type as those being tested. There are several certified reference materials for soils and rocks, which should be used. Failing this, the laboratory can use its true typical samples that will be checked by various protocols in AAS and with other techniques (*see* Chapter 18).

3.4 Principal Operating Conditions in FAAS

Table 10.4, taken from Voïnovitch (1988), is a compilation of the principal operating conditions. In actual practice the manufacturer of the instrument and the literature on analytical chemistry will be reported in the data.

3.5 Operating Conditions in EAAS

Table 10.5 gives certain matrix modifiers and recommended decomposition and atomization temperatures for various elements. Rather than those given in Table 10.4, it is advisable that the values in Table 10.5 be used, which are only indicative, and to reoptimize the settings of the instruments actually used and the analytical media (*see* § 2.3 above).

Table 10.5 Analytical conditions in EAAS (Hoenig and Kersabiec, 1990)

Element	Modifier	Decompn T, °C	Atomizn T, °C
Al	$Mg(NO_3)_2$	1700	2600
Ag	$Mg(NO_3)_2$	950	2200
As	$Ni(NO_3)_2$	1400	2500
	$Pd(NO_3)_2$	1500	
Be	$Mg(NO_3)_2$	1500	2700
	NH_4OH	1600	
Bi	$Ni(NO_3)_2$	900	2200
Cd	$Mg(NO_3)_2 + NH_4H_2PO_4$	900	2200
	$Pd(NO_3)_2$	800	
Cr	$Mg(NO_3)_2$	1650	2600
Co	$Pd(NO_3)_2$	1100	2600
Cu	$Pd(NO_3)_2$	1100	2500
Fe	$Mg(NO_3)_2$	1400	2600
Mn	$Mg(NO_3)_2$	1400	2600

(*Contd. p. 202*)

Table 10.4 Principal analytical conditions in AAS (after Voïnovitch, 1988).

D.L.—detection limit; U.L.—upper limit of range of determination; λ—wavelength.

Element	λ, nm	Slit width	D.L. ng mL^{-1}	U.L. μg mL^{-1}	Flame	Buffer
Ag	328.1	0.7	2	4	Oxidizing air-C$_2$H$_2$; danger of precipitation and/or adsorption	0.2% KCl
	338.3		10			
Al	369.3	0.6-0.8	20	50	Reducing N$_2$O-C$_2$H$_2$; determine optimal height with single-element lamp	1% La, 0.3 mol L^{-1} HCl
As	193.7	0.5	200	50	Oxidizing air-C$_2$H$_2$; slightly reducing N$_2$O-C$_2$H$_2$ (UV-sensitive PM)	
B	249.8	0.7	700	500	Slightly reducing air-C$_2$H$_2$	0.5% KCl
Ba	553.6	0.2-0.3	8	25	Reducing N$_2$O-C$_2$H$_2$; air-C$_2$H$_2$ (10 times less sensitive)	
Ca	422.7	0.7	0.5	5	Slightly reducing N$_2$O-C$_2$H$_2$; air-C$_2$H$_2$;	0.2% KCl
		0.2	5			
Cd	228.8	0.2-0.3	1	2	Air-C$_2$H$_2$; N$_2$O-C$_2$H$_2$	3% (NH$_4$)$_2$HPO$_4$
	326.2					
Co	240.7	0.2	10	5	Oxidizing air-C$_2$H$_2$	1-2% NH$_4$Cl or
Cr	357.9	0.7	3	5-10	Reducing air-C$_2$H$_2$	0.5 mol L^{-1} oxine
	425.4		10			
	427.5					
Cu	324.8	0.7	0.2	5	Oxidizing air-C$_2$H$_2$	
	327.5		20			
	249.2		200			
Fe	248.3	0.2	5	5	Oxidizing air-C$_2$H$_2$	0.03-0.04 mol L^{-1} Ba, Sr or Ca
	253.3					
	372					
	386					
Hg	253.7	0.5	250	300	Oxidizing air-C$_2$H$_2$; cold vapour	
	253.7	0.2	0.01	0.5		
K	766.5	1-2	2	2	Oxidizing air-C$_2$H$_2$	1% Cs or Na
	769.9		5	2		
	766.5		0.5	2		

(Contd.)

Table 10.4 (contd.)

Element	λ, nm	Slit width	D.L. ng mL^{-1}	U.L. μg mL^{-1}	Flame	Buffer
Li	670.8	0.7	0.3	2-3	Oxidizing air-C$_2$H$_2$	0.1% KCl
	323.3		10			
	610.40		1000			
Mg	285.2	0.7	0.1	0.5	Stoichiometric air-C$_2$H$_2$	0.2% KCl; 0.1–0.5% La
	202.6		50			
Mn	279.5	0.2	2	3	Oxidizing air-C$_2$H$_2$	0.2% La CaCl$_2$
	403.1		50			
Mo	313.3	0.7	20	40	Reducing N$_2$O-C$_2$H$_2$	0.2% Al in 0.01 mol L^{-1} HCl
	315.9		10		Highly reducing air-C$_2$H$_2$	0.01 mol L^{-1} Cs + NH$_4$F 2% NH$_4$Cl
	320.9		200			0.5% Na$_2$SO$_4$
Na	589.0	0.7	0.2	1	Oxidizing air-C$_2$H$_2$	0.1% K or Cs
	589.6					
	330.2				Three-slot air-propane burner	
Ni	232.0	0.1–0.2	40	5	Highly oxidizing air-C$_2$H$_2$	
	341.5		10	10		
Pb	283.3		10	20	Oxidizing air-C$_2$H$_2$	
Si	251.6	0.2	50	100	Reducing N$_2$O-C$_2$H$_2$	
Sn	235.5	0.2	70	300	N$_2$O-C$_2$H$_2$	
	286.3				Ar-H$_2$	
	224.6	0.2	50	100		
SnH$_4$			0.2		Hydrides	
Sr	460.7	0.7	20	5	Reducing air-C$_2$H$_2$	0.2–1% La; 0.3% KCl
Ti	365.4	0.2	40	200	Reducing N$_2$O-C$_2$H$_2$	0.2% KCl
	364.3	0.5	200			
	399.0	0.5	200			
V	318.5	0.7	40	150	Reducing N$_2$O-C$_2$H$_2$	0.1–0.5% AlCl$_3$
	318.4					
	318.3					
Zn	213.9	0.7–2	1	1	Highly oxidizing air-C$_2$H$_2$	
	307.0	0.1	10	1		

Table 10.5 (*Contd. from p. 199*)

Element	Modifier	Decompn T, °C	Atomizn T, °C
Mo	—	—	—
Ni	$Mg(NO_3)_2$	1400	2500
Pb	$Mg(NO_3)_2 + NH_4H_2PO_4$	800	2400
	$Pd(NO_3)_2$	1200	
Se	$Ni(NO_3)_2$	900	2600
	$Pd(NO_3)_2$	1200	
Te	$Ni(NO_3)_2$	900	2400
	$Pd(NO_3)_2$	1300	
Tl	H_2SO_4	600	2200
Zn	$Mg(NO_3)_2$	600	2200

References

Bois N., Louvrier J. and Voïnovitch I.A. 1984. Examen théorique de certaines interférences interéléments, sur binaires, en spectrométrie d'absorption atomique avec flammes. *Analusis*, **12**: 396-403.

Hoenig M. and de Kersabiec A.M. 1990. *L'atomisation électrothermique en spectrométrie d'absorption atomique*. Masson, Paris, 296 pp.

Pansu M. 1985. OPTIMI: optimisation des mesures par utilization de plans d'expériences et méthodologie des surfaces de réponse. *Fiches techniques Agence Nationale du Logiciel*. Vandoeuvre les Nancy, France.

Pinta M. 1978. *Modern Methods for Trace Element Analysis*. Ann Arbor Science, Ann Arbor MI, USA.

Pinta M. 1979. Atomisation dans les flammes et dans les fours. Recherches des conditions analytiques. In (M. Pinta, ed.), *Spectrométrie d'absorption atomique*. Masson, vol. I, pp. 179-249.

Pinta M. and Laporte J. 1979. Appareillage. In (M. Pinta, ed.), *Spectrométrie d'absorption atomique*. Masson, vol. I, pp. 1-29.

Pinta M. 1980. Roches, sols et minérais. In (M. Pinta, ed.), *Spectrométrie d'absorption atomique*. Masson, vol. II, pp. 263-305.

Pinta M. 1985. Spectrométrie d'absorption atomique. *Techniques de l'ingénieur* **10**, P 2825, 24 pp.

Pinta M., de Kersabiec A.M. and Richard M.L. 1982. Possibilités de l'exploitation de l'effet Zeeman pour la correction d'absorptions non spécifiques en absorption atomique. *Analusis*, **10**: 207-215.

Riandey C., 1979. Les perturbations. In (M. Pinta, ed.), *Spectrométrie d'absorption atomique*. Masson, vol. I, pp. 82-78.

Ure A.M. 1991. Atomic absorption and flame emission spectrometry. In (K.A. Smith, ed.) *Soil Analysis. Modern Instrumental Techniques*. Marcel Dekker, 2nd ed., pp. 1-62.

Voïnovitch I.A. 1988. *Analyse des sols, roches et sédiments. Méthodes choisies*. Masson, Paris, pp. 14-98.

Bibliography

Butcher D.J. and Sneddon J. 1998. *A Practical Guide to Graphite Furnace Atomic Absorption Spectrometry* (Chemical Analysis, vol. 149). Wiley-Interscience.

Dean J.R. and Ando D.J. 1998. *Atomic Absorption and Plasma Spectroscopy*. (Analytical Chemistry by Open Learning Series), John Wiley and Sons.

Dedina J. and Tsalev D.L. 1995. *Hydride Generation Atomic Absorption Spectrometry* (Chemical Analysis, vol. 150). John Wiley and Sons, 544 pp.

Haswell S.J. (ed.). 1991. *Atomic Absorption Spectrometry: Theory, Design and Applications* (Analytical Spectroscopy Library, vol. 5). Elsevier Science.

Kurfurst U. (ed.). 1998. *Solid Sample Analysis: Direct and Slurry Sampling Using GF-AAS and ETV-ICP*. Springer Verlag.

Tsalev D.L. (ed.). 1995. *Atomic Absorption Spectrometry in Occupational and Environmental Health Practice: Progress in Analytical Methodology*. CRC Press, Boca Raton FL, USA.

Varma, Asha. 1990. *Handbook of Furnace Atomic Absorption Spectroscopy*. CRC Press, Boca Raton FL, USA.

Emission Spectrometry[1]

[1] Text by Francis Sondag and Florence le Cornec, IRD, 32 Avenue Henri Varagnat, 93 140 Bondy Cedex, France (with the assistance of the authors).

1. Principle

Methods of analysis by emission spectrometry are based on measurement of the intensity of photon emission by atoms or molecules placed in an energy source. When a chemical substance, for example, in aqueous solution, is introduced in a heat source, the energy provided causes dissociation of the molecular combinations; the atoms formed pass for a very short time (of the order of 10^{-8} s) to an excited level of energy E_m before falling back to a less excited level E_0. The difference in energy $E_m - E_0$ is released in the form of photons of frequency v:

$$E_m - E_0 = hv \qquad (1)$$

where h is Planck's constant.

As orbital transitions are defined in atomic structure, the emission lines are characteristic of the latter and the determination of their wavelength enables identification of the element emitting them. Furthermore, there is a proportionality between the intensity of the lines and the concentration of atoms. Concentration of the element in the emitting medium can be determined by measuring the light flux.

The earliest applications of emission spectrometry go back to the work of Kirchoff and Bunsen on rubidium and caesium in 1859. The first determinations by this method were done for sodium in 1873 and the technique was considerably developed in the 1930s. From 1960 on, development of atomic absorption eclipsed the emission methods. However, the invention of high-energy excitation sources such as nitrous oxide-acetylene flames and, more recently, plasma torches, has given rise to a veritable revival of these methods. The reader will find broader treatments of their theoretical aspects in the works of Dean and Rains (1975), Pinta (1979; 1987).

Emission spectrometry can be done in different modes grouped according to the source enabling emission. They enable analysis of increasingly refractory elements by increasing the excitation temperature from the butane-air flame to plasma. In most cases the sample is introduced into the energy source as an aerosol (Fig. 11.1). The energy supplied first causes *desolvation*, whereby the sample is transformed into solid microparticles. The next step consists of decomposition to a gas of free molecules *(vaporization)*, which will then dissociate into atoms *(atomization)*. Lastly, if the energy is sufficient, the atoms are *excited* and/or *ionized*, either by radiation or by collision.

The rest of this chapter explains the two principal emission methods currently in use: flame-emission spectrometry and plasma-emission spectrometry. Arc spectrography, rarely used nowadays, is done by passing an alternating or direct current between two electrodes, most often of graphite. A very hot (about 3600°C) emission source is thereby obtained, which can excite a larger number of elements. Arc-emission spectrography

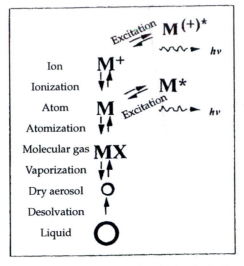

Fig. 11.1 Various steps in the process of excitation-ionization of a sample.

also has the advantage of enabling analysis of solid samples, provided they are conductors (or made conductive by mixing with graphite for example). It is also a technique whereby a rapid qualitative or semiquantitative survey can be done of all the elements in a substrate prior to more precise analysis, without all the precautions required for a truly quantitative analysis. However, progress in this type of instrument has been interrupted to the gain of plasma-emission spectrometry.

2. Flame Spectrometry

2.1 Generalities

The radiation emitted in the flame is not monochromatic but, rather, composed of a mixture of wavelengths. To measure its intensity it is necessary to separate these lines; selective optical filters or a monochromator fulfil this function.

The instruments are fitted with two kinds of measuring devices:

—Flame photometers: the wavelength is selected by means of optical filters: these are cheap devices, but the number of elements that can be analysed is small, mainly K, Na, Li and Ca. These instruments are satisfactory for the serial analyses most common in soils.

—Spectrophotometers: the selection of wavelength is done with a monochromator; analysis of many elements is possible (among others, magnesium, which is difficult to determine with optical-filter instruments) through reduction of spectral interferences.

Filter instruments are very robust; because of the large area available for illumination, the photon flux is large and can be recorded by simple measuring devices, which results in cheap, easily serviced instruments. But this solution is valid only for determination of elements with sufficiently distinct radiation wavelengths. For study of spectra rich in lines, often close to each other, it is necessary to use instruments fitted with a monochromator; the quality of the monochromator depends on many parameters, of which one is resolution. Commercially available instruments today combine an emission spectrometer with an atomic absorption spectrometer (*see* Chapter 10) because they have similar excitation and detection devices.

2.2 Atomization Source: The Flame

In most cases the test solution is introduced into the flame in the form of a solution nebulized by the fuel gas. The purpose of the flame is to decompose the sample in order to take it to the atomic or molecular vapour state and to excite the largest possible number of atoms or molecules, whose emission line is to be measured.

Two kinds of flames are distinguished according to the manner in which the fuel and the oxidant reach the combustion zone:

—if the fuel and oxidant are mixed, *premixed flame* (air-acetylene, nitrous oxide-acetylene, etc.);

—if the gas flows are separate (for example, fuel jet entering air with it is mixed while burning), *diffused flame* (oxygen-acetylene, etc.).

Most flames are a compromise between these two types, with one or the other dominating according to burner configuration and flow rate and nature of the gases. The upper part of the flame, where background emission is weaker and combustion reactions are more complete, is mostly used.

Different gas mixtures can be used; the air-acetylene flame is easy to operate and free from risks, and enables excitation of many elements with weak background spectrum but very distinct emission lines. This flame is hotter than that with butane, propane and town gas (Table 11.1).

Air can be supplied economically from a high-output (for example, 70 L min^{-1}) industrial compressor with a large (100 L) tank, ensuring regular flow without sudden fluctuation or stoppage. The compressor should have an oil filter to purify the air before it leaves the reservoir.

Table 11.1 Principal mixtures used in flame-emission spectrometry.

Fuel	Oxidant	Temperature, K respectively
Town gas	Air, Oxygen	2115, 3015
Butane	Air, Oxygen	2175, 3175
Propane	Air, Oxygen	2200, 3125
Hydrogen	Air, Oxygen, Nitrous oxide	2300, 2935, 2925
Acetylene	Air, Oxygen, Nitrous oxide	2500, 3400, 3175

Acetylene is supplied from commercial cylinders. The two gases are introduced into the burner through a set of two-stage regulators; the final regulation is obtained by a needle-valve regulator. Such a set-up enables stabilization of the flows with $\pm 1.5\%$ relative variation.

2.3 Emission Spectra

2.3.1 Characteristics

According to whether the emitting bodies are molecules, atoms or ions, band or line spectra are obtained, with positions characterized by a wavelength (Table 11.2). In addition to the spectra produced by the elements present in the injected solution, the flame too has its own emission. Air-acetylene flames give a spectrum containing OH bands and a continuous background between 300 and 500 nm caused by dissociation of CO molecules, with a maximum around 450 nm.

2.3.2 Interferences

The principal interferences are physical and chemical, related to the matrix analysed:

—formation in the flame of stable combinations of the type Ca-P-O or Ca-Al-O for example (*see* Chapter 10);

—differences in salt content or viscosity of the solutions influencing nebulization;

—ionization interferences in matrices containing elements with low ionization potential: for the alkali metals in an air-acetylene flame, for example, the degree of ionization ranges from 5% for Li to 82% for Cs, with that for K around 50% (*see* Chapter 10).

Table 11.2   Principal wavelengths excited in the air-acetylene flame for the commonly analysed elements (after Dean and Rains, 1975).

Element	Wavelength emitted, nm	Relative intensity
Ba	553.55	
Ca	422.67	
Cs	852.11	10,000
	894.35	5,000
K	766.49	400,000
	769.90	200,000
Li	670.79	700,000
Mg	285.21	
Mn	403.08	
Na	589.00	800,000
	589.59	400,000
Rb	780.02	50,000
	794.47	25,000
Sr	460.73	

These interferences apply chiefly to the alkali and alkaline-earth metals:

—for alkali metals the interferences are relatively low and it is possible to make direct determinations, except in the case where one alkali metal is present in large amounts relative to another; there could be an exaggerating effect, for example, when sodium is determined in presence of a large amount of potassium; addition of calcium when much aluminium is present enables a better determination of Na, K and Li;

—in the case of alkaline-earth metals, the element that suffers greatest interference is calcium; this very large interference is caused chiefly by aluminium and anions such as phosphate.

In all cases, for a determination it is necessary to obtain the calibration curve by first using data from pure solutions. Comparison of the curve obtained with interfering ions present will give information on the magnitude of the interferences.

The method of 'standard additions' (*see* Chapter 18) may also be used. Two aliquots, A and B, of the test solution are taken; to aliquot B a known concentration C_e of the analyte element is added and the intensities I_a and I_b of the two solutions are measured. The ratio $C_e/(I_b - I_a)$ enables drawing of the standard interference curve.

2.4 Example of Analysis and Calibration

2.4.1 Generalities

This example pertains to the determination of exchangeable K and Na in soils by the barium chloride method (Gillman, 1979). The exchangeable bases are extracted with a 0.1 mol L^{-1} $BaCl_2$ solution containing 1% $CsCl_2$; they are determined in the solution directly.

The emission spectra of the alkali metals (Li, Na, K, Rb and Cs) are particularly intense and thus are very suitable for quantitative determination of these metals in aqueous solution. The emission corresponding to the transition between ground state and the first excitation level often gives rise to a doublet (Table 11.2), the shorter wavelength having approximately twice the intensity of the longer. In very hot flames, other lines are emitted but they are not useful for analysis; they actually cause spectral interferences for other elements.

2.4.2 Operating conditions and calibration

The operating conditions naturally are related to the quality of the instrument, but they also depend on the settings: burner position, regulation of supply of gases and solution to the flame, etc.

In the present example, the air-acetylene flame is used with the burner parallel to the detector. Determinations are done in three replications, the mean being used for drawing the calibration curves. A 'blank' is run with each set of analyses and its intensity is subtracted from that of the standards and test solutions. The instrumental settings are presented in Table 11.3.

Table 11.3   Instrumental settings for determination of K and Na by flame emission.

	K	Na
Wavelength, nm	766.49	589.00
Slit width, mm	1.00	0.50
Air pressure, bar	4.00	4.00
Acetylene pressure, bar	0.80	0.80
Integration time, s	3	3

For potassium, the intensity (I) determined for four standards of varying concentration C (Table 11.4) can be fitted to a straight line of equation

$$I = 0.145 + 0.133\ C \qquad (2)$$

with coefficient of determination of 99.6% and an acceptable percentage standard error [$100 \times (C_{estimated} - C_{standard})/C_{standard}$] because it is less than 2% for the values estimated for contents greater than 3 mg L^{-1}.

Careful examination of the calibration, however, brings out certain limitations in the instrument used. For zero potassium the signal is not totally negligible ($I = 0.145$). This indicates a background noise (probably not caused by a specific potassium signal), which has a deleterious influence on the detection limit when the element is present in trace quantities. The confidence interval (*see* Chapter 18 for details of calculation) at the concentration corresponding to the signal with an unknown solution is also large (Table 11.4 and Fig. 11.2), despite the high value of the coefficient of determination. Actually the fitting can be further improved by taking into account a slight curvature as in the case of sodium below, where it is more evident.

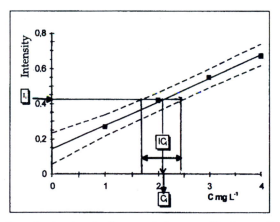

Fig. 11.2   Calibration of potassium by flame emission according to equation 2
Points and continuous line - actual values and those predicted from standards, respectively; dashed lines - confidence intervals at 95% for calibration error;
I_i - intensity of signal from an unknown solution; C_i - content corresponding to this signal; IC_i - confidence interval at 95% for C_i.

Table 11.4   Emission intensities and error of estimate for K (eqn 2).

Standards C, mg L^{-1}	Reading I	Estimated C mg L^{-1}	Error of estimate, %	Confidence interval at 95%
1.0	0.27	0.94	−6	±0.46
2.0	0.42	2.07	3.5	±0.39
3.0	0.55	3.05	1.7	±0.40
4.0	0.67	3.95	−1.3	±0.45

For sodium the intensity of the four standards (Table 11.5) actually gives a curve (Fig. 11.3) with an equation that can be approximated by a polynomial of the second degree:

$$I = 0.17 + 0.048C - 0.00084C^2 \tag{3}$$

with a coefficient of determination of 0.9995; the error of estimate is less than 2% over the entire range of standards.

Table 11.5   Emission intensities and error of estimate for Na.

Standard mg L^{-1}	Reading	Estimated standard mg L^{-1}	Error of estimate, %
5.0	0.385	4.96	−0.80
10.0	0.568	10.14	1.40
15.0	0.694	14.80	−1.31
20.0	0.793	20.10	0.52

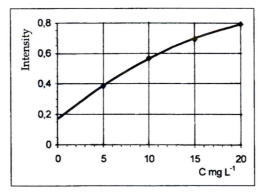

Fig. 11.3   Calibration curve for sodium (eqn 3).

3. Inductively Coupled Plasma Emission Spectrometry

3.1 Introduction

Atomic emission spectrometry using an *argon plasma* induced by high-frequency as an ionization and excitation source (Inductively Coupled Plasma—Atomic Emission Spectrometry: *ICP-AES)* has received considerable impetus in the last thirty years. Today this multi-element analytical method is very popular and routinely used in many laboratories in greatly varied fields, especially environment and geology (Riandey *et al.*, 1982; Cresser *et al.*, 1994). Many papers describe in detail the principle and instrumentation, and give examples of application of this technique (Trassy and Mermet, 1984; Boumans, 1987; Thompson and Walsh, 1989; Sharp *et al.*, 1994).

The term 'plasma' denotes gases that are partially ionized, but are electrically neutral macroscopically. The *ICP* source appears as a bright and very intense white flame dangerous to view without protective glasses. Its temperature (>5000 K), higher than that of chemical flames (3000 K), is the reason behind the interest it has sparked in analytical work. Actually, at these temperatures, not only are most molecular species totally dissociated, but the production of excited and/or ionized atoms is also very high. This allows analysis of a large number of elements with a high degree of sensitivity, but also entails two hazards: (1) low performance with regard to detection limit of volatile elements (alkali metals) and (2) risk of error caused by spectral interferences because of the large number of species in the atomic or ionized state.

3.2 Characteristics and Interferences in Analysis by ICP-AES

3.2.1 Generalities

ICP-AES analysis, like all emission methods, is a *multi-element method*. It allows *quantitative determination* of a large number of elements of the Periodic Table (about 70 for sequential instruments; for simultaneous-analysis instruments a number defined at the time of manufacture) and gives *detection limits in water of the order of $\mu g\ L^{-1}$* for many of these elements.

Like most physicochemical methods, this is a *comparative method*: the elemental concentrations in the test sample are determined by using calibration curves prepared every day with solutions containing known quantities of the elements to be determined. It is imperative that the standards also be prepared in the same medium or same matrix as the samples. Although there are few chemical interferences (*see* §3.2.2) in analysis by *ICP-AES*, the matrix can, because of the nature and concentration of its constituents, modify passage of the sample into the plasma and thus change the excitation and ionization states of the element studied. Lastly, *these calibration curves are linear over many orders of magnitude* (0.01-1000, 0.1-10,000 mg L^{-1}, etc.).

3.2.2 Interferences

Chemical interferences

Because of the *high temperature* (*>5000 K*) of *plasmas*, not only is the efficiency of excitation and ionization improved, but also all chemical associations, even the most refractory, are decomposed with very great reduction and even practically total elimination of the interelement effects observed in flames or graphite furnaces. In the case of argon plasma, the temperature in the detection zone is about 7000 K. One of the characteristics of *ICP* is that the sample is introduced into the centre of the plasma. It stays in the high temperature zones for a relatively long time, approximately 2 ms. This residence time of the aerosol in the centre of the discharge also explains the low matrix effects encountered in analysis.

Spectral interferences

Practically speaking, a spectral line can never be represented by a point value of wavelength, but presents a profile covering a range of wavelengths. The various causes of broadening of lines are chiefly related to movements and collisions of electrons and atoms in the plasma. The profile of a line is shown in Fig. 11.4

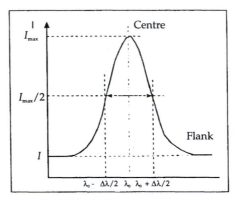

Fig. 11.4   Profile of a spectral line.

Also, the use of plasma as excitation source gives spectra richer in lines than other sources. The principal interferences encountered in *ICP-AES* are therefore very numerous spectral interferences that make selection of the analytical wavelength critical in order to minimize these problems. They may be grouped as follows:

(1) Continuous background emitted by plasma

The continuous radiation, essentially caused by radiating recombination $(Ar^+ + e^- \rightarrow Ar^* + h\nu_{continuum})$, is superposed on the line emission. It is proportional to the electron density. In practice, the background varies according to the operating conditions of the plasma: introduction of the sample (viscosity and acid concentration), flow of the gases, observation height, etc. Lastly, the longer the wavelength, the more intense the background. To eliminate these variations, a background correction is generally necessary, especially for determination of trace quantities. Many corrective methods are available but it should always be verified that the one selected is suitable for the line being measured.

(2) Interfering radiation

This is the ensemble of wavelengths other than those caused by relaxation of the elements in the test solution; this radiation reaches the detector in the same manner. It is mainly due to optical imperfections and particle diffusion in the monochromator. Nowadays, with the use of holographic gratings, such radiation has been greatly reduced.

(3) Total or partial masking of the analytical line by a molecular band or another spectral line

This masking can originate from the following causes.

—Lines of argon are systematically present in the plasma but are relatively few and well catalogued, and there is no line between 175 and 300 nm. The most intense are located between 420 and 440 nm.

—Presence of water leads to the formation of hydrogen and OH radicals not destroyed by a high-frequency plasma. The hydrogen lines are few, but can be broad as well. The principal OH band is located at 306.4 nm.

—Presence of air also leads to spectral radiation, especially emission bands of N_2, N_2^+ (329-590 nm) and NO (200-300 nm). The last can be excited only in the periphery of the plasma as the NO bonds do not survive residence in the plasma.

—With organic solutions, the spectra are still more complex. Band spectra of C_2, CH, N_2, N_2^+ and NH, which extend practically over the entire spectral domain, are seen.

—Presence of any other element than that studied, even at low concentration, can obstruct the measurement of certain lines (Fig. 11.5: interference of iron on Boron; Table 11.6: example of interfering lines emitted around the erbium line at 337.276 nm). Determination of trace quantities of elements can also be hampered by the foot of the spectral line of other elements present in large quantities even if the wavelength is not very close (example: Ca 393.4-396.7 nm).

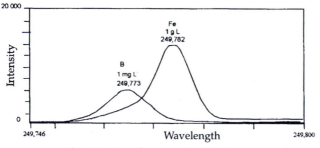

Fig. 11.5 Interference of iron on boron.

The largest compilations of spectral lines in *ICP-AES* are presented in the publications of Winge *et al.*(1985) and Boumans (1980).

Table 11.6 Example of interfering lines emitted around the Er line at 337.276 nm. Presence of Ti in the solution renders this line unusable without correction. The other elements give emission lines less intense than or far from those of erbium and are therefore less troublesome in the measurement.

Element	Wavelength, nm	Intensity
Ni	337.199	35
U	337.201	6
Sc	337.215	4000
Ti	337.221	85
Rh	337.225	16
Nb	337.256	28
Th	337.270	7
Er	**337.276**	**1600**
Ti	337.280	6800
Pd	337.300	60
Zr	337.342	120
Ce	337.346	90
Ar	337.348	11
Ce	337.373	0
U	337.410	6

Selection of the analytical line is extremely important and depends on the matrix of the sample analysed. In soils, the presence of major elements can hamper the selection of spectral lines for analysis of trace elements, and sometimes makes their direct determination impossible without pretreatment of the sample. Lastly, it must be borne in mind that today all instruments are equipped with computers that enable correction of certain spectral interferences, such as background correction or interelement correction, provided these interferences have been identified beforehand, a delicate operation if the major-element composition of the solution is not known and if the masking of the analytical wavelength by interfering radiation is almost total. It is always preferable to use an unaffected line if at all possible, rather than embark on numerous corrections. Lastly, if a correction is used, it is always necessary to verify that it properly matches the sample.

3.2.3 Sensitivity

The detection limits for many elements are of the order of μg L^{-1} or tens of μg L^{-1} in pure water. They may however be greatly augmented in very concentrated or complex media. For most elements *ICP-AES* is a more sensitive method than flame atomic absorption but less sensitive than electrothermal atomic absorption and mass spectrometry with ionization by high frequency induced plasma (ICP-MS). Thus, for determination of trace elements such as lead or cadmium in unpolluted soils, the sensitivity of ICP-AES with a classic nebulizer will be inadequate and the analysis of these elements should preferably be done by electrothermal atomic adsorption or ICP-MS. Similarly, direct determination of arsenic and selenium in waters by ICP-AES is impossible and should be done by the hydride method coupled with ICP-AES or by flame atomic absorption spectrometry.

3.3 Instrumentation

The principal application of atomic emission spectrometry is the analysis of liquid samples that are aspirated and converted to aerosols (assemblage of very fine liquid particles in a gaseous flux) by nebulization. This aerosol is then transported into the plasma where it is desolvated, vapourized, atomized, excited and/or ionized. The excited atoms and ions emit characteristic radiations of the elements present, which are separated by means of a spectrometer. These radiations are detected and converted to data for the analyst. A classic ICP-AES instrument thus will comprise the following components (Fig. 11.6):
—a system for introducing the sample into the plasma,
—a torch,
—a high-frequency generator,
—an optical system to analyse the emitted spectrum, and
—systems for detection and measurement of the signal, which enable qualitative and quantitative determinations from the emitted radiation.

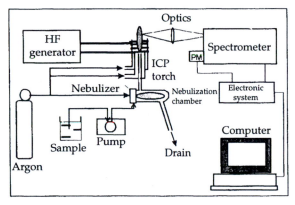

Fig. 11.6   Principal components of an ICP-AES instrument.

3.3.1 System for introducing the sample

The most commonly used system for introducing liquid samples is nebulization. The ideal nebulizer should have high efficiency and should produce a fine aerosol in which the size of droplets is the smallest possible, independent of the physical characteristics of the sample (such as viscosity, density, etc.), very stable over time. However, although techniques for introduction of the sample have occupied a prime position in research and development of the atomic emission spectrometer, they still remain the weak and limiting link in analysis. The traditional techniques for liquids used a pneumatic or ultrasonic nebulizer.

3.3.1.1 PNEUMATIC NEBULIZERS

Three major groups of pneumatic nebulizers are encountered in plasma analysis:

—*Concentric nebulizers*: the gas and liquid flows are coaxial (Fig. 11.7a). The liquid flows through a capillary of very small diameter (< 0.5 mm) into a nozzle in which a constriction speeds up the gas and creates the aerosol. A peristaltic pump should be used to impel the sample to avoid any variation in the quantity of liquid aspirated, especially in the case of solutions with different viscosities. These nebulizers give excellent sensitivity and high stability. The major inconvenience is that they are easily blocked when a suspended particle is introduced from the solution or when solutions very rich in salts are aspirated. The most popular nebulizer is the Meinhard type, made entirely of glass, which, of course, cannot be used for solutions containing hydrofluoric acid.

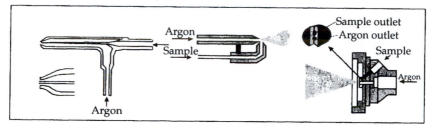

Fig. 11.7 Pneumatic nebulizers: a - concentric nebulizer; b - cross-flow nebulizer; c - V-shaped nebulizer.

—*Cross-flow nebulizers*: the axes of the tubes for gas and liquid are at 90° to one another (Fig. 11.7b). The liquid forced by a peristaltic pump is introduced into a vertical capillary tube while the gas passes through a horizontal capillary, the outlets of the two capillaries are close to one another. The performance of these nebulizers depends largely on the alignment of the two capillaries, for which reason they are often fixed by the manufacturer.

—*V-shaped nebulizers (based on the Babington nebulizer)*: the liquid is introduced into the bottom of the V by means of a peristaltic pump (Fig. 11.7c). The nebulizing gas is forced in through a hole drilled in the centre of the V. These nebulizers are available in glass and various plastic materials resistant to hydrofluoric acid. They are much less susceptible to blocking than concentric nebulizers because the liquid can be introduced through a wide-bore capillary (generally 1 mm). They also work well with very viscous solutions and are recommended for all salt-rich solutions containing suspended particles. Lastly, their performance is similar to that of other nebulizers.

All these nebulizers should be connected to a *nebulization chamber*, in which the droplets of the aerosol are sorted according to size. Actually, the diameter of many droplets is often too large to let them be injected directly into the plasma without causing loss of sensitivity or reproducibility. The function of the chamber is therefore to hold back the largest drops and to allow only a fine, uniform aerosol to flow into the plasma. This chamber also allows smoothening of the pulses often caused by the peristaltic pump. In general, a nebulization chamber allows only drops smaller than 10 μm in diameter to enter the plasma. Then *the output of the nebulizer-chamber assembly is less than 5% for sample flow of the order of mL min⁻¹*, the remaining 95% being discharged into the drain. Sample volumes required for a determination by ICP-AES are quite large, for many elements larger than for ICP-MS and, more so, for EAAS. A minimum sample volume of 5 mL is required for determining 10 elements. This technique is therefore less suitable for microanalysis than EAAS.

3.3.1.2 ULTRASONIC NEBULIZER

An ultrasonic nebulization system (Fig. 11.8) contains a piezoelectric crystal excited by a generator with the resonance frequency of the crystal. The ultrasonic energy creates an aerosol. This nebulizer has the advantage of producing a very fine (mean droplet size 1.5 μm) uniform aerosol with yield much higher than pneumatic systems. However, a desolvation unit has to be introduced between the nebulizer and the torch to eliminate nebulization of solvent or water in large quantity along with the sample. Such systems improve detection systems by a factor of 5 to 50 for simple aqueous media compared to pneumatic nebulizers. They are not easy to use, however, and furthermore are quite expensive.

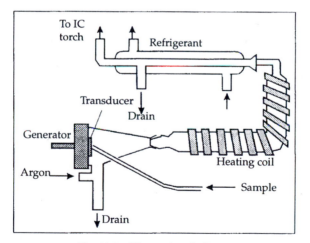

Fig. 11.8 Ultrasonic nebulizer.

3.3.2 *The torch*

The torch consists of three coaxial tubes (Fig. 11.9):
   —an outer quartz tube,
   —a middle quartz tube, and
   —an injection tube of quartz or alumina.

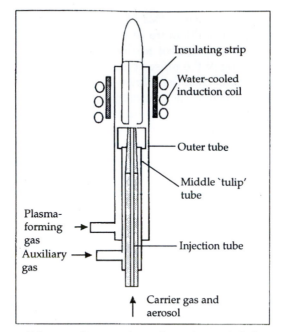

Fig. 11.9 Schematic diagram of an inductively coupled plasma torch.

The three tubes provide three passages for the argon.

—The very narrow channel between the outer and middle tubes, which the argon reaches tangentially, receives the gas designed for feeding the plasma, called the plasma-forming gas. The flow of this gas is generally between 8 and 17 L min^{-1}.

—The channel between the middle and injection tubes receives the auxiliary gas (argon), which enables raising the plasma tongue above the injection and middle tubes. The flow is usually between 0 and 3 L min^{-1}.

—The central channel through which the sample passes in the form of an aerosol in the carrier gas (argon).

Monobloc or demountable torches are available, with injection tubes of different diameters depending on the type of sample to be analysed.

3.3.3 High-frequency generator

The high-frequency generator has the function of supplying the energy necessary for obtaining and maintaining the plasma. The power of the generator, generally between 600 and 1500 watts, is transmitted to the plasma-forming gas through an induction coil surrounding the top of the torch. The generators used in ICP-AES work at a frequency between 27 and 56 MHz, although an increasing number of instruments work at 40 MHz with better analytical performance.

3.3.4 *Optical system*

The energy that the atoms have acquired from the plasma is re-emitted in the form of electromagnetic radiation. This radiation consists of wavelengths corresponding to the elements present. Measurement of the various lines can be done simultaneously or sequentially, with dispersion of the rays accomplished by means of a diffraction grating. The optical system comprises two parts:

—the system for focussing the emission source,

—the monochromator or polychromator and measuring device.

3.3.4.1 OPTICAL SYSTEM

This transmits the radiation emitted by the plasma to the entrance slit of the spectrometer. It is composed of lenses and/or mirrors that project the optical image of the plasma on the entrance slit of the dispersing device. In some instruments a supplementary mirror facilitates measurement of the radiation emitted at different heights in the plasma.

3.3.4.2 MONOCHROMATOR

The two monochromator systems most commonly used are the Czerny-Turner and Ebert mountings (Fig. 11.10). This spectrometer comprises an entrance slit, one (Ebert) or two (Czerny-Turner) collimating and focussing mirrors, a plane grating and an exit slit. Multielement analysis is done successively by sequential scanning at different wavelengths by rotation of the grating. A monochromator enables analysis of close to 70 elements in sequence whatever the sample matrix.

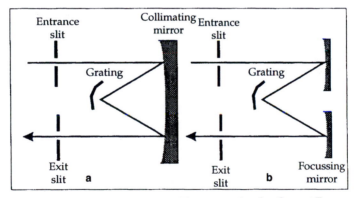

Fig. 11.10   Two types of monochromators: a - Ebert mounting; b - Czerny-Turner mounting.

An instrument fitted with a monochromator is also called a sequential instrument.

3.3.4.3 POLYCHROMATOR

The most classic mounting is the Pashen-Runge mounting (Fig. 11.11). It comprises an entrance slit, a concave grating and exit slits; all these elements

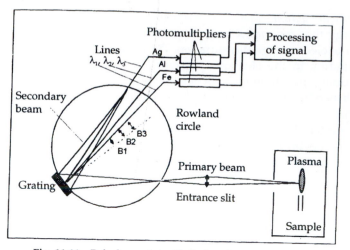

Fig. 11.11   Polychromator with Pashen-Runge mounting.

are arranged in a circle called the Rowland circle, whose diameter is equal to the radius of curvature of the grating. The grating is fixed and as many slits and detectors are provided as the number of lines to be analysed, all these having been defined when the instrument was designed. The polychromator therefore enables simultaneous measurement of only a certain predetermined number of lines; the other lines are inaccessible. Such an instrument, because of its lack of flexibility, is designed more specifically for routine analyses; it is not suitable for the analytical work of a research laboratory, particularly for trace elements (unsatisfactory detection limit). An instrument so equipped is termed a simultaneous instrument. Its major feature is great speed in analysis. It is possible to combine the two types of instrument—sequential and simultaneous—in the same torch.

Gratings may be engraved, holographic or blazed (echelette). They have 600 to 4200 lines per millimetre. The spectral range covered by these devices extends from 160 to 900 nm. To be able to detect lines below 190 nm, the spectrometer should be evacuated or purged with argon or nitrogen eliminate oxygen molecules that absorb these radiations.

3.3.5 Detection and processing of the signal

The diffracted radiation is converted to electric current by a photomultiplier, this current giving a relative measure of the intensity of the radiation. In a monochromator a photomultiplier is placed behind the exit slit. In a polychromator one photomultiplier is placed behind each exit slit to receive the diffracted rays directly or from a mirror or fibre-optic cable. The type of photomultiplier depends on the spectral domain and intensity expected of the signal. Photomultipliers are slowly being replaced by charge-coupled devices (CCD) or charge-injection devices (CID). The current is then transformed to a voltage that is itself converted to digital data that can be used by a computer.

Today all these instruments are connected to a computer that controls all the functions of the ICP-AES: ignition, regulation of gases, study of wavelengths, management of methods and results, etc.

3.4 Example of Analysis and Calibration

3.4.1 Operating conditions

The major elements determined in soils are silicon, aluminium, calcium, iron, magnesium, sodium, potassium, phosphorus and titanium. For this application a fusion with lithium metaborate was done to retain the silica of the samples in solution.

Operating conditions depend on the type of instrument used. Following dissolution, a torch, nebulizer and nebulization chamber compatible with the final medium of the samples should be selected. In the present case, a monochromator instrument with a demountable torch and alumina injector, a V-shaped nebulizer and the standard nebulization chamber were used. It is necessary to determine and optimize the following conditions for each element analysed:

—wavelength,
—power of the generator,
—flow of plasma-forming, auxiliary and nebulization gases,
—observation height in the plasma,
—time and mode of integration of the peak,
—mode of background correction, etc.

The conditions used for these determinations are presented in Table 11.7. Silicon is separately determined, while all the other elements are determined in one single programme.

Table. 11.7   Instrumental parameters for quantitative analysis of major elements in soils.

Element	Si	K	Al	Na	Ca	Ti	Fe	Mg	Mn	P
Wavelength, nm	288.16	769.90	396.15	588.99	317.93	337.28	259.94	285.21	257.61	213.62
Observation height, mm	8	10	11	11	8	8	11	8	12	8
Integration time, s	3	5	2	5	5	5	1	5	5	3
Power, kW	1.50	0.70	1.00	1.00	1.00	1.50	1.50	1.50	1.50	1.50
Plasma gas flow, L min^{-1}	10.5	10.5	10.5	10.5	10.5	10.5	10.5	10.5	12.0	16.5
Auxiliary gas flow, L min^{-1}	1.5	2.25	0.75	0.75	1.5	0.75	0.75	0.75	0.75	2.25
Background correction	yes	yes	yes	yes	yes	yes	yes	yes	yes	yes

3.4.2 Calibration

The multielement standards are prepared in the same medium as the samples, that is, with the same concentration of lithium metaborate. In *ICP-AES* the calibration curves representing, for each element, the intensity of the signal *I* as a function of the concentration *C* are always straight lines with the general equation:

$$I = a + bC \tag{4}$$

with the coefficient 'a' not significantly different from zero if spectral background correction has been done, as in our example. Table 11.8 reports the values of 'a' and 'b' and also the value of the coefficient of determination r^2 for the line fitted for 10 elements of the example. The table also shows the results of the test of significance of the constant 'a' at 5% level of significance. The values obtained for 'a' and 'b' are not transposable from one determination to another because they depend on instrument settings. However, a high value of the slope 'b' indicates good sensitivity of the method of determination of the corresponding element. A value of 'a' not significantly different from zero is favourable for a lower detection limit.

Table 11.8   Values of the calibration parameters according to equation 4 for the ten elements of the example. The values of 'a' and 'b' are in arbitrary units depending on instrumental settings.
N.S. - not significantly different from zero (at 5% significance).

Element	Slope 'b'	Intercept 'a'	Coefficient of determination r^2, %	Significance of 'a'
K	819	120	100.00	N.S.
Al	9450	750	100.00	N.S.
Fe	13830	1200	100.00	N.S.
Ca	4520	1400	99.96	N.S.
Ti	54310	1600	99.99	N.S.
Mg	7240	260	100.00	N.S.
Mn	93800	1000	100.00	N.S.
Na	42000	6900	99.99	N.S.
Si	335	90	99.80	N.S.
P	750	30	99.33	N.S.

Table 11.9 shows the errors of estimate between the actual contents of the standards and the contents calculated using equation 4. The relative error of estimate is always larger for low contents. However, in every case, it is lower than 2% except for determination of phosphorus.

Figure 11.12 shows an example of the calibration curve in the case of aluminium. We have not reported here the related curves of confidence interval (Fig. 11.2) because they almost coincide with the prediction curve.

Table 11.9 Errors of estimate in % for the indicated contents of 10 standards.

Content	K	Al	Fe	Ca	Ti	Mg	Mn	Na	Si	P
0.76										−11.84
1.33							0.83			
1.67					1.74					
1.91										10.47
2.67							0.00			
3.33					−0.60					
3.82										−2.36
4.17								1.99		
5.33							0.00			
6.67				1.94	0.00	1.20				
8.33	−1.50	1.68	0.83					0.36		
13.33				1.73		−0.60				
16.66	0.60	−0.72	−0.60					−0.24		
26.66				−0.56		0.08				
33.33	−0.06	0.00	0.09							
38.70									−1.21	
55.75									−0.38	
68.49									−2.47	
71.94									−1.35	

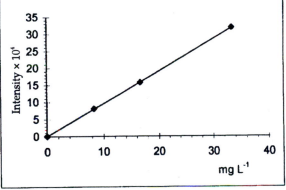

Fig. 11.12   Calibration curve for Al_2O_3.

3.5 Conclusion

Today many manufacturers offer only one range of more and more automated sequential and simultaneous instruments. This technique is tending to progressively replace flame atomic absorption spectrometry in laboratories. On the other hand, electrothermal atomic absorption still remains the method of choice for analysis of elements present in trace quantities and for microanalysis. More recently a new technique, inductively coupled plasma mass spectrometry (ICP-MS) has been introduced. The ICP-MS combines a plasma for producing ions with a mass spectrometer as detector. It thus provides data on each atomic-mass entity and enables quantitative determination of various isotopes of an element.

4. Comparison of ICP-AES, AAS and ICP-MS

A user may be faced with the problem of choosing between the ICP-AES and atomic absorption methods, which are today commonly used in physicochemical analysis laboratories. Several parameters have to be taken into account for selecting the one or the other of these techniques. The major parameters are:

—sensitivity and detection limits,
—analytical accuracy,
—usable concentration range,
—whether monoelement analysis or multielement analysis,
—quantity of analyte available (microanalysis),
—time taken for the analysis,
—interferences encountered in the proposed application,
—degree of automation necessary,
—price and operational costs.

A major difference between atomic absorption and ICP emission techniques lies in the fact that the latter enables rapid determination of several elements while atomic absorption allows a single element because of the need for a light source producing the radiation of each element to be analysed. Flame emission is essentially applicable to the alkali metals.

Table 11.10   Comparison of detection limits ($\mu g\ L^{-1}$) in pure water between plasma emission (ICP-AES and ICP-MS) and atomic absorption (flame and electrothermal) spectrometry.

Element	Flame AAS	EAAS	ICP-AES	ICP-MS
As	500	1.00	20	0.050
Al	50	0.50	3	0.010
Ba	50	1.50	0.2	0.005
Be	5	0.05	0.5	0.050
Bi	100	1.00	20	0.005
Cd	5	0.03	3	0.010
Ce	200,000		15	0.005
Co	10	0.50	10	0.005
Cr	10	0.15	10	0.005
Cu	5	0.50	5	0.010
Gd	4000		5	0.005
Ho	80		1	0.005
In	80	0.50	30	0.010
La	4000		0.05	0.005
Li	5	0.50	1	0.020
Mn	5	0.06	0.5	0.005
Ni	20	0.50	10	0.005
Pb	20	0.50	20	0.005
Se	1000	1.00	50	0.100
Tl	40	1.50	30	0.010
U	100,000		30	0.010
Y	500		0.5	0.005
Zn	2	0.01	1	0.020

The detection limits of each of these techniques should be appraised in relation to the problem posed (amount of soluble salt in the matrix, presence of interfering elements, etc.). Table 11.10 presents the approximate detection limits in pure water that can be attained with each technique including inductively coupled plasma mass spectrometry (ICPMS).

Application of one technique or another should also take into account human capabilities (necessary knowledge), equipment (physical plant, availability of gases, etc.) and availability of finance. Table 11.11 summarizes a simplified comparison of different elements to be taken into consideration while evaluating which technique is better suited for solving a given analytical problem.

Table 11.11 Simplified comparison of the characteristics of plasma emission spectrometers (ICP-AES and ICP-MS) and atomic absorption spectrometry (flame and electrothermal)

	Flame AAS	Electrothermal AAS	ICP-AES	ICP-MS
Detection limit	Very good for some elements	Excellent for most elements	Very good for most elements	Excellent for most elements
Speed of analysis	15 s/el./ sample	4 min/el./ sample	2-30 el./min/ sample	All elements in 2-6 min/sample
Linear range	10^3	10^2	10^5	10^6
Short-term precision	0.1-1%	1-5%	0.3-2%	1-3%
Long-term (4h) precision			<5%	<5%
Spectral interferences	Practically none	Few	Common	Few
Chemical interferences	Many	Many	Practically none	Moderate
Ionization interferences	Some	Minimal	Minimal	Minimal
Isotope interferences	No	No	No	Yes
Max. dissolved solids	0.5-3%	>20%	2-25%	0.1-0.4%
Number of elements	>68	>50	>73	>75
Sample volume	Very large	Very small	Large	Small
Semi-quantitative analysis	No	No	Yes	Yes
Isotopic analysis	No	No	No	Yes
Routine operation	Easy	Easy	Easy	Easy
Development of method	Easy	Experience necessary	Experience necessary	Experience necessary
Supervised operation	No	Yes	Yes	Yes
Fuel gas	Yes	No	No	No
Operating cost	Low	Moderate	High	High
Price	Low	Moderate/high	High	Very high

References

Boumans P.W.J.M. 1980. *Line Coincidence Tables for Inductively Coupled Plasma Atomic Emission Spectrometry.* vols. 1 and 2. Pergamon Press, Oxford, UK.

Boumans P.W.J.M. 1987. *Inductively Coupled Plasma Emission Spectroscopy. Part I: Methodology, Instrumentation and Performance; Part II: Applications and Fundamentals.* John Wiley and Sons, New York.

Cresser M.S., Armstrong J., Cook J., Dean J.R., Watkins P. and Cave M. 1994. Atomic spectrometry update-environmental analysis. *J. Analyt. Atomic Spectrometry*, **9**: 25R-85R.

Dean J.A. and Rains T.C. 1975. *Flame Emission and Atomic Spectrometry*. vols. 1 to 3. Marcel Dekker, New York.

Gillman G.P.. 1979. A proposed method for the measurement of exchange properties of highly weathered soils. *Aust. J. Soil Res.* **17**: 129-139.

Pinta M. 1979. *Spectrométrie d'absorption atomique*. vols. 1 and 2. Masson, Paris.

Pinta M. 1987. Spectrométrie des vibrations et des particules: spectrométrie d'émission de flamme. *Techniques de l'ingénieur*, **2815**, 12 pp.

Riandey C., Alphonse P., Gavinelli R. and Pinta M. 1982. Détermination des éléments majeurs des sols et des roches par spectrométrie d'émission de plasma et spectrométrie d'absorption atomique. *Analusis*, **10**: 323-332.

Sharp B.L., Chenery S., Jowitt R., Sparkes S.T. and Fisher A. 1994. Atomic Spectrometry update-atomic emission spectrometry. *J. Analyt. Atomic Spectrometry*, **9**: 171R-200R.

Thompson M. and Walsh J.N. 1989. *Handbook of Inductively Coupled Plasma Spectrometry*. Blackie and Son Ltd, New York, 2nd ed.

Trassy C. and Mermet J.-M. 1984. *Les applications analytiques des plasmas haute fréquence*, Technique et Documentation, Paris.

Winge R.K., Fassel V.A., Peterson V.J. and Floyd M.A. 1985. *Inductively Coupled Plasma Atomic Emission Spectroscopy. An Atlas of Spectral Information*. Elsevier, Amsterdam.

Bibliography

Hill S.J. (ed.), 1998. *ICP Spectrometry and Its Applications* (Sheffield Analytical Chemistry). Sheffield Academic Press, 320 pp.

Jarvis K.E., Gray A.L. and Houk R.S. 1991. *Handbook of Inductively Coupled Mass Spectrometry*. Chapman and Hall, London.

Montaser E. (ed.). 1998. *Inductively Coupled Plasma Mass Spectrometry*. VCH Publications.

Montaser E. and Golightly D.W. (eds.). 1992. *Inductively Coupled Plasmas in Analytical Atomic Spectrometry*. John Wiley and Sons, 2nd ed., 1017 pp.

Ionometry

1. Principle and Definitions

1.1 Introduction

Ionometry by means of ion-selective electrodes (ISE) designates a long-known measurement technique, used mostly for the measurement of pH (glass electrode). Following observations on the response of the pH electrode in solutions with high salt concentration, an electrode sensitive to sodium ions was perfected. Later, the work of Ross (1967) resulted in an electrode sensitive to calcium ions (solid-membrane electrodes). Progress then accelerated, especially under the influence of the companies Orion and Fluka[1] (polymer-membrane electrodes).

Selective electrodes render *in situ measurement* very easy. They allow unusual determinations, those of *ionization state and activity* of elements, in place of just their concentration. They thus are a valuable tool, perhaps irreplaceable for certain investigations on the functioning of soils. Nevertheless their use for purely analytical purposes requires knowledge of where they can be applied. Determinations without dilution can be done in a wide range of concentrations (usually 1 to 10^{-4} mol L^{-1}, and to less than 10^{-6} mol L^{-1}). The wide variety of shapes that can be given to these electrodes allows measurement without disturbance in sites difficult to reach. Amongst the numerous electrodes now available, those suitable for a very restricted range of media, perfected for one particular case, are distinguished from electrodes that can work over a wide range, which are the ones most often marketed. These electrodes can also be made in laboratories and synthesis of the active compounds is relatively easy (Susini, 1986). Also, active products are available for many electrodes (Fluka[1] 'ionophores', etc). Solvents, plasticizers and accessories used for making membranes are also marketed.

It must be emphasized that the quality of data given by the electrodes depends on that of the instrument used. Progress in electronics has led to development of this method well suited to continuous measurement, quality control and recording.

1.2 Theory

The reaction of ions with an electrode is directly influenced by other ions present in the solution. The state of the solution is expressed by a characteristic, the activity a of ions, which is related to concentration by the equation:

$$a_x = \gamma_x C_x \tag{1}$$

where C_x denotes the molar concentration of the ion x and γ_x is its activity coefficient, which is close to unity in very dilute solutions but different in

---

[1] *See* Appendix 6 for addresses.

concentrated solutions. It is a criterion of the behaviour of the ion x in solution, which may be calculated approximately from the equation established in 1937 by Kielland after Debye-Huckel for concentrations up to 0.1 mol L^{-1}:

$$- \log \gamma_x = \frac{A Z^2 J^{\frac{1}{2}}}{1 + B K J^{\frac{1}{2}}} \qquad (2)$$

where γ_x is the activity coefficient, A and B are constants depending on the solvent (*see* Appendix II to this chapter), Z is the ionic charge, J is the ionic strength and K is a characteristic related to the size of the hydrated ion (*see* Appendix II to this chapter).

The ionic strength J is calculated for all the ions present using the relation (at 25°C):

$$J = 0.5 \sum_x Z_x^2 C_x \qquad (3)$$

where Z_x is the ionic charge and C_x the concentration in mol L^{-1}.

The activity coefficient thus does not vary linearly with ionic strength. For low ionic strengths $J \leq 10^{-3}$ mol L^{-1}, the activity coefficient is close to unity. The activity is then almost equal to the concentration. It is also possible to compute the activity coefficient from measurements made with an ion-specific electrode.

1.3 Principle of Measurements

An ion electrode comprises, like a pH electrode, a solid or liquid membrane separating two solutions of an electrolyte. A potential is created at the interface, which is related to the activities a_x of the ion in the two solutions.

The variously shaped electrodes are generally composed of a tube, to one end of which is fixed the membrane sensitive to the ion under study. The interior of the tube contains a solution of the same ion. A metallic electrode attached to a coaxial wire is dipped in this solution to ensure electrical contact.

For measurement in the simplest case, this electrode is dipped in the solution containing the corresponding ion x (Fig. 12.1). The potential difference that appears between it and a reference electrode dipped in the same solution is measured. This potential difference, which is the sum of all component potentials, can be expressed by the Nernst equation (eqn 4 below).

1.4 Electrode Potential and Slope

The potential of an electrode depends, over a certain range, on the logarithm of the activity of free ions to which it is sensitive (Fig. 12.2) according to the relation:

$$U = U_0 + 2.303 \, (RT/nF) \log a_x \qquad (4)$$

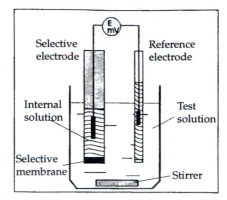

Fig. 12.1 Diagram of a measurement with an ion-selective electrode.

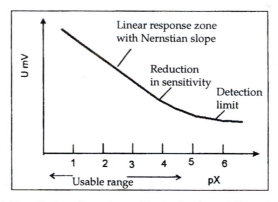

Fig. 12.2 Potential in mV of an electrode sensitive to the element X in relation to the activity a_x of the element in solution ($pX = -\log a_x$).

where U is the potential in mV developed compared to a reference; U_0 is the standard potential in mV of the ion for $a_x = 1$; R is the gas constant, 8.314 J K^{-1} mol^{-1}; T is the temperature in degrees Kelvin; F is the Faraday constant, 96,500 C mol^{-1}; n is the ionic charge with its sign; 2.303 is the factor for converting common logarithms to natural logarithms; a_x is the activity of the ion in the solution.

Sometimes the slope of the electrode is defined in mV by the Nernst factor $S = 2.303RT/nF$. This slope, a characteristic of the electrode, depends on the temperature θ in °C, the ionic charge and the sign of the charge (Figs. 12.2 and 12.3). For an increasing activity of cations the electrode potential increases positively (negatively for anions). For divalent ions the slope is half that for monovalent ions (Fig. 12.3).

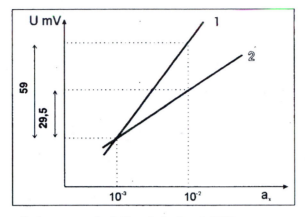

Fig. 12.3   Theoretical response of selective electrodes at 25°C:
1 - monovalent cation;
2 - divalent cation.

The slope S is 59.16 mV at 25°C for a monovalent cation. In other cases, if θ is expressed in °C, the slope is easily calculated from the relation:

$$S = 0.1984 \, (273.16 + \theta)/n \qquad (5)$$

At low ionic concentration the slope is no longer proportional to the activity because of the slight solubility of the substance constituting the membrane, and it is necessary to take this influence into account. Actually, many factors modify the response of the electrode, but the techniques of measurement described below enable elimination of these effects most of the time.

1.5 Selectivity and Detection Limit

In most cases, the test solution contains ions other than those to be measured. This can result in modification of the potential corresponding to the activity of the ion by formation of complexes or by precipitation. In other cases, the interfering ions can penetrate the sensitive membrane, which is what happens with polymer membranes. Considering these influences, the potential of the sequence is modified. According to Nikolsky:

$$U = U_0 + 2{,}303 \, \frac{RT}{n_A F} \, \log \left(a_A + k_{AB} \, a_B^{\frac{n_A}{n_B}} \right) \qquad (6)$$

where a_A is the activity of the analyte ion; a_B is the activity of the interfering ion; n_A is the charge of the analyte ion; n_B is the charge of the interfering ion; k_{AB} is the selectivity coefficient.

The selectivity coefficient k_{AB} expresses the preference of the detector for ion 'A' compared to ion 'B' (interference of B). It is possible to determine the value of k_{AB} by measuring the potential of the electrode in a solution containing only the ion to be determined and in another solution containing only the interfering ion:

$$U_1 = U_0 + S \log a_A$$

$$U_2 = U_0 + S \log (a_B^{n_A/n_B} k_{AB})$$

with $$\Delta U = U_1 - U_2$$

$$K_{AB} = \frac{a_A}{a_B^{n_a/n_B}} 10^{S/\Delta U}$$

For $n_A = n_B$, by making $a_A = a_B$,

$$k_{AB} = 10^{S/\Delta U}$$

It is also possible to use a graphical procedure. When the straight lines PQ and RS (Fig. 12.4) are extended, at the intersection point T, $U_1 = U_2$ or

$$a_A = K_{AB} \, a_B^{n_A/n_B}$$

and equation (6) becomes

$$U = U_0 + (2.303RT/n_AF) \log (2a_A)$$

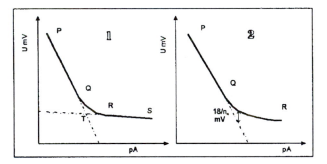

Fig. 12.4   Determination of selectivity of an electrode.

Compared to the ideal straight line (pure A), the potential difference arising from the interference can then be expressed by

$$\Delta U = 2.303(RT/n_AF)[\log (2a_A) - \log a_A] \cong 18/n_A \text{ (in mV at 25°C)}$$

In the most common situation the prominent flattening of the RS portion of the curve (Fig. 12.4) is not seen. The Nernstian response curve is extended by extrapolation of the straight-line portion PQ to the height of a deviation of $18/n_A$ mV compared to the actual curve. The detection limit of element A in the medium is obtained on the x-axis.

For electrodes with crystalline membrane, the interfering ion can form soluble complexes or nearly insoluble precipitates. In this case, k_{AB} is given by the ratio of the solubility products of the material of the electrode and the difficultly soluble combination in the solution. Taking the example of the influence of chlorides on a bromide-selective electrode:

$$k_{AB} = L_{AgBr}/L_{AgCl} = 3.62 \times 10^{-3}$$

where L_{AgBr} is the solubility product of silver bromide = 3.98×10^{-13}, and L_{AgCl} solubility product of silver chloride = 1.10×10^{-10}.

This relation is used to define the maximum concentration of Cl⁻ ions permitted without altering the sensing crystal:

$$a_B = (1/k_{AB})a_A.$$

For a bromide-ion activity of $a_A = 10^{-4}$,

$$a_B = \text{maximum activity of the Cl}^- \text{ ion} = 2.8 \times 10^{-2}.$$

2. Apparatus

2.1 Measuring Electrodes

Ion electrodes are grouped according to type of detector into
    —glass-membrane electrodes: Na^+, H^+
    —crystalline-membrane electrodes: Cu^{2+}, Ca^{2+}, Pb^{2+}, F^-, Cl^-
    —polymer-membrane electrodes: Ca^{2+}, K^+, NO_3^-, BF_4^-
    —gas-diffusion electrodes: NH_3, CO_2 (are of pH type)
    —optical detectors ('Optodes'): NH_3, etc.
Specialized firms (Orion, Metrohm, Ingold[1], etc.) commonly distribute many models of ion electrodes with varying shape more or less suitable for particular situations. Among these electrodes, the most useful for soils are of the membrane type (K^+, Ca^{++}, Ca^{++}+ Mg^{++}, NH_4^+, NO_3^-, etc.). Models with interchangeable sensor should preferably be selected.

It is possible to make sensors in the laboratory from the reagents for fabricating membranes (for example, Ionophores-Selectrophores from Fluka). As an illustration may be cited the technique of preparing a calcium electrode (Fig. 12.5) described by Susini (1986) according to a principle conceived by Griffith *et al.* (1972).

2.2 Reference Electrode

The role of this electrode is to provide a potential that is as constant as possible and is independent of the composition of the test solution.

It is composed of a metallic conductor in contact with a sparingly soluble salt of the same metal and a solution of constant composition (reference electrolyte). It is this electrolyte that is placed in contact with the solution in

---

[1] *See* Appendix 6 for addresses.

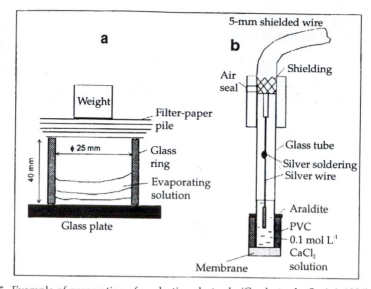

Fig. 12.5 Example of preparation of a selective electrode (Ca electrode, Susini, 1986):
a - preparation of a Ca-sensitive membrane by evaporation of 2.5 mL of a solution of a mixture of 5 mL tetrahydrofuran + 0.170 g polyvinyl chloride Breon 110 + 0.320 g di-n-octylphenylphosphonate + 0.033 g di-4 (1,1,3,3 tetramethylbutyl) phenylphosphoric acid; the membrane is detached 24 hours after evaporation; it is sufficient for making 4 electrodes (b);
b - mounting of the electrode.

order to establish electrical connection. There should be no free mixing, and this condition is most usually fulfilled by using a porous plug.

A potential difference, termed the 'diffusion potential', develops at the interface of the two solutions due to the difference in mobility of the anions and cations. Also, the choice of reference electrolyte plays an important part in obtaining a potential difference as stable as possible. To enable free access between the reference electrolyte and the test solution, a double-junction reference electrode is used.

Arrangements with a separate reference electrode are preferable to combined electrodes. That many anomalies are caused by defects in the reference electrode should not be lost sight of. It is appropriate to take into account in the computations the true potential of the electrode expressed relative to the hydrogen electrode (Table 12.1).

2.3 Ionometers

Conditions required to be fulfilled for use of ion electrodes are very similar to those for measurement of pH. The instruments are similar to pH instruments but the measured value is expressed in mV. It is therefore necessary to ensure that the instrument selected has facility for pH-mV conversion, with a scale displaying millivolts.

Table 12.1 Reference potentials relative to the hydrogen electrode.

Reference system	Potential difference U (mV) vs the hydrogen electrode					
	0°C	25°C	40°C	60°C	80°C	95°C
Silver chloride:						
Ag, AgCl with KCl (sat.)	+220	+197	+181	+160	+138	+121
Ag, AgCl with KCl (3.0 mol L^{-1})	+226	+208	+195	+178	+160	+147
Ag, AgCl with LiCl (sat. in ethanol)		+143				
Ag, AgCl with KNO_3 (sat.)		+467				
Ag, AgCl with $LiClO_4$ (1.0 mol L^{-1} in glacial acetic acid)		+350				
Calomel						
Hg, Hg_2Cl_2 with KCl (sat.)	+260	+244	+234	220		
Mercurous sulphate						
Hg, Hg_2SO_4 with K_2SO_4 (sat.)		+656 (22°C)				
Thalamide®						
Pt Hg, Tl, TlCl with KCl (sat.)		−577	−592	−608	−624	

However, this pH meter should have more refined characteristics than the usual pH meters: the impedance should not be smaller than 10^{12} ohms and the resolution should be 0.1 mV with precision of ± 0.1 mV (0.001 pH unit).

Thus, for a slope S of 59.16 mV, for an activity measurement with 1% repeatability, the potential difference will be

$$\Delta U = 59.16 \log (100/100 \pm 1) = \pm 0.25 \text{ mV}.$$

It is therefore imperative that the measurement be done with an instrument having a resolution of at least this value.

The instrument should, of course, carry the usual controls of a pH meter: temperature setting, slope correction (70-100%), compensation potentiometer and a display. It is not possible, however, to read concentrations directly from such an ionometer; they are obtained from a calibration curve drawn with standard solutions or by the method of standard additions.

There is a higher-priced category of ionometers with, in addition to the above features, an integral calculation system that enables direct reading of the results from various analytical techniques.

It is useful to supplement the instrument with an automatic or manual selector that allows the same measuring device to be switched between 4 to 6 different electrodes. Such an instrument will have settings for eliminating drift in potential of each electrode and thus obtaining the original potentials using a single value, which greatly simplifies display of data and computations.

For direct measurements *in situ* in soils, the measuring sites should be more numerous and provided with independent power supply (Loyer and Susini, 1978).

3. Calibration and Measurement

3.1 Response Time of an Electrode

Ion electrodes respond to the activity of the ion and it is only if the ionic strength of the medium is low that activity is equal to the concentration. Before starting determinations it is necessary to know the response time of the electrode. Actually, when the electrode is dipped in the test solution, the response is not instantaneous; the time taken for equilibrium to be attained is long or short depending on the nature of the medium. Equilibrium is considered to have been attained when the deviation in potential between two measurements is of the order of 1 mV (Fig. 12.6). Therefore an arbitrary time is fixed, to be followed in the entire series of analyses.

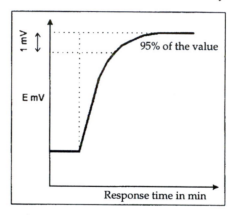

Fig. 12.6   Response time of an electrode.

3.2 Calibration Curve for 'Activity'

A series of standard solutions of known ionic strength should be carefully prepared using distilled water and very pure compounds, in order to make dilute solutions of given concentration.

Calibration is always begun with low concentrations. The ion electrode and reference electrode are dipped in each standard solution at 25°C. After stability has been achieved the potential difference is recorded and a curve is drawn from the values obtained using semilogarithmic paper or a computer. The calibration should be repeated once or twice in a day. If deviation from an earlier curve is noticed (very often, the new curve is parallel to the earlier one), the compensator setting of the instrument can be adjusted to establish coincidence of the two curves. Some instruments have a scale permitting direct reading of activities in mol L^{-1}.

In many determinations concerning ionic equilibria in solutions, knowledge of the activity is more pertinent than that of the concentration. Ionometry is also a technique of choice for following the evolution of a complexed or insoluble compound (masking phenomenon). The technique can also be used for measuring ionic concentrations.

3.3 Determination of Ionic Concentration

Concentration can be calculated from the activity a_x according to equation 1. In most cases, unfortunately, the activity coefficient γ and the ionic strength of the medium are ignored. The analyst has two solutions available: (1) adjusting the ionic strength of the medium; (2) using the method of standard additions (*see* Chapter 18).

3.3.1 Method using constant ionic strength

The principle consists of adding to the sample a certain volume of a solution of high ionic strength whereby the ionic strength of the test solution is brought to a high value. This results in the activity γ remaining unvarying. These buffers are called 'total ionic strength adjustment buffers' (TISAB) and an entire range of them has been studied for various ions (*see* Appendix I to this chapter). There are also special formulae for certain particular techniques using a complexing action for masking the highly interfering ions (Susini, 1986).

A series of standards is prepared to give the chosen range of concentrations. These solutions are added to a volume of buffer (TISAB) to give an ionic strength close to 2 (*see* Appendix IV to this chapter), and the corresponding potential differences are determined with due compensation for temperature. Using semilogarithmic paper a standard curve is drawn from the values obtained. The unknown samples are taken through the same steps, strictly following the procedure used for the calibration. The concentrations are determined from the standard curve (*see* Chapter 18) with the logarithmic scale on the abscissa.

3.3.2 Method of standard additions

This method (*see* Chapter 18) is successfully applied to solutions of high ionic strength. The ions present need not be known, except for highly interfering ions that require addition of special complexing buffers. Figure 12.7 illustrates the accuracy and reproducibility of the method of standard additions in the case of determination of the calcium ion.

If lower precision can be tolerated, the method may be simplified as follows. To a volume V_0 of the sample (at one point in the range) a volume V_a of a solution of the analyte ion is added in such a way that the volume V_a does not exceed 1% of the volume V_0. Also, the concentration C_a of the added solution should be about 100 times the expected concentration in the sample. Under these conditions the change in volume caused can be disregarded. The change in concentration ΔC caused by addition of V_a is

$$\Delta C = V_a C_a / V_0 \tag{8}$$

and the concentration C_x of the sample is given by

$$C_x = A \cdot \Delta C \text{ where } A = 1/(10^{\Delta U/S} - 1) \tag{9}$$

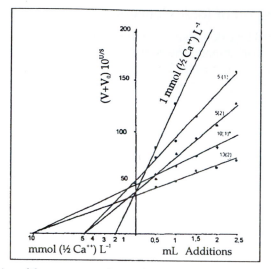

Fig. 12.7   Illustration of the accuracy and reproducibility of the method of standard additions with a Ca electrode (Susini, 1986): C_x = 2, 5, 10 mmol (1/2 Ca^{2+}) L^{-1}; V_0 = 50 mL, 25 mL triethanolamine buffer pH 8.4 + 25 mL H$_2$O, added solution 100 mmol (1/2 Ca^{++}) L^{-1}.

C_x being the concentration of the sample (same units as C_a), A the increment factor for T = 25°C, ΔU the difference in potential ($U_2 - U_1$) in mV (U_1 and U_2 being the potentials before and after addition, respectively), S the value of the slope of the electrode in mV.

For a temperature different from 25°C ΔU will be corrected according to equation 5.

Another mode of calculation taking into account the variation in volume can also be adopted. It then becomes possible to chose any volume whatsoever for V_a:

$$C_x = C_a Q \quad \text{avec} \quad Q = \frac{V_a}{(V_0 + V_a)(10^{\Delta U/S} - V_0)} \tag{10}$$

4. Optodes (Ion-Selective Photodiodes)

This set-up is an extension of determinations with ion-selective electrodes operating not by ionometry, but by molecular spectrometry (*see* Chapter 9). The membrane containing the reagent sensitive to the analyte ion plays the part of selective filter for the radiation of wavelength corresponding to the compound measured, the absorbance being a function of the concentration (Fig. 12.8).

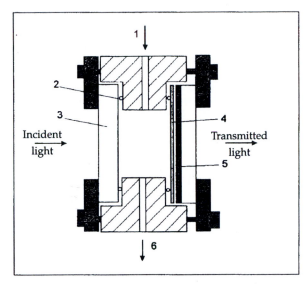

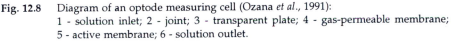

Fig. 12.8   Diagram of an optode measuring cell (Ozana *et al.*, 1991):
1 - solution inlet; 2 - joint; 3 - transparent plate; 4 - gas-permeable membrane;
5 - active membrane; 6 - solution outlet.

This system lends itself to many arrangements that can be easily miniaturized. For example, the membrane can be directly deposited on a photodiode, the light being led through an optical fibre. Therefore, these set-ups can be installed in sites hard to access.

This technique does not require a reference electrode. Response is quick, the set-up can be immersed in the reaction medium and the technique lends itself well to continuous automated measurements. An example of the technique of preparation of a membrane for NH_3 gas can be found in Seiler *et al.* (1989).

References

Griffiths G.H., Moody G.S. and Thomas J.D.R. 1972. An investigation of the optimum composition of polyvinyl chloride matrix membranes used for selective calcium electrodes. *Analyst*, **97**: 420-427.

Loyer J.Y. and Susini J. 1978. Réalisation et utilisation d'un ensemble automatique pour la mesure en continu et *in situ* du pH, du Eh et du pNa du sol. *Cah. ORSTOM Sér. Pédol.* **16**: 425-437.

Ozanna S., Hauser P.C. Seiler K., Tan S.S.S. and Morf W.E. 1991. Ammonia-gas selective optical sensor based on neutral ionophores. *Anal. Chem.* **63**: 640-644.

Seiler K., Morf W.E. and Rusterholz B. 1989. Design and characterization of a novel ammonium ion selective optical sensor. *Anal. Sci.* **5**: 557-561.

Susini J. 1986. Électrode ionique sensible aux ions calcium. Sa fabrication, son utilisation dans les eaux et les suspensions de sol. *Cah. ORSTOM Sér. Pédol.* **22**: 87-104.

Bibliography

Ari Ivaska. 1980. Linear titration plots with ion selective electrodes. *Talanta*, **27**: 161-164.

Bailey P.L. 1979. Industrial applications for ion selective electrodes. *Ion-Selective Electrode Rev.* **1**: 81-133.

Cheng K.L., Hung Jui-Chi and Da Prager. 1973. Determination of exchangeable calcium and magnesium in soil by ion selective electrode method. *Microchem. J.* **18**: 256-261.

Doiron E.B. and Chamberland E. 1973. Rapid determination of exchangeable calcium in soil with the calcium electrode. *Comm. Soil Sci. Pl. Analysis*, **4**: 205-209.

Ebdon L., Ellis A.T. and Corfield G.C. 1979. Ion selective polymeric membrane electrodes with immobilized ion exchange sites. *Analyst*, **104**: 730-738.

Govington A.K. and Rebolo M.J.F. 1983. Reference electrodes and liquid junction effects in ion-selective electrode potentiometry. *Ion-Selective Electrode Rev.* **5**: 93-128.

Graggs A., Doyle B., Hassan S.K.A.G.G., Moody G.S. and Thomas J.D.R. 1980. PVC ion-selective electrodes based on calcium bis dialkyl and di (4-alkylphenyl) phosphates and mixed solvent mediators. *Talanta*, **27**: 277-280.

Hansen E.H., Ruzicha J. and Ghose A.K. 1978. Flow injection analysis for calcium in serum, water and waste waters by spectrophotometry and by ion selective electrodes. *Anal. Chem.* **100**: 151-165.

Horval G., Toth K. and Pungor E. 1989. Theoretical models of ion-selective membranes. *Analytica Chimica Acta*, **216**: 163-176.

Koryta, J. 1990. Theory and applications of ion-selective electrodes. *Analytica Chimica Acta*, **233**: 130 (770 references).

Kovda V.A. and Materova Ye. A. 1977. Experiment in the use of ion-selective electrodes in agrochemical soil investigations. *Soviet Soil Sci.* **235** (1).

Moody G.J. and Thomas J.D.R. 1978. Developments in coated-wire ion-selective electrodes. *Laboratory Practice*, pp. 285-289.

Thomas J.D.R., 1978. Design of calcium ion selective electrodes. *Laboratory Practice*, pp. 857-861.

Zykina G.K. 1985. Determination of the ion composition of soil solutions by means of ion-selective electrodes. *Pochvovedeniye*, **4**: 104-107.

Evans A. and James A.M. 1987. *Potentiometry and Ion Selective Electrodes* (Analytical Chemistry for Open Learning). John Wiley and Sons.

Ion-selective Electrode Reviews. Pergamon Press.

Koryta J. and Dvorak J. 1987. Ion-selective electrodes. In: *Principles of Electrochemistry*. Wiley.

Moody G.J. and Thomas J.D.R. 1971. *Selective Ion Sensitive Electrodes*. Merrow Technical Library, England.

Mort W.E. 1985. *The Principle of Ion-selective Electrodes and of Membrane Transport*. Elsevier-Mir.

Pungor E. 1978. *Ion-selective Electrodes*. Elsevier Scientific Publishing Company.

Pungor E. 1988. *Dynamic Characteristics of Ion-selective Electrodes*. CRC Press, Boca Raton FL, USA.

Seiler K. 1991. *Ionenselective Optodenmembranen*. Éditions Fluka.

Thomas R.C. 1978. *Ion-sensitive Intracellular Microelectrodes*. Academic Press.

Umezawa Y. 1990. *CRC Handbook of Ion-selective Electrodes: Selectivity Coefficients*. CRC Press, Boca Raton FL, USA.

Yu T.R. and Ji G.L. 1993. *Electrochemical Methods in Soil and Water Research*. Pergamon Press, 477 pp.

Appendix I. Formulae of buffers for adjusting ionic strength

Ion	Reagent	Weight for 100 mL
Na^+	Trihydroxymethyl aminomethane (TRIS) [$(HOH_2C)_3CNH_2$]	12.11 g; adjust pH to 8-10 with HNO_3
K^+	Sodium chloride NaCl, 1 mol L^{-1} to 0.1 mol L^{-1} according to K concentration	5.84 g to 0.584 g
Ca^{2+}	Potassium chloride KCl, 1 mol L^{-1}	7.46 g
Cu^{2+}	Potassium nitrate KNO_3, 1 mol L^{-1}	10.11 g
Ag^+	Potassium nitrate KNO_3, 2 mol L^{-1}	20.22 g
Cd^{2+}	Potassium nitrate KNO_3, 5 mol L^{-1}	50.55 g
Pb^{2+}	Sodium perchlorate $NaClO_4.H_2O$, 1 mol L^{-1}	14.05 g, adjust pH to 5-9
F^-	Water	50 mL
	Sodium chloride	5.84 g
	Acetic acid	5.75 mL
	EDTA Komplexone IV	0.5 g
Cl^-	Potassium nitrate, 2 mol L^{-1}	20.22 g
Br^- I^-	Sodium nitrate, 2 mol L^{-1}	17 g
CN^-	Sodium hydroxide, 0.1 mol NaOH L^{-1}	0.40 g; eliminate sulphides
SCN^-	Potassium nitrate, 1 mol KNO_3 L^{-1}	10.11 g
NO_3^-	Ammonium sulphate, 0.1 mol $(NH_4)_2SO_4$ L^{-1}	1.32 g
	Aluminium sulphate, 0.1 mol $Al_2(SO_4)_3$ L^{-1}	3.42 g
BF_4^-	Ammonium sulphate, 2 mol $(NH_4)_2SO_4$ L^{-1}	26.42 g
S^{2-}	Sodium hydroxide, 2 mol NaOH L^{-1}	8 g; pH $\geq$ 13

Appendix II. Tables for calculation of activity coefficient

Table II.1   Constants A and B (eqn 2) for water at different temperatures.

Temperature °C	Constant A		Constant B	
	Molarity scale	Molality scale	Molarity scale	Molality scale
0	0.488	0.488	0.324	0.324
10	0.496	0.496	0.325	0.325
20	0.505	0.504	0.328	0.327
25	0.509	0.509	0.328	0.328
30	0.514	0.513	0.330	0.329
40	0.524	0.522	0.332	0.331
50	0.535	0.532	0.334	0.332
60	0.547	0.542	0.337	0.334
70	0.560	0.554	0.339	0.335
80	0.574	0.566	0.342	0.337
90	0.598	0.579	0.345	0.339
100	0.606	0.593	0.346	0.341

Table II.2   Values of K (eqn 2) for various hydrated ions.

K	Monovalent ions
9	H^+
6	L^{i+}, $(C_2H_5)_4N^+$
	$C_6H_5COO^-$, $C_6H_5CH_2COO^-$
5	$(C_2H_5)_3NH^+$
	$CHCl_2COO^-$, CCl_3COO^-
4	Na^+, $(CH_3)_4N^+$, $(CH_3)_3NH^+$, $(C_2H_5)_2NH_2^+$, $C_2H_5NH_3^+$
	HCO_3^-, $H_2PO_4^-$, CH_3COO^-
3	K^+, Rb^+, Cs^+, Tl^+, Ag^+, NH_4^+, $(CH_3)_2NH_2^+$, $CH_3NH_3^+$
	OH^-, F^-, Cl^-, Br^-, I^-, CN^-, SCN^-, ClO_4^-, H_2 (citrate)$^-$, $HCOO^-$

K	Divalent ions
8	Be^{2+}, Mg^{2+}
6	Ca^{2+}, Cu^{2+}, Zn^{2+}, Sn^{2+}, Mn^{2+}, Fe^{2+}, Co^{2+}, Ni^{2+}
5	Sr^{2+}, Ba^{2+}, Cd^{2+}, Hg^{2+}, Pb^{2+}
	CO_3^{2-}, $(COO)_2^{2-}$, $CH_2(COO)_2^{2-}$, $(CHOHCOO)_2^{2-}$, $H(citrate)^{2-}$
4	Hg_2^{2+}, SO_4^{2-}, $S_2O_3^{2-}$, CrO_4^{2-}, HPO_4^{2-}

K	Trivalent ions
9	Al^{3+}, Fe^{3+}, Cr^{3+}, La^{3+}, Ce_2^{3+}
5	$(citrate)^{3-}$
4	PO_4^{3-}

K	Quadrivalent ions
11	Th^{4+}, Ce^{4+}, Sn^{4+}

Appendix III. Activity coefficient γ of aqueous solutions at 25°C

	Ionic strength J (eqn 2 and 3)						
	0.001	0.0025	0.005	0.01	0.025	0.05	0.1
	Activity coefficient γ_i						
K	Monovalent ions						
9	0.967	0.950	0.934	0.913	0.881	0.854	0.826
6	0.966	0.948	0.930	0.907	0.868	0.834	0.796
5	0.965	0.947	0.928	0.904	0.863	0.826	0.783
4	0.965	0.946	0.927	0.902	0.858	0.817	0.770
3	0.965	0.946	0.925	0.899	0.852	0.807	0.754
K	Divalent ions						
8	0.872	0.813	0.756	0.690	0.592	0.517	0.445
6	0.870	0.808	0.748	0.676	0.568	0.483	0.401
5	0.869	0.805	0.743	0.668	0.555	0.464	0.377
4	0.867	0.803	0.738	0.661	0.541	0.445	0.351
K	Trivalent ions						
9	0.737	0.632	0.540	0.443	0.321	0.242	0.178
5	0.728	0.614	0.513	0.404	0.266	0.178	0.111
4	0.726	0.610	0.505	0.394	0.251	0.161	0.095
K	Quadrivalent ions						
11	0.587	0.452	0.348	0.525	0.151	0.098	0.063

Appendix IV. Suggested Adjustment of Ionic Strength for Various Solutions.

Test ion	Sample solution mol L^{-1}	mL buffer per 50 mL sample	Ionic strength mol L^{-1}
Na^+	10^{-5} to 10^{-5}	50	2
F^-	≤ 0.1	50	
	0.1 to 0.5	450	2
Ca^{2+}	10^{-6} to 10^{-4}	0.5 ISAB diluted 1:3	10^{-2}
	10^{-4} to 10^{-1}	1	8×10^{-2}
	0.1 to 1	17	1
K^+	10^{-6} to 10^{-4}	0.5 ISAB diluted 1:4	1.2×10^{-2}
	10^{-4} to 10^{-1}	1	1.2×10^{-1}
	0.1 to 1	10	1
NO_3^-	10^{-6} to 10^{-4}	0.5 ISAB diluted 1:4	4×10^{-3}
	10^{-4} to 10^{-1}	1	4×10^{-2}
	0.1 to 1	50	1
BF_4^-	7×10^{-6} to 1	1	1.2×10^{-1}

13

Chromatographic Techniques

1. Definitions and Principles

1.1 General Definition

There are great similarities in principle between the various chromatographic techniques. Chromatography comprises analytical methods that enable separation and then identification and quantification of constituents of mixtures. Separation is achieved by using differences in equilibrium constants of components between a mobile phase that tends to transport them and a stationary phase that tends to retain them. An illustration of the phenomenon can be obtained from a highly tannic stain of wine or concentrated coffee on a white shirt: when one tries to remove this stain by pouring water little by little on it, the shirt gets wetted (migration of the solvent front on the stationary phase) and aureoles will be formed and grow larger around the original stain. These aureoles are the manifestation of chromatographic separation of the coloured components of the stain between the mobile phase (water) and the stationary phase (cloth). One has perhaps done chromatography unknowingly.

1.2 Classification of Chromatographic Methods

The methods are grouped (Fig. 13.1) essentially according to the nature of the stationary phase and the nature of the mobile phase. If the latter is a gas,

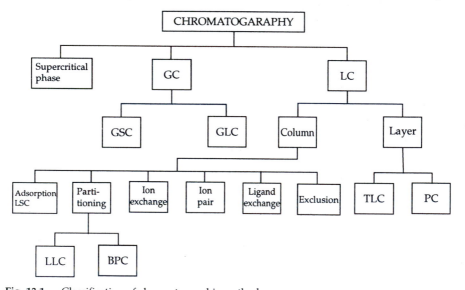

Fig. 13.1   Classification of chromatographic methods
GC - gas chromatography; LC - liquid chromatography; GSC - gas-solid chromatography; GLC - gas-liquid chromatography; LSC - liquid-solid chromatography; LLC - liquid-liquid chromatography; BPC - bonded-phase chromatography; TLC - thin-layer chromatography; PC - paper chromatography.

the technique is gas chromatography (GC) and if liquid, the technique is liquid chromatography (LC). Supercritical phase chromatographic techniques may be considered intermediate between GC and LC.

The stationary phase can be an adsorbent solid or a liquid impregnating a solid. Gas chromatography is called GSC (gas-solid chromatography) if the stationary phase is an adsorbent solid and GLC (gas-liquid chromatography) if the stationary phase is a liquid. Similarly LC becomes LSC (liquid-solid chromatography) or LLC (liquid-liquid chromatography). If the stationary phase is intermediate between a liquid and a solid, we have LGC (liquid gel chromatography). The adsorbent can also be a solid on which the appropriate molecules are bonded: BPC (bonded-phase chromatography).

Another grouping is based on the technology applied. In GC the chromatographic separation is always done in a column. In LC, sheet and column methods are distinguished. A chromatographic column is a tube of very varied diameter and shape, which holds the adsorbent as a filling or an impregnation. Sheet methods can use a solid material directly as adsorbent, generally paper (cloth in the example in §1.1) or more often a material deposited on an inert plate (glass, plastic, etc.). The theory presented below pertains mostly to column methods; its relation to sheet methods is presented in Chapter 15.

Gas chromatography uses interactions between the solute to be separated and the stationary phase. In liquid chromatography, the interactions are more complex because they concern the stationary phase and the mobile phase concomitantly. Another classification of LC is based on the nature of the phenomena put to use (Fig. 13.1): adsorption chromatography (LSC), partition chromatography (LLC and BPC), ion-exchange chromatography, ion-pair chromatography, ligand-exchange chromatography and exclusion chromatography. These diverse techniques are explained in Chapter 15.

2. Elementary Parameters of Chromatography

2.1 Similarities Between Techniques

There are great similarities in principle between chromatographic techniques since they are all based on the partition of solutes in a mixture between a mobile phase and a stationary phase (Fig. 13.2).

For a given chromatographic system, the distribution of each solute can be characterized by a distribution (or partition) coefficient K such that

$$K = C_s / C_m \tag{1}$$

where C_s and C_m respectively denote the equilibrium concentrations of the solute in the stationary and mobile phases. When the solutes have different partition coefficients, they are transported at different speeds by the mobile

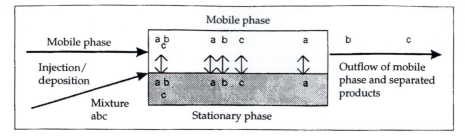

Fig. 13.2   Diagram of the principle of chromatographic separation of the components of a ternary mixture *abc*.

phase and will be separately obtained at the outlet of the system (Fig. 13.2). In sheet chromatography, separation is manifested by a spot corresponding to each component of the mixture (aureoles in the example given in §1.1). In column chromatography, a detector provides a peak of more or less Gaussian shape on the recorder or on the monitor screen (Fig. 13.3).

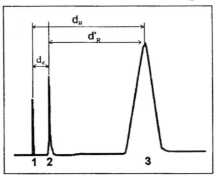

Fig. 13.3   Elementary retention parameters of a chromatogram:
1 - instant of injection; 2 - peak of unretained substance (air or methane in the case of GC); 3 - summit of peak of component to be characterized; d_d - retention distance corresponding to the dead volume of the column (volume of mobile phase); d_R and d_R' - original and corrected retention distances of the component.

2.2 Retention Parameters

The formulae below pertain to gas or liquid column chromatography; the parameters used for sheet chromatography are slightly different. Figure 13.3 gives a schematic presentation of the principal parameters pertaining to the chromatographic separation of a solute.

The first experimental parameter in chromatography is *retention time, t_R*: this is the time that elapses between the injection (instant of injection) and the outflow at the summit of the peak (presumed to be symmetrical) of the solute under consideration.

Retention time depends on:
—nature of the solute,
—nature of the stationary phase and its packing density in the column,

—rate of flow of the mobile phase,
—column temperature,
—nature of the mobile phase in the case of LC.
However, t_R is, in principle, independent of
—quantity of solute injected (provided it is small),
—nature and abundance of other constituents of the mixture,
—nature and pressure of the carrier gas in the case of GC.

With a column under defined experimental conditions, t_R is a qualitative characteristic of the solute studied.

Computerized instruments can give the retention time in seconds. With a recorder, the retention distance d_R between the instant of injection and the summit of the peak is measured on the paper. Then,

$$t_R = d_R/V_E \tag{2}$$

where V_E is the speed of the recorder.

A more precise characteristic of a solute under given experimental conditions is the corrected retention time t_R' given by

$$t_R' = t_R - t_d \tag{3}$$

This parameter is very closely related to the phenomenon of retention proper. Actually, t_d, being the retention time of the peak of the component presumed not to be retained by the stationary phase, is supposed to represent the time taken by the solute molecules to pass through the voids in the column, during which time they are not in direct contact with the stationary phase for participating in the exchange process.

By the same logic, parameters more and more independent of the experimental conditions and therefore able to better characterize the solutes have been proposed. When the pressure-gradient in the column is neglected, the parameter V_R may be defined by

$$V_R = t_R D_S \tag{4}$$

where V_R, termed retention volume, no longer depends on the flux of the mobile phase D_S (volumetric flow rate of the carrier gas in GC). At constant temperature this is the volume of the mobile phase that should pass through the column from the instant of injection to the summit of the outflow peak. The retention volume of the unretained peak and V_R', the corrected retention volume, are defined just as in the case of t_R.

In GC, the pressure-gradient in the column can be taken into account to calculate the absolute retention volume V_N from the retention volume reduced by a correction factor established in 1952 by James and Martin, the inventors of the technique (*see* Chapter 14). In the case of gas-liquid chromatography, the specific retention volume V_g related to the mass of the stationary liquid phase can be computed.

The retention volume V_R (eqn 4) is related to the partition coefficient K (eqn 1) as follows:

$$V_R = V_d + KV_S \tag{5}$$

where V_d denotes the volume of the mobile phase contained in the column and V_S the volume occupied by the stationary phase. This relation is valid only in the range over which K is independent of the concentration of the solute in the eluent phase.

In order to be independent of the geometric parameters of the column, the retention of a component can be characterized by a capacity factor k' such that

$$k' = C_S V_S / C_d V_d = KV_S / V_d \tag{6}$$

For a given column, the ratio V_S / V_d is constant and k' depends only on the capacity factor K, which is itself directly related to the exchange process by an equation of the type

$$\ln K = -\Delta G° / RT \tag{7}$$

where $\Delta G°$ denotes the free enthalpy of dissolution of the solute between the phases, R the gas constant and T the absolute temperature. The capacity factor can be experimentally determined by using the equation

$$k' = (V_R - V_d)/V_d = (t_R - t_d)/t_d \tag{8}$$

Retention time is thus related to the capacity factor by the equation

$$t_R = t_d(1 + k') = L(1 + k')/u \tag{9}$$

where L is the length of the column and u the speed of the mobile phase.

2.3 Relative-Retention Parameters

Characterization of solutes by a relative-retention value allows freedom from auxiliary measurements (flow rates of gases in particular) necessary for the conversion of retention distance d'_R to specific retention volume V_g. These parameters are therefore popular in chromatography for qualitative analysis. For a given mixture, it is necessary to select a reference component 'R' and compute the relative-retention parameters of the other components 'i' using the ratio d'_i / d'_R.

2.4 Efficiency of a Column

This is a measure of the sharpness of a peak compared to its retention distance. It is expressed by the number of theoretical plateaux by reference to the old theory of distillation. The number n of theoretical plateaux is given by

$$n = 16(d_R/\omega)^2 = (d_R/\sigma)^2 = 5.54(d_R/l)^2 \tag{10}$$

where ω is the segment intercepted on the base line by the tangents at the inflection points of the peak (Fig. 13.4), σ the standard deviation of the Gaussian peak and I the width of the peak at half-height.

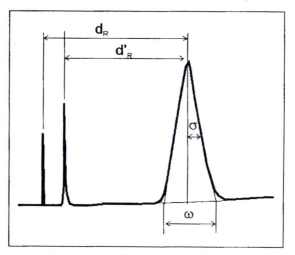

Fig. 13.4 Experimental parameters of a presumably Gaussian chromatographic peak.

The equivalent height h of a theoretical plateau is related to n and the length of the column by

$$h = L/n \qquad (11)$$

Efficient columns are characterized by a large number of theoretical plateaux, the order of the value of n varying little with solute. A classic packed GC column may have 2000 plateaux per metre; the efficiency per metre is of the same order for the ordinary capillary columns, but these can be much longer.

Similarly, in the case of capillary GC columns (d_d relatively greater than for packed columns) the number of effective theoretical plateaux is defined by

$$n_{eff} = 16(d_R'/\omega)^2 \qquad (12)$$

2.5 Resolution of Two Components

This is a parameter for estimating the quality of a chromatographic separation between two peaks of a chromatogram. It depends on the sharpness of the peaks (efficiency of the column) and also on the relative separation of the summits of the peaks (relative-retention coefficient). The parameters to be taken into account are indicated in Fig. 13.5, the resolution R of the two peaks A and B in that figure being expressed by

$$R_{A,B} = \frac{2(d_{R_B} - d_{R_A})}{\omega_B + \omega_A} \qquad (13)$$

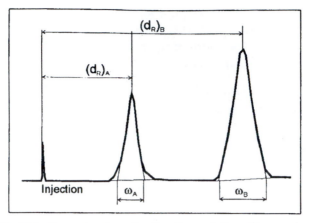

Fig. 13.5 Parameters for estimating the resolution of two peaks

For qualitative analysis a value of R between 0.5 and 1 suffices, but for precise quantitative work, values of R of at least 1 or larger are desirable. In computerized instruments, signal-processing programmes capable of artificially recreating two poorly resolved peaks are available, but this aid does not obviate the need for improving resolution by classic means; accuracy and precision of the analysis will always be improved.

3. Practical Chromatography

3.1 Sample Preparation

Sample preparation is always a much longer operation than chromatographic fractionation itself. This statement, valid for all chromatographic techniques, is particularly true for gas chromatography, in which components to be separated must be introduced in a relatively volatile and not excessively polar form to avoid irreversible adsorption on the column. The different preparation techniques vary widely according to the groups of substances to be identified and possibly quantified. In general, the substances of solid or liquid substrates should be converted to liquid form for conducting operations such as extraction, purification of extracts, derivative preparation, etc. on natural substrates. The level of complexity depends also on the equipment used; thus, purification in GC should be done more carefully for an electron-capture detector than for a flame-ionization detector. In high-pressure liquid chromatography, it is particularly necessary to ensure purity of the solvents and absence of particles and dissolved gases.

3.2 Qualitative Analysis

3.2.1 *Comparison of retention distances*

Qualitative analysis under given chromatographic conditions is generally done by comparing the retention time or retention distance of the unknown substance with the retention time of known standards injected under the same conditions. This identification is not always absolutely certain, as two different compounds may have similar retention times. The confidence level can be greatly enhanced by:

—simultaneous injection in the column of the mixture to be identified and a small quantity of the standard; increase in peak height without noticeable effect on peak width signifies a coincidence of thermodynamic properties of the compounds compared;

—repetition of the comparison on one or two columns of different characteristics or polarity.

Thus the identification becomes almost certain. The problem is that the number of organic compounds is very large and comparison among them all is impossible. Supplementary techniques must be used to reduce the number of standards needed; the following steps will be of help:

—preliminary sample-preparation operations (see §3.1 above) by which the great families of substances (peptides, hydrocarbons, carbohydrates, fatty acids, etc.) can be separated;

—use of detectors selective for broad groups;

—use of methods based on certain chromatographic techniques such as bi-directional flow of solvent in sheet chromatography and use of retention coefficients (*see* Chapter 14).

3.2.2 *Combination of chromatographic and spectroscopic methods*

These instruments provide the most certain means of qualitative identification by analysis of the spectra of compounds as they get eluted. In this manner chromatography is combined with most spectroscopic methods (infrared, NMR, atomic absorption, ICP, etc.) but the most well-known combination is that of gas chromatography and mass spectrometry. The use of coupling techniques is mostly limited to difficult cases of identification of eluted compounds or to confirmation of observations made using the classic techniques described here.

3.3 Quantitative Analysis

3.3.1 *Principle*

Chromatography enables quantitative analysis of a given solute 'i' by application of the proportionality

$$m_i = K_i A_i \qquad (14)$$

linking the mass m_i of the solute 'i' to the area A_i of the peak it gives on the chromatogram and the proportionality coefficient K_i.

Various manual methods are available for measuring the area of a peak when only analogue recording on paper is available (triangulation, weighing, planimetry). When the peaks are narrow and symmetrical, measurement of the peak heights, which are proportional to the areas, is also satisfactory. However, the highest precision is given by integrators. Originally mechanical with a system attached to the recorder pen to give the integrated curve, today practically all integrators are electronic. Electronic integrating recorders are now often replaced by computer systems with more comprehensive functions of spectrum processing; they can also be integrated with a computer network.

In principle, knowing the exact mass of an injected standard compound, it should be possible to calculate its proportionality coefficient K_i directly using equation 14, after measuring the area of the peak. In practice, this determination is not easy because although it is possible to inject relatively reproducible volumes, it is difficult to ascertain the exact absolute volume of a liquid injected by syringe. Various other methods of standardization have been applied: internal normalization, external standardization, standard additions, internal standardization.

3.3.2 Internal normalization

This method gives the relative concentrations of the solutes among themselves. The area on the chromatogram of each peak pertaining to the solutes 1, 2, i, . . ., n is measured and the mass percentage C_m of any one of the solutes 'm' is obtained using the equation

$$C_m\,(\%) = 100\,\frac{K_m A_m}{\sum\limits_1^n K_i A_i} \tag{15}$$

As a first approximation, in the same family of substances, all the K_i may be considered equal, transforming equation 15 to

$$C_m = 100\,\frac{A_m}{\sum\limits_1^n A_i} \tag{16}$$

It is also possible to formulate a standard mixture containing all the compounds 'i' with concentrations close to those of the components in the unknown mixture. The proportionality factor of any component 'm' of this mixture can be calculated by using the equation

$$K_m = \frac{C_m \sum\limits_1^n m_i}{100\,A_m} \tag{17}$$

3.3.3 External standardization

The area of the peak of solute 'm' in the mixture of unknown concentration is compared with the area of the same peak in a standard mixture with known concentration. It is the classic standardization curve technique widely used in analytical chemistry. In principle, since equation 14 shows a linear relationship between area and concentration, it is sufficient to proceed with the comparison using a single concentration in the standard mixture, especially if the concentration selected is not too different from that in the unknown mixture. Let $C_{m,i}$ and $C_{s,i}$ be the respective concentrations of the compound 'i' in the mixture and in the standard solution, and $A_{m,i}$ and $A_{s,i}$ the corresponding areas of the peaks of the same compound. Then $C_{m,i}$ is given by

$$C_{m,i} = C_{s,i} \frac{A_{m,i}}{A_{s,i}} \tag{18}$$

Nevertheless, it is more precise to confirm linearity by tracing at least once the calibration curve $A = f(C)$ for a range of concentration. This method requires great reproducibility in the volumes injected and is not the most precise in chromatography. Its chief use is to provide an absolute measure of the contents (and not relative as with internal normalization, § 3.3.2) without addition and the attendant risk of contamination of the mixture (standard addition in § 3.3.4 and internal standardization in § 3.3.5). Its use is often limited to analysis of compounds or elements present in traces.

3.3.4 Method of additions

A first analysis is done by injecting the sample with unknown concentration $x_{i,s}$ of compound 'i'. A second analysis is done after addition of a reference mixture with known concentration $x_{i,r}$ of the same compound 'i'. If A_i and A_i' denote respectively the peak areas for the compound in the first and second chromatograms, the concentration $x_{i,s}$ may be obtained from

$$x_{i,s} = \frac{A_i f x_{i,r}}{(A_i' - A_i)} \tag{19}$$

where f is the factor of dilution due to the addition of the reference compound. This method is useful when there is another peak t close to the peak i in the unknown mixture (and not in the added solution). The ratio A_t/A_t' of the areas of this peak before and after addition gives an estimate of the factor f.

3.3.5 Internal standardization

This method is popular because it is practicable and precise in GC whenever the conditions permit its use. For this it is necessary to find a compound that is definitely not present in the test mixture and whose retention time is close to that of the components of the mixture without causing interference. A calibration mixture is prepared containing known concentrations of all the components of the test mixture and the selected compound, termed the internal standard. A calibration chromatogram then enables determination of the relative response coefficient of each component 'i' (area A_i, mass m_i) compared to the internal standard (area A_s, mass m_s), or

$$K_{i/s} = K_i / K_s = m_i A_s / m_s A_i \qquad (20)$$

A chromatogram of the unknown mixture containing the same quantity of the internal standard as the calibration mixture is then run; the concentration x_i of the compound 'i' in the unknown mixture is given by

$$x_i = \frac{K_{i/s} x_s A_i}{A_s} \qquad (21)$$

It is not necessary to repeat the determination of $K_{i/s}$ (eqn 20) for each new sample in a set of determinations. The technique also facilitates qualitative identification of compounds by means of relative-retention times compared to the internal standard.

Bibliography

Barcelo D. (ed.). 1997. *Advanced Chromatographic and Electromigration Methods (Journal of Chromatography Library,* **60**). Elsevier Science Ltd.

Brown P.R. (ed.), 1998. *Advances in Chromatography,* **39**. Marcel Dekker.

Guiochon G., Golshan-Shirazi S. and Katti A.M. 1994. *Fundamentals of Preparative and Nonlinear Chromatography.* Academic Press, 701 pp.

Gas Chromatography

1. Definitions and Principles

1.1 General Description of a Chromatograph

Gas chromatography (GC) was invented by James and Martin in 1952. As in other chromatographic systems (Chapter 13), its aim is to fractionate the components of a mixture using the principle of partition between a mobile phase and a stationary phase. In this technique, the mobile phase is a gas and the mixture injected must be in the gaseous form, which involves:

—a system for regulating pressure and/or flow rate of the gas of the mobile phase (called carrier gas);

—an injection system for introducing the mixture at the head of the column at the operating pressure and temperature, and causing total vaporization of the mixture, if it is a liquid;

—a furnace with regulated temperature to maintain the column at the proper temperature.

Lastly, as in all chromatographic instruments, there should be a detector at the outlet of the column for identifying and quantifying the components of the injected mixture. The general arrangement of a chromatograph is presented in Fig. 14.1.

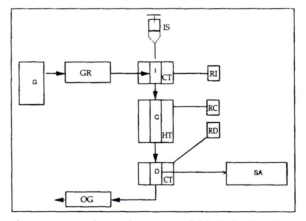

Fig. 14.1   General arrangement of a gas chromatograph labelled with reference to the sections describing the various elements:

G - carrier gas (§2.2.1); GR - regulator for pressure and flow of the carrier gas (§2.2.2); I - injector; IS - injection system (liquid syringe, gas syringe, injection rings, solid injector, automated systems); C - chromatographic column (§2.1); D - detector (§2.4); RI, RC, RD - temperature controls for the injector, the column furnace and the detector; CT, HT - circuits (injector, detector) and thermostated furnace; SA - system for amplification of signal (electrometer) and data acquisition (recorder, integrator and/or computer); OG - outflow of gases and effluents from the column with, if necessary, measurement of flow rate, capture of compounds, preparative-column chromatography collector, interface with spectrometry system.

1.2 Elementary Parameters

1.2.1 Retention parameters

The retention parameters defined in Chapter 13 are applicable to GC as well. This particular technique is based on two major hypotheses: compliance of the gaseous phase to the gas laws (hypothesis of ideality) and proportionality of the concentrations of the constituents in the liquid and gaseous phases (hypothesis of linearity). The retention volume V_R and the corrected retention volume V_R' (eqns 3 and 4 in Chapter 13) should in this case be corrected by taking into account the compressibility of the mobile phase. Thus a net or absolute retention volume V_N is calculated by

$$V_N = jV_R' \tag{1}$$

where j is the coefficient for pressure-gradient proposed by James and Martin (1952) according to the equation

$$j = (3P^2 - 1)/(2P^3 - 1) \tag{2}$$

with $P = P_e/P_o$, where P_e and P_o are the entry and outflow pressures of the carrier gas in the column. The specific retention volume V_g can also be computed by dividing V_N by the weight of the active filling m_f at 0°C as follows:

$$V_g = (V_N/m_f)(273/T_c) \tag{3}$$

T_c being the temperature of the column in K.

1.2.2 Variation of retention parameters with temperature

The retention volume generally decreases rapidly when the temperature rises and variation follows a Clapeyron type law applicable to a condensed phase:

$$\log V_g = \frac{\Delta H}{2.3\,RT} + Cte \tag{4}$$

where ΔH is the enthalpy of dissolution or adsorption of the solute on the stationary phase, T the absolute temperature and R the gas constant. A graph of the logarithm of retention volume for each component of the mixture against the reciprocal of the temperature will help in choosing the most appropriate temperature for chromatographic separation.

1.2.3 Variation of retention parameters with chain length in a homologous series

By a homologous series is meant a family of compounds of analogous developed formula differentiated only by the number n of carbon atoms: the series of saturated hydrocarbons, unsaturated hydrocarbons, methyl esters of saturated fatty acids, etc. At constant temperature (isothermal chromatography) variation in retention volume mostly conforms to the equation

$$\log V_g = an + b \tag{5}$$

This law often helps in identifying peaks when all the comparison standards (*see* §3 below) are not available. The chief consequence of this law lies.in the fact that it is difficult to separate the components of a series when the molecular masses of the end-members differ greatly. Choice of a high column temperature gives poor separation of the lowest peaks; a very low column temperature results in excessive retention times for determination of the heaviest components. The solution to this problem is to work with programmed increase in column temperature. Programming of temperature enables transformation to regular spacings the inter-peak spacings that increase exponentially with molecular size. Resolution (*see* §2.5 in Chapter 13) is improved and more constant for light and heavy compounds. Detection limit for the heaviest compounds is also improved because of diminution of peak width when the retention volume is increased.

2. Apparatus and Equipment

2.1 Columns and Materials for Chromatographic Separation

2.1.1 Chromatographic columns

The GC column has been defined (*see* §1.1 above) as the heart of the system, all the rest of the equipment serving only to make this column operate under optimum conditions and to identify the gases at its outlet. Two broad types of gas-chromatography columns are distinguished:

—*Packed columns*: these are generally made of glass (Pyrex type borosilicate) or stainless steel and have internal diameter of 2 to 6 mm, corresponding to an external diameter often indicated in British units (for joining Swagelock type fittings: 1/16″, 1/8″, 1/4″). Their length ranges from 0.5 to 10 m according to the features of the separation sought (*see* §1.2 above). They should be carefully packed with an adsorbent in the case of gas-solid chromatography or an inert support impregnated with a thin film of a stationary phase in the case of gas-liquid chromatography.

—*Capillary columns*: these are made from very thin, very long tubes fabricated from various materials. The most popular not so long ago were made from tubes of Pyrex type glass drawn out into capillaries. The disadvantage of these systems was evidently their extreme fragility, and which chromatographer has not pulled out a few hairs while contemplating all her high resolves shattered at the base of the furnace? Today, this disadvantage can be avoided because the technique has made great progress by indirectly benefiting from technologies opened up in the area of research on optical fibres: nowadays capillary columns are made of fused silica coated externally with a polyimide polymer film and the whole structure has great strength. The internal diameter of these columns ranges from 0.2 to 0.75 mm and the length from 10 to 150 m.

2.1.2 Chromatographic supports

Used in packed columns, chromatographic supports have the function of fixing on their surface a film of more or less viscous polymer that attains the liquid state at the operating temperature (gas-liquid chromatography, GLC). It is this liquid film that will accomplish the chromatographic separations. The supports should have two major qualities: large specific surface area and great inertness *vis-à-vis* the compounds to be separated. Various kinds of materials can be used: infusorial earth, diatomaceous earth and balls made of metal, silica, polymers, etc. The supports should be carefully sieved to a particle size of 0.2 to 0.3 mm mean diameter (80-100 mesh, 100-120 mesh). They are deactivated by various treatments, especially acid washing, followed if necessary by treatment with dimethylchlorosilane (for blocking polar functional groups). The adsorbents for gas-solid chromatography (GSC) are available in powder form similar to GLC supports, but are used as stationary phases without impregnation.

2.1.3 Stationary phases

These more or less viscous products should attain the liquid state at the operating temperature and should also have a very low vapour pressure to avoid the hazard of being entrained by the carrier gas. They are deposited in a thin layer on the support grains or on the inner walls of the tube in capillary columns.

There are many stationary phases that, apart from the properties mentioned above, have great chemical inertness with regard to the mixture to be analysed. Also, they are characterized by many parameters to be taken into account for the separation of a given mixture.

—*Minimum operating temperature*: the process of gas-liquid chromatography involves, as the name indicates, exchange of solutes between a gaseous phase and a liquid phase; when the temperature drops too low, certain stationary phases may crystallize and the efficiency of the column may be considerably reduced.

—*Maximum operating temperature*: this is the temperature above which the volatility and decomposition of the stationary phase is not negligible; this temperature is also given by instrument suppliers but it can vary with several factors: sensitivity of the detector used, quantity impregnated, flow rate of the carrier gas, chemical inertness of the support, acceptable background noise, tolerance of the detector for contaminants.

—*Polarity*: when there is no interaction of the solute with the stationary phase other than the laws of liquid-gas exchange, the theory of GC states that the components of a mixture are eluted in the order of increasing boiling point. This is true for non-polar solutes on non-polar stationary phases. In actual practice the order can be totally upset because of dipole moments of the solutes and the solvents. This is why investigators have long tended to classify various stationary phases according to their polarity. This research culminated in 1970 with the McReynolds constants that have been used most of the time since then.

McReynolds constants

Continuing the work of Rohrschneider, McReynolds (1970) proposed characterization of the polarity of stationary phases in relation to 5 principal reference standards representative of the major organic groups (Table 14.1). These constants are based on using the retention indices (*see* §1.2 above and §3.3.2 below) by taking squalane as non-polar reference standard (relative polarity = 0). For each reference substance a column is characterized by the difference in retention index of the substance on that column and its index on squalane at the same temperature. For example, for characterizing the behaviour of a phase in relation to aromatic substances:

$X' = \Delta I$ of benzene on the phase

$= I$ of benzene on the phase-I of benzene on squalane.

Table 14.1 Principal reference substances used by McReynolds for characterizing polarity of stationary phases.

Symbol	Reference substance	Principal groups addressed
X'	Benzene	Aromatics, olefins
Y'	1-butanol	Alcohols, nitriles, acids
Z'	Methylpropylketone	Ketones, aldehydes, esters, ethers, epoxides, derivatives with dimethylamine groups
U'	Nitropropane	Nitriles, derivatives with nitro groups
S'	Pyridine	Pyridine

The sum $X' + Y' + \ldots + S'$ expresses the total polarity of the column and each individual term more or less its polarity with respect to the corresponding compounds: high value for a particular constant signifies that the corresponding phase selectively slows down the elution of compounds with chemical functions of that group. Most suppliers of chromatographic products have propagated use of these constants, particularly the firm Supelco Inc. (Bellefonte PA, USA), which was the first to give the McReynolds constants for all the stationary phases in its catalogue.

2.1.4 Preparation of a packed column

Impregnation of support

—Select the support and its preparation (sieving, treatment, etc.). Choose the appropriate stationary phase and the impregnation quantity. Low impregnation levels generally give low retention times; they can be disastrous for some separations but will reduce the time taken for certain analyses. High impregnation levels allow better retention of volatile compounds; they are, on the other hand, inappropriate for the separation of slightly volatile compounds and may favour contamination of the detectors by carrying in the carrier gas. By convention, the impregnation level is expressed in weight of stationary phase per 100 g support.

—For 20 g support impregnated at 2%: weigh out 400 mg of stationary phase and dissolve it in the appropriate solvent (indicated in the manufacturer's catalogue) taking care to avoid using excess solvent.

—Keep the solution in contact with the support for at least 30 min, if necessary under vacuum to eliminate free air from the support.

—Eliminate the solvent in a flash evaporator (or by other evaporation or filtration process) at low rotation speed to avoid damage to particles of the support. Resieving is recommended at the end of the operation.

Packing the column

—Plug with a wad of glass wool the end designed to be connected to the detector side.

—Connect the same end to a vacuum pump and aspirate.

—Connect the other end to a funnel (Fig. 14.2) and slowly pour in the dry, impregnated support, tapping the column with a spatula, or using an electric vibrator, to achieve uniform packing. If necessary, leave a dead volume at the funnel end according to the geometry of the injector and length of the injection needle, the tip of which should be level with the

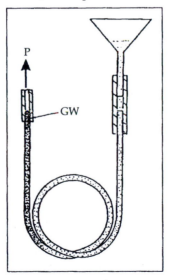

Fig. 14.2 Schematic diagram of the filling of a column
P - connection to water pump; GW - glass-wool plug

support during injection. Disconnect the funnel and stopper the top of the column with a plug of glass wool.

Gas-solid chromatography (GSC)
The step of packing the column is the same but no impregnation is done. The column is simply filled with the chosen adsorbent. The column can also be conditioned as given below but the drying time could be shorter.

Conditioning the column before use
  —Connect the column with the injector side to the chromatograph, not the detector side, to avoid contamination.
  —Pass a gentle gas flow for an hour when cold, then for half-a-day at 50°C.
  —Raise the temperature gradually ($1°C$ min^{-1} if a furnace-temperature programmer is available, otherwise in steps) up to about 10°C lower than the maximum operating temperature of the stationary phase, but at least 10°C above the maximum anticipated working temperature.
  —Let stabilize for at least two days.
  —Columns once dismantled should be reconditioned at least overnight before the next use.

Conditioning can cause a slight reduction in the volume of the filling; in such a case, the column can be freshly packed under vacuum with vibration as in the filling operation and topped up before proceeding with a shorter conditioning step. Used columns sometimes become loaded with carbonized substances at the injector and simultaneously a reduction in efficiency can be observed. A treatment enabling some regeneration consists of removing the glass-wool plug at the injector end, replacing a few centimetres of the filling at the beginning of the column and proceeding with a fresh drying of shorter duration before use. Similarly, used capillary columns can regain some efficiency if a small piece is cut off from the injector end.

2.1.5 Capillary columns
When these columns were invented in 1958 by Marcel Golay, laboratories used to make capillaries themselves. Some still do, but the operations are very delicate. The inner wall of the tube should be subjected to attack to increase its specific surface area. Then a passivation treatment is given to eliminate all irreversible adsorption of solutes on the support before the adsorbent film is deposited and dried. Techniques of fabrication of these columns will not be described here; the reader is referred to specialized publications (Prévôt, 1982).

It has been indicated earlier that the material of choice supplanting all others today for capillary columns is fused silica reinforced with an outer plastic coating. Apart from its great strength, silica is a pure material that does not have active sites capable of inducing irreversible adsorption of solutes. In specialized catalogues today there is a wide choice of capillary columns of this type and selection of a column truly appropriate to the needs may be problematical. Some manufacturers' reviews (*The Supelco Reporter* and *Biotext* distributed by Supelco and Biorad[1]) regularly report experiments done with these columns on various groups of substances. Some general information is given below for selection of a capillary column.

Influence of column diameter and thickness of the stationary-phase film
Theoretical calculations in capillary chromatography are simple because of the simple geometry of the carrier gas-adsorbent interface. According to equations 4 and 5 in Chapter 13, the retention time t_R of a compound R related to the non-retained peak of retention time t_m (peak for air, methane, etc., according to the separation being done and the apparatus) can be expressed by

$$t_R = t_m + (K/\beta)t_m \qquad (6)$$

where K denotes the partition coefficient of the solute between the gaseous phase and the stationary liquid phase, and β the ratio between the volumes of these phases, which can be expressed as follows:

$$\beta = V_m/V_s = \pi r^2 L / 2\pi r e_s L = r/2e_s \qquad (7)$$

in which V_m and V_s are respectively the volumes of the mobile phase (gas) and the stationary phase (liquid), r and L are the internal radius and length of the column, e_s is the thickness of the stationary-phase film.

From equations 6 and 7,

$$t_R = t_a(1 + 2e_s/r) \qquad (8)$$

Therefore it is with the thinnest films and widest columns that the most rapid analyses are achieved. But the efficiency may be unacceptably reduced by decrease in t_R (eqn 10 in Chapter 13), unless the peaks are extremely narrow. The same holds true for resolution (eqn 13 in Chapter 13) because two peaks with short retention time will necessarily be very close.

Selection of column diameter
In practice, capillary columns of small diameter (0.2 and 0.25 mm internal diameter ID) ensure maximum resolution and are ideal for difficult analyses and complex samples. Their chief disadvantage is their low capacity; they can hold only very small amounts of sample (<100 ng) and it is necessary to use splitter-injectors (*see* §2.3 below, injection). The quantity of solute passing through the detector is reduced, whereby the detection limit is lowered and these columns are not recommended for trace analysis.

Columns of 0.32 mm ID have a slightly lower resolution but almost equal to that of 0.25 mm columns and hold up to 500 ng of each constituent of the sample. Higher optimum flows make these columns ideal for 'splitless' or 'cold-column' type injections (*see* §2.3.3 below).

---

[1] *See* Appendix 6 for addresses.

Columns of larger diameter are also called 'wide-bore capillary columns':
—those of 0.53 mm ID have a capacity of 2000 ng per constituent; they can use the injectors for packed columns;
—those of 0.75 mm ID accept samples almost as large as done by packed columns (up to 15,000 ng), often with greater efficiency than the latter.

Loss of efficiency due to increase in diameter can also be compensated for by lengthening the column.

Choice of thickness of adsorbent-phase film
Standard films are 0.2 to 0.4 µm thick and are suitable for many mixtures. Thicker films (0.5 to 5 µm) can be used for separation of complex samples with low boiling point. A column with a thick film can reduce costs by ensuring a resolution similar to that of a longer column with standard film, but may have the disadvantage of loss of adsorbent phase when operated close to its maximum working temperature.

Bonded or unbonded phase
Bonded phases represent the latest development in the technology of fused-silica capillary columns. The adsorbent phase is chemically bonded to the support, thereby giving these columns great thermal stability and high resistance to displacement of the adsorbent by the carrier gas and injected products. In case of contamination they can even be washed internally to eliminate soluble contaminants. Nowadays many bonded-phase columns are available with the same polarity as unbonded-phase columns.

2.2 Circulation of the Carrier Gas

2.2.1 The carrier gas
Carrier gases are generally neutral or reducing gases used to preclude oxidation of injected products and columns and to preserve the detectors. Choice of carrier gas depends greatly on the type of detector used (*see* §2.4 below), the response of which may vary with the gas used. Hydrogen or helium is chosen for a catharometer, nitrogen or helium for a flame-ionization detector and nitrogen or an argon-methane mixture for an electron-capture detector.

Two important physical properties of the carrier gas are viscosity and diffusion coefficient of a compound in the gas (Table 14.2). From this point of view the ideal carrier gas is hydrogen, but it is scarcely used for this purpose because of the risk of leakage, bringing on its accumulation in the hot furnace.

Another important parameter is the purity of the carrier gas. In most cases, gases of 'U' quality can be used but for work with very sensitive detectors such as electron-capture detectors, gases of 'N60' quality should be used. It is also recommended that gas-purification systems be attached to the system: activated charcoal for organic compounds, 5-Å molecular sieves for trapping water and a system for catalytic elimination of oxygen.

Table 14.2 Two important physical constants of four carrier gases.

Carrier gas	Viscosity at 100°C 10^{-5} poise	Diffusion coefficient of butane at 25°C under 1 bar in cm^2 s^{-1}
Argon	270	0.10
Nitrogen	212	0.10
Helium	234	0.36
Hydrogen	105	0.40

2.2.2 Carrier-gas circuits and regulation

A two-stage regulator should be used at the outlet of the cylinders if they are beside the chromatograph. If the cylinder is outside the laboratory, it will suffice to have a single-stage regulator on it, but it is imperative to have a regulator adjustable at scale 0-5 bars beside the chromatograph. For work with sensitive detectors stainless regulators are recommended. The gas paths are 1/16" or 1/8" tubes of plastic, copper or stainless steel with 'Swagelock'-type connectors with tapered copper or Teflon sealing rings. It is not advisable to use plastic tubes with sensitive detectors.

At the regulator outlet the carrier gas is led through a flow regulator ahead of the injector. The flow is regulated to the chosen value when cold, before a chromatographic separation, by means of measurement at the outlet with a soap-bubble flowmeter generally graduated from 0 to 20 mL. The value chosen is increased with column diameter to maintain constant speed of gas within. For packed columns the flow rate is usually between 10 mL and a few tens of mL per minute. For standard small-diameter capillary columns the pressure at the head of the column is kept constant and the flow regulator may be short-circuited or fully opened at the base. Regulation of inlet pressure is then more precise and is achieved by measurement of retention time t_m (eqn 6) of a peak not retained on the column (air, methane, etc.). The ratio 'column length/t_m' gives the speed of the carrier gas, which can be adjusted through the inlet pressure (generally low, 0.1 to 1 bar).

It is important to frequently check all gas joints for absence of leaks, using soap solution. To be a good 'chromatographer', one should be a good plumber, because leaks can greatly perturb background noise and the quality of analyses.

2.3 Injectors

The injector achieves interfacing of the column with the exterior to enable introduction of the sample into the chromatograph; it should therefore be airtight. It also has the function of converting liquid or solid samples to gaseous form and transferring them to the column. It should therefore be heated but it is not imperative to exceed the boiling point of the components of the mixture to achieve complete vaporization; the quantity injected is small enough to produce a volume of vapour that does not cause overshooting the saturated vapour pressure of the various components at the operating temperature.

2.3.1 Injector with elastomer plug (septum) for packed columns

This·type of injector is the most popular for laboratory apparatus with packed columns. Injection is done by means of a special syringe through an elastomer plug called the *septum*, held by an adjustable nut (Fig. 14.3) in a vaporization chamber. This chamber is heated and its geometry should be studied to ensure rapid transport of the vaporized solute to the column.

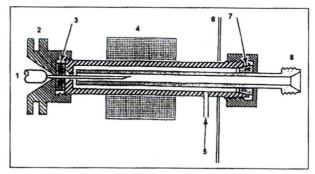

Fig. 14.3   Injector with septum for packed columns:
1 - injection needle; 2 - adjustable nut with cooling fins; 3 - septum; 4 - thermostatic block; 5 - carrier-gas inlet; 6 - furnace area; 7 - airtight cone of brass or Teflon; 8 - connection to column

The plug or septum should be easy to pierce, retain airtightness when the needle is withdrawn, have high resistance to temperature and not release volatile substances. It might well be said that the ideal material for making it does not exist; there are always hazards of extraneous peaks caused by the plug when the injection temperature used is high. Also it is necessary to replace the plug after a few tens of injections when tightening of the clamping nut is no longer sufficient to stop leaks.

The volume injected is usually in the range 0.5 to 5 μL. The length of the injector needle should match the geometry of the injector; training is necessary in how to give a proper injection, especially without bending the needle.

2.3.2 Injection of solids and gases

Injection of a gas can be done as in the preceding section using an injector with septum, but a larger syringe is required. As an indication, 1 μL of liquid hexane corresponds to 260 μL gaseous hexane under standard conditions. However, most often injection rings are used that are capable of containing a given quantity of gas and controlled by multiple valves. This type of injector may also be used for liquids and they are generally used in HPLC (*see* Chapter 15).

Injection of solids is less common but is done in special set-ups that permit taking a solid sample in an injector. There are also pyrolytic injectors for identifying heavy substances according to the chromatographic signature of the decomposition products released when the sample is heated.

2.3.3 Injectors for capillary columns

Standard capillary columns support only very small injections of the mixture and the quantities used for packed columns (0.5-5 μL) will result in their saturation. It is necessary to use about 100 times smaller amounts, and syringes of adequate precision for such volumes are not readily available. It is then the injector that has to reduce the volume of mixture injected by representatively dividing it or by selective elimination of the solvent.

Splitter-injectors

This device is as old as the invention of capillary columns by Golay in 1958. Injection of a volume of 0.5 to 2 μL is accomplished through a septum as described earlier for packed columns. The vaporized product is then transported by the carrier gas to the column, but a system of controlled bleed (split) with a microvalve (Fig. 14.4) allows only a small fraction (say 1%) to enter the column.

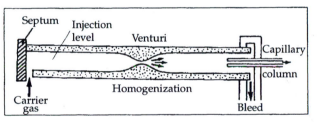

Fig. 14.4   Schematic diagram of a splitter-injector with gas-homogenizing Venturi tubes for capillary columns.

The chief disadvantage of the splitter-divider is loss of sensitivity caused by the division of the mixture; this renders the device poorly suited in principle to analysis of trace components.

Injection without division-'splitless' technique

Injection can be done with a splitter-injector of the same type as described above, but following a more complicated procedure with a cooled column (Table 14.3).

Sophisticated injectors of 'split-splitless' type are also available that enable working in both modes with automation of the operations.

Glass-needle injector

Reduction in the injected· volume does not use a system of division but concentration from the solvent by evaporation. The injection is always done through a septum but using an unheated glass needle, held in place by a magnet (Fig. 14.5). Passing the carrier gas for the chosen time (15 to 45

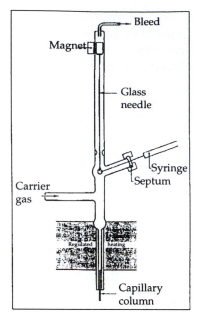

Fig. 14.5 Glass-needle injector.

seconds) will evaporate the microdrop of solvent deposited in the needle, leaving only less volatile compounds. The point of the needle is then pushed by means of the magnet into the heated injection chamber before the column.

Table 14.3 Splitless technique or injection without division (Prévôt, 1982).

Procedure	Effect
Close the bleed	Carrier gas passes only through the column.
Open furnace door	Column cooled to about 30°C.
Inject 1 μL of a very dilute solution in hexane	The vaporized sample and hexane slowly penetrate the column.
Wait about 30 seconds	As the column is cold, hexane is recondensed in the first few cm and itself acts temporarily as a stationary phase for the components of the sample (solvent effect). β decreases and t_R is increased (*see* eqn 6). A large part of the sample is transferred from the injector to the beginning of the column through the gaseous phase.
Open the bleed and close furnace door	The injector is swept with what is left of the solvent and sample
Start temperature programming	The solvent evaporates and passes through the entire column; it gives a very large peak that quickly drops to the baseline; β increases, the components of the sample are moved and sorted turn by turn in the column.

Theoretically the glass-needle injector is ideal for analysis of components present in traces such as pesticide residues in soils. Also, it concentrates the injected products instead of splitting them, but allows working with higher amplification by maintaining a reasonable background noise because of

—absence or great reduction of the solvent peak;

—great reduction in risk of contamination by volatile products from the septum, because it remains cold;

—reduction in contamination by possible heavy non-volatile substances that remain in the needle if they are not decomposed in the injector.

On the other hand, this injector has two major disadvantages:

—its use is limited to relatively less volatile substances;

—manipulation during injection is particularly tricky; use is recommended of suitable illumination and a mounted lens to properly locate the tip of the syringe on that of the glass needle of the injector and to confirm the transfer from the one to the other of the injected microdrop; this type of injector lends itself least to automation.

Direct injection into the column (on-column injector)

In this type of injector, the sample is led directly in the form of a liquid into the cooled capillary column by means of a very thin needle. Figure 14.6 presents a schematic diagram of a commercial injector. This injector has no septum, the needle passing through the hole of a Teflon stopcock, which is opened at the time of injection. This apparatus is advantageous for mixtures with components of widely varying volatility and for substances with low volatility and low stability that should not be overheated in the injector. However, danger of contamination of the column is maximum and the first few cm are stripped from the stationary phase.

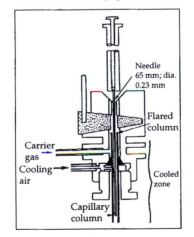

Fig. 14.6 On-column injector (Fisons Instruments[1]).

2.4 Detectors

2.4.1 Detection

On leaving the column, the effluents should be identified and quantified according to the electrical signals recorded by an analogue recorder or

---

[1] *See* Appendix 6 for addresses

digitized in a computer. These signals are produced by numerous types of detectors, which may be grouped as follows:

—those that give a signal proportional to the total quantity of each of the components of the sorted mixture (integral detector);

—those that give a signal according to the change in gravimetric or volumetric concentration of the solute in the gaseous mixture (differential detector).

Mostly differential detectors are employed in practice, to provide recordings or chromatograms composed of a sequence of peaks, ideally well separated in time and appearing Gaussian in shape.

The detection systems are based on any change in the physical or chemical properties of the gaseous mixture caused by the presence of the solute to be determined (Table 14.4). The following detectors are distinguished:

—universal detectors that respond to most compounds;

—selective detectors for one or several groups of compounds;

—specific detectors that can be set to one property of a particular compound. The most commonly used specific detector is the mass spectrometer, which is very practicable for confirming the products identified, but not always satisfactory in detection limit.

Table 14.4   Chief gas chromatography detectors (after Untz, 1982). MDQ - minimum detectable quantity (peak with height twice that of the background).

Detector	Linearity	MDQ	Selectivity	Chief applications
Catharometer	10^5	1 to 10 ng	Non-selective	All compounds
Flame-ionization	10^7	20 to 100 pg	Non-selective	Combustible organic compounds
Electron-capture	10^4	0.1 pg	Variable selectivity	Halogenated compounds
Thermo-ionic	10^4	1 pg P	10^4 to 10^5 compared to hydrocarbons	Organophosphorus, organonitrogen
Flame photometry	10^3 to 10^4 (P) 10^3 (S)	10 pg P 1 ng S	10^3 to 10^4 compared to hydrocarbons	Sulphur and phosphorus compounds
Electrolytic conductivity	10^2 to 10^4	0.1 ng	10^5 compared to hydrocarbons	Compounds with N, S, Cl
Gas-density balance	10^5	100 ng	Non-selective	All compounds
He-ionization	10^4	10 pg	Non-selective	Permanent gases
H.F. plasma	10^3	10 to 1000 pg	Non-selective	Permanent gases
Microwave plasma	10^3 to 10^4	10 to 500 pg	Selective	All compounds
Infrared		10 ng to 1 μg	Selective	All compounds
Mass spectrometry	10^3	<1 pg	Specific	All compounds
Photoionization	10^7	1 pg	Non-selective	All ionizable compounds

2.4.2 Thermal-conductivity detector: the catharometer

Universally used, this detector has been the most popular in gas chromatography, particularly in the early days of use of this technique. It is now often substituted by the flame-ionization detector, less universally used however, but more sensitive for combustible organic compounds. It is today still very widely used for non-combustible substances such as, most soil gases, for example.

The principle of this detector uses the change in electrical resistance under the effect of variation in thermal conductivity of the carrier gas when a solute appears at the outlet of the column. In practice, a resistance enclosed in a constant-temperature heated block and set up in a Wheatstone bridge is used (Fig. 14.7). This resistance receives the gas from the effluents of the column; another resistance of the bridge is bathed by the pure carrier gas. The bridge is balanced and the equilibrium point corresponds to the base-line of the recorder; when a solute appears in the carrier gas the imbalance is recorded at the point corresponding to the voltage Δe defined by

$$\Delta e = \left[\frac{1}{G}\right]\left[\frac{aR_0 EI^2}{4J}\right]\left[X\left(\frac{1}{\lambda_s} - \frac{1}{\lambda_g}\right)\right]$$

(9)

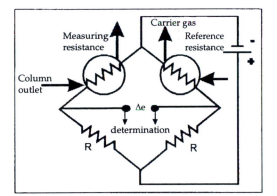

Fig. 14.7   Theoretical scheme of a catharometer detector.

The first term between brackets in equation 9 corresponds to the effect of the geometry of the detection cell, the second to the electrical effect and the third to the thermal-conductivity effect of the gaseous mixture, with

G = cell constant,
a = temperature coefficient of resistance of the wire,
R_0 = resistance of the wire at 0°C,
E = voltage applied between the ends of the bridge,
I = current in the bridge,
J = Joule's equivalent (4.19 J cal^{-1}),
λ_s = thermal conductivity of the solute,
λ_g = thermal conductivity of the carrier gas,
X = molar fraction of the solute in the carrier gas.

For a given setting, the imbalance of the bridge is directly proportional to the molar fraction of the solute and the signal is the larger the greater the difference between the thermal conductivities of the solute and the carrier gas. Generally hydrogen or helium is used as the carrier gas because of their high thermal conductivity.

2.4.3 Flame-ionization detector

Operation of this detector requires burning the effluent from the column in an air-hydrogen flame. Combustible organic compounds produce free radicals that are ionized in the flame at high temperature with application of an electrical field on all sides (Fig. 14.8). The ions formed cause an ionization current between the electrodes, which is then amplified by an electrometer. The signal generated is proportional to the gravimetric concentration of the solute in the carrier gas; it is the larger the more the combustible atoms C and H but the less the oxidizing atoms (O, Cl, etc.).

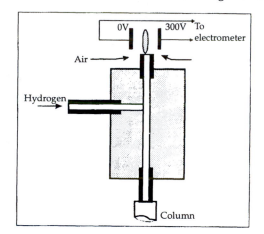

Fig. 14.8 Schematic diagram of a flame-ionization detector.

This type of detector is about 100 times as sensitive as the catharometer for combustible organic compounds. However, it is not universally used because it does not respond to inorganic compounds, non-combustible gases (such as soil CO_2) and fully oxidized organic compounds such as oxalic acid. Compared to the catharometer, this detector also requires a cylinder of hydrogen in addition to one of the carrier gas and compressed-air supply to ensure combustion of the hydrogen. However, the carrier gas can be nitrogen, which is cheaper than helium.

2.4.4 Selective electron-capture detector

This detector uses a β-ray source such as ^{63}Ni or tritium. The source is placed between two electrodes (Fig. 14.9); when operating in the rest mode, the background current generated by the emitted electrons is measured.

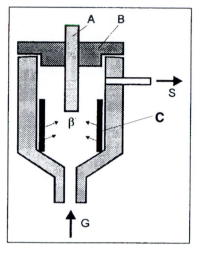

Fig. 14.9   Theoretical scheme of an electron-capture detector (A - collecting anode; B - insulating material, C - radioactive cathode emitting β-rays; G and S - entry and exit of effluents in the column).

When the detector is bathed by a substance having affinity for free electrons, these are captured and a sudden drop is caused in the background current. Thus an inverted peak is obtained compared to the flame-ionization detector and the peak is righted by reversing polarity.

The response of this detector is thus limited to compounds having affinity for free electrons. So, it is almost insensitive to most organic compounds except if they contain electronegative groups such as halogenated, nitrated, etc. groups. Specifically this detector is very sensitive to halogenated compounds. It is therefore the tool of choice for analysis of chlorinated-pesticide residues and polychlorobiphenyl (PCB) pollutants. It is mostly through use of this detector that it has been possible to confirm that the earth is polluted with organochlorine compounds up to the ocean deep of polar regions. Perhaps today the entire earth is also polluted by other groups of substances for which means of detection are not available, none of the other known detectors apparently attaining the sensitivity that this detector has for halogenated compounds. Besides, although classic chlorinated pesticides are no longer widely used, new, less toxic halogenated compounds are commonly used and interest in this detector has not waned. For example, we ourselves have had the occasion to observe a response of this detector to Deltamethrine (Decis) about 100 times greater than that with mass spectrometry.

Like the catharometer, this detector does not require supplementary gas installation, particularly if nitrogen is used as the carrier gas. If another carrier gas is used, a supply of nitrogen is required in addition, as the electron-capture detector prefers this gas for its functioning; also, there should be sufficient flow of nitrogen and if capillary columns are used, the detector needs additional gas supply (quenching gas).

The major disadvantage of this detector lies in its high susceptibility to contamination and oxidation; without adequate precautions the signal rapidly becomes unstable and cannot be interpreted. Very pure gases should be used and it is recommended that they be further purified of oxygen, water and organic traces by passage through suitable absorbers. Extreme rigour should also be observed in the quality of solvents and reagents and in the cleanliness of the glassware used.

Radioactive sources should be declared to the competent regulatory authority; a contaminated source should not be dismantled and cleaned by the user except if the laboratory has been approved for handling radionuclides.

2.4.5 Thermo-ionic detector

The thermo-ionic detector resembles the flame-ionization detector and the effluent from the column passes through an air-hydrogen flame as in the latter. The principal difference lies in the presence in the detector of a pellet of an alkali salt. This results in the release of alkali atoms or ions, which are responsible for the selectivity of the detector.

Compared to the flame-ionization detector, the sensitivity of this detector is actually much higher for organonitrogen and, more so, organophosphorus compounds. This detector is therefore the tool of choice for study of residues of organophosphorus insecticides and nitrated herbicides, compounds widely used in agriculture and which can be found in soils, surface waters and groundwaters.

2.4.6 Other types of detectors

A wide variety of detectors is available and the major ones have been listed in Table 14.4. The usefulness of the selective flame-photometric detector should be underscored for its useful sensitivity to sulphur-bearing and phosphorus-bearing compounds. Though its sensitivity is low (of the same order of magnitude as in the catharometer) this detector, which is not affected by gas density, has the advantage of providing a response calculable as a function of the molecular mass of the solute compared to carrier gas. Lastly, the combination of gas chromatography and mass spectrometry is also widely used in several variants, from use of the mass spectrometer as a simple detector to sequential recording of complete mass spectra of the eluted compounds to identify them.

2.5 Temperature Controls

A gas chromatograph carries three temperature controls for:
    injector temperature;
    temperature of the furnace containing the column;
    detector temperature.

The regulators are at least of the proportional action type controlled by platinum sensors. There generally is also a dial for temperature control working in parallel through thermocouples. This control is only necessary in our view for detecting large breakdowns. In normal operation it is preferable to rely on the temperature registered by the regulators, platinum probes being more precise than thermocouples. Regulation of furnace temperature should be most carefully done; it is better if a blower is fitted to homogenize the air and hence the furnace temperature. For well-mastered separations or those that present no particular difficulty, a simple regulator for the furnace temperature is adequate whereas for more complex analyses, it is imperative that a programmer be installed. There are more or less sophisticated programmers but models with a single programme allow most of the work to be done under satisfactory conditions.

3. Practice

3.1 Sample Preparation

Sample preparation is always the most time-consuming operation in chromatography (*see* §3.1 in Chapter 13). In GC the preparation can be particularly tricky. It often requires prefractionation of the broad groups of compounds, purifying them, and dissolving them in solvents in the least polar and adequately volatile form (derivatization reactions). This preparation is sometimes made easier by progress in technology of chromatographic columns. Thus light organic acids earlier used to be esterified and dissolved in non-polar solvents; today they can be injected directly in the form of acid in aqueous medium into some columns.

3.2 Injections

Injection is most often done with suitable syringes through a septum (*see* §2.3 above) and should be very carefully done. For liquids syringes of 1 to 10 μL capacity are generally used for injecting volumes of 0.5 to 5 μL. For each injection it is necessary to

—wash the syringe 5 to 6 times with pure solvent;

—rinse 5 to 6 times with the mixture to be injected, emptying the syringe each time outside the flask;

—finish the rinsing without emptying the syringe with rapid suction and blowing out to eliminate air bubbles;

—slowly aspirate the volume to be injected, ensuring absence of air bubbles, if bubbles are present, repeat the preceding operation;

—wipe the needle dry with a small piece of clean, thin filter paper without removing the liquid within by capillary action;

—aspirate a little air (1 μL) to prevent the injection from starting before the injector is positioned in the desired location in the injector;

—insert the needle through the septum to the desired position;

—inject the mixture and withdraw the needle while holding, if necessary, the injector body in the other hand.

The listing of these operations might appear fastidious, but with a little practice the injection can be rapidly done with satisfactory reproducibility. Some authors recommend enclosing the substance to be injected between small amounts of pure solvent: upstream for pushing the compound out of the syringe and downstream to fill the needle; most of the time these additional precautions can be overlooked. The manual operation sequence above can be programmed into robots imitating manual injectors. These robots are sometimes also called 'injectors (automatic)', but this term might lead to confusion with the injector described above (*see* §2.3): the interface between the chromatographic column and the exterior.

3.3 Qualitative Analysis

3.3.1 Comparison of retention distances

This technique is common to all chromatographic methods (*see* §3.2.1 in Chapter 13). In isothermal GC, supplementary assistance is given by use of formula 5 above for identification of the peaks of compounds of a homologous series. Identification can also be facilitated by use of retention indices described below.

3.3.2 Use of retention indices

The relative retention values allow bypassing the auxiliary characterization of specific retention volume. Generally a known reference compound 'R' is chosen, injected along with the unknown compound 'i' and the relative retention of 'i' is expressed relative to 'R' (*see* §2.3 in Chapter 13). These values can be compared with those in standard tables established for certain stationary phases at various temperatures.

However, choice of standards is not always obvious and it is often preferable to refer to a more universal scale, that of retention indices proposed by Kovàtz (1958). Use of these indices eliminates for the first time preparation of standards other than a homologous series of hydrocarbons. Two hydrocarbons of the series are chosen to bracket the peak of the unknown solute 'x' and its retention index I_x is calculated using the formula

$$I_x = 100 \frac{\log \dfrac{V_{g\,x}}{V_{g\,z}}}{\log \dfrac{V_{g\,z+1}}{V_{g\,z}}} + 100z \tag{10}$$

established with the specific retention volumes V_g of the compounds: the unknown 'x', the hydrocarbon with z carbon atoms and the hydrocarbons with $z + 1$ carbon atoms. Formula 10 is equivalent to

$$I_x = \frac{100 \log (d_R^2)_x - \log (d_R^2)_z}{\log (d_R^2)_{z+1} - \log (d_R^2)_z} + 100z \tag{11}$$

The retention indices have been established with great precision for a large number of compounds and tabulated. They are also used for establishing the McReynolds constants for grouping stationary phases used in GC (*see* §2.1.3 above). The value can also be deduced from the stereochemical structure of the analysed products by the method of increments ΔI so that

$$\Delta I = I_p - I_a \tag{12}$$

in which I_p is the index with a polar column and I_a the index with a nonpolar column.

2.3.3 Other analytical identifications
The most reliable identification techniques use the combination of GC with the usual structural-chemical spectrometric techniques, but these combinations are not only for the gaseous phase (*see* §3.2.2 of Chapter 13) even if the best known combination is GC-mass spectrometry.

Quantitative analytical techniques by GC are strictly the same as those by other chromatographic methods (*see* §3.3 in Chapter 13).

3.4 Applications
It is not intended here to draw up a catalogue but to highlight the broad areas in which this technique has been or could be applied in soil analysis.

The soil is a complex mineral or organic medium containing a very wide variety of molecules of natural or anthropic origin (organic amendments, fertilizers, pesticides). Its precise study is therefore necessarily done through methods of fractionation among which chromatographic techniques will be useful, particularly for organic molecules.

The principal advantages of GC compared to other liquid-phase methods (Chapter 15) are essentially founded on

—*its high resolution*: it enables much finer separations than liquid-phase techniques and is therefore often irreplaceable, especially with the further improvement in resolution given by capillary columns; the GC techniques enable resolution especially of many mixtures of isomers including optical isomers; they are often used alone but can also be complementary to liquid-chromatography techniques for completing the separation of compounds prefractionated into broad groups by those methods or for analysing constituent chains of larger molecules;

—its great sensitivity and possible use of numerous *specific detectors*, which make it the premier tool for trace analysis of organic molecules.

The chief disadvantage of GC is the need for converting the molecules to gaseous form to separate them; it follows that:

—it is not suitable for fractionation of very heavy molecules such as most humic acids;

—it is not suitable for fractionation of fragile molecules.

However, GC is used, among other methods, for the study of humic acids when it is combined with pyrolytic techniques (also mentioned in §2.3.2 above). Furthermore, the decomposition of labile compounds can be limited by suitable operating conditions, as shown in a study of pyrethrinoid residues (Pansu *et al.*, 1981).

The following can be indicated among the various applications known or possible in the study of soils:

—analysis of soil gases: CO_2 from respiratory processes, nitrogen from denitrification, methane, other light hydrocarbons form anaerobic decomposition and various rare gases trapped in the soil. Gas chromatography is the only tool for fractionation as well as determination of these gases, even if there are other techniques of direct measurement of some of them (CO_2 in particular);

—analysis of fundamental biochemical constituents of soils: gas chromatography rivals liquid-phase methods in this area. Its major handicap often lies in the great difficulty of preparing samples that have to be introduced in the injector in sufficiently pure, not too polar form, etc. However, it is unrivalled for the study of carbohydrates (Pansu, 1992), of soil lipids (especially fatty acids) and of light organic acids (sulphate-reducing decomposition, etc.). It enables efficient fractionation of amino acids but is often placed second in this area behind LC, unless finer separations are desired, such as of the optical isomers of these acids;

—analysis of pesticide residues and pollutants in soils where, as indicated above, GC is an indispensable tool because of its sensitivity as well as its numerous specific detectors;

—lastly, despite the strong competition of other techniques in this area, GC is also a sensitive analytical tool for mineral elements in the form of various complexes, generally organometallic (Guiochon and Pommier, 1971).

References

Golay M.J.E. 1958. *Théories de la chromatographie dans des colonnes tubulaires ouvertes et enduites à sections droite, circulaire et rectangulaire.* Butterworths Scientific Publications, London, 13 pp.

Guiochon G. and Pommier C. 1971. *La chromatographie en phase gazeuse en chimie inorganique.* Gauthier-Villars, 355 pp.

James A.T. and Martin A.J.P. 1952. Gas-liquid partition chromatography: the separation and microestimation of fatty acids from formic acid to dodecanoic acid. *Biochem. J.* **50**: 679.

Kovàtz V.E. 1958. Gas-chromatographische Charakterisierung organischer Verbindungen. Teil I. Retentionindiçes aliphatischer Halogenide, Alkohol, Aldehyde und Ketone. *Helv. Chim. Acta*, **41:** 1915-1932.

McReynolds, W.O. 1970. Characterization of some liquid phases. *J. Chrom. Sci.* **8:** 685.

Pansu, M. 1992. *Les sucres neutres dans les sols: opportunité et tentatives d'amélioration de leur détermination.* Document IRD, Montpellier, France, No. **4,** 25 pp.

Pansu M., Dhouïbi M.H. and Pinta M. 1981. Détermination des traces de pyréthrinoïdes dans les substrats biologiques par chromatographie en phase gazeuse. *Analusis*, **9:** 55-59.

Prévôt A. 1982. Les colonnes capillaires. In (J. Tranchant, ed.). *Manuel pratique de chromatographie en phase gazeuse.* Masson, Paris.

Untz G. 1982. Appareillage. In (J. Tranchant, ed.). *Manuel pratique de chromatographie en phase gazeuse.* Masson, Paris.

Bibliography

Grant D.W. 1996. *Capillary Gas Chromatography* (Separation Science Series). John Wiley and Sons, 304 pp.

Grob R.L. 1995. *Modern Practice of Gas Chromatography.* Wiley-Interscience, 2nd ed., 897 pp.

Jennings W. and Mittlefehldt E. 1997. *Analytical Gas Chromatography.* Academic Press, 2nd ed., 315 pp.

Lavoue G. 1976. *La chromatographie en phase gazeuse et ses applications.* Labo France, Paris, 225 pp.

McMaster M. and McMaster C. 1998. *GC-MAS: A Practical User's Guide.* John Wiley and Sons.

McNair H.M. and Miller J.M. 1997. *Basic Gas Chromatography* (Techniques in Analytical Chemistry). John Wiley and Sons.

Moye H.A. (ed.), 1990. *Analysis of Pesticide Residues.* Krieger Publishing Company.

Tranchant J. 1982. Séparations par distribution entre phases: Chromatographie en phase gazeuse. *Techniques de l'Ingénieur,* Paris, **P-1485** and **P-1485-1,** 30 pp.

Tranchant J. 1995. *Manuel pratique de chromatographie en phase gazeuse.* Masson, 3rd ed., 720 pp.

Liquid Chromatography

1. Definition and Principle

1.1 Introduction

Historically, liquid chromatography (LC) made its appearance before gas chromatography (GC): chronologically, partition chromatography and paper chromatography in the 1940s, GC and thin layer chromatography (TLC) in 1950, exclusion chromatography in the 1960s and modern LC in 1969.

However, column liquid chromatography, though one of the oldest techniques, has remained scarcely used principally because of the slowness of separations and lack of a detector. The slowness of separation was due to the slow elution rates in low-efficiency columns. Today, the efficiency of columns (*see* §2.4 in Chapter 13) has been greatly improved and work at high pressure has multiplied a hundred-fold the speed of the mobile phase, improving in the same proportion the time and quality of analyses. Nowadays the technique is often designated 'high-pressure liquid chromatography' or 'high-performance liquid chromatography' (HPLC). We shall restrict ourselves in this book to this modern form of LC.

The technique of HPLC does not present the limitations of GC in regard to compounds of low volatility or low thermal stability. Separation mechanisms in LC are more complex than in GC. In the latter technique, there are no interactions between the injected solutes and the stationary phase, whereas in LC there are interactions between solutes and the mobile phase. The variety of fractionations possible is thus wider in LC, with additional possibilities of optimization of separations according to the composition of the mobile phase. However, the resolution and sensitivity of GC are quite often irreplaceable for compounds of low molecular weight; the variety and more universal character of GC detectors is an added advantage in that technique. Actually the two techniques are not competitive but complementary. In some instances LC is even used as a technique for prefractionation prior to GC.

1.2 General Design of a HPLC Apparatus

The general design of an apparatus for high-pressure liquid chromatography (Fig. 15.1) resembles the design of a GC apparatus as described in Chapter 14. The essential differences arise from the nature of the mobile phase. The simple flow/pressure regulator for the carrier gas is here replaced by a high-pressure pump with optionally a system for maintaining a solvent gradient. On the other hand, the temperature-regulations systems of GC are not required here, except when a column temperature regulation system is opted for.

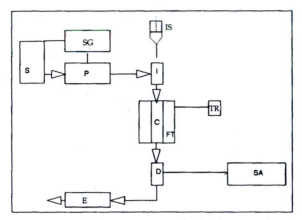

Fig. 15.1   General design of a liquid chromatograph (with reference to descriptive paragraphs
in the text):
S - reservoir(s) for solvent(s); P - high-pressure pump (§2.2); SG - optional solvent-
gradient system (§2.3); I - injector; IS - injection system (syringe, injection ring, etc.)
(§2.4); C - chromatographic column (§2.1); D - detector (§2.5); TR - optional
temperature regulator for column furnace; TF - optional thermostatic furnace; SA -
system for signal amplification and recording (display, recorder and/or computer);
E - effluent exit of the column with optional fraction collector or interface with
spectrometry system.

1.3 Theory

The fundamental parameters pertaining to chromatography (*see* §2 in Chapter
13) are certainly applicable here too. However, from the theoretical viewpoint,
three major differences between LC and GC should be considered:
   —diffusion coefficients in liquids are 10^4 to 10^5 times smaller than in
gases,
   —viscosity of liquids is about 100 times greater than that of gases,
   —gases are compressible, whereas liquids are incompressible up to about
300 bars.

1.3.1 Diffusion coefficient
The slowness of diffusion in the stagnant mobile phase in pores of the
stationary phase leads to a low rate of exchange between these two phases.
With older LC columns it was necessary to operate with very low speeds of
the mobile phase and separations were very time-consuming. Increase in
the speed of exchange could be obtained by modifying the texture of the
stationary phase in two different ways:
   —*stationary phases termed 'coated'*: spherical support impermeable to the
liquid phase (diameter of particles 30 to 50 µm) on which is deposited a thin
(about 1 µm) film of the stationary phase in which the exchange processes
take place; these phases have a low hydraulic pressure gradient but have
the disadvantage of low capacity (small amounts of sample to be injected),
which explains their present lack of popularity;

—porous stationary phases of uniform fine particle size: as the depth of pores decreases with particle size, reduction in this size has been studied; the particle diameters most commonly used today range from 3 to 5 μm; attaining this homogeneous particle size is delicate and filling the columns is equally difficult, restricting their length to about 30 cm.

1.3.2 Viscosity and hydraulic pressure gradient

The hydraulic pressure gradient is given, in a first approximation, by Darcy's law:

$$\Delta P = \varnothing \, \frac{\eta L u}{d_p^2} \tag{1}$$

in which

ΔP = hydraulic pressure gradient (dimension $ML^{-1}T^{-2}$, Pascal in S.I. units),

$\varnothing$ = coefficient denoting resistance to flow, depending on sphericity of the stationary phase and the quality of the packing (dimensionless),

η = dynamic viscosity of the mobile phase ($ML^{-1}T^{-1}$, Pa s = 10^3 cP),

L = length of column (L, m in S.I. units),

u = speed of flow of mobile phase (LT^{-1}, m s^{-1} in S.I. units),

d_p = diameter of particles of the stationary phase (L, m in S.I. units).

As the viscosity of liquids is about 100 times higher than that of gases, the pressure of the mobile phase required to be applied at the head of the column for satisfactory LC should also be 100 times higher than in GC. Use of stationary phases with small diameter to compensate for low diffusion coefficient of liquids (*see* §1.3.1) has a negative effect on hydraulic pressure gradient; as this term appears as its square, a reduction of particle size by a factor of 2 necessitates increasing the pressure at the column inlet by a factor of 4. However, since short (3-25 cm) columns are used, with low speeds of the mobile phase (0.1-1 cm s^{-1}), the pressures necessary rarely exceed a hundred bars (10^7 Pa) or so. Also packing arrangements giving the smallest values of ϕ (sphericity and uniformity of particles, etc.) can be looked for.

1.4 Comparison of Thin Layer Chromatography (TLC) and Column Liquid Chromatography (CLC)

Techniques of TLC are rather simple to execute with inexpensive equipment (Fig. 15.2). However column chromatography, especially in its current form (HPLC), has many advantages compared to TLC: the time taken for separations is often much shorter, it enables precise quantitative analysis (*see* §3.3 in Chapter 13) and lends itself well to automation with recovery of various solutes (preparative chromatography). Layer chromatography is used, however, when a column apparatus is not available or, in earlier times, for roughly studying the fractionation conditions for a mixture prior to column liquid chromatography.

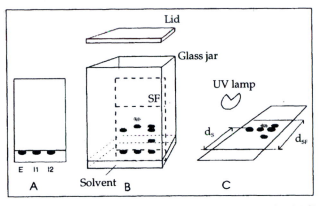

Fig. 15.2 Schematic diagram of (ascending) thin layer chromatography (TLC):
A - deposition of mixtures to be analysed (11, 12) and of a standard mixture (E) for comparison on a TLC plate;
B - development of an ascending chromatogram in a glass jar (containing a small amount of solvent) to the level of the solvent front (SF);
C - reading of the plate (visually for coloured compounds, by spraying of reagent, under a UV lamp, etc.) and measurement of chromatographic parameters (d_{SF} = distance between the line of deposition and the solvent front, d_S = migration distance of solutes).

In TLC the migration of a solute is characterized by the parameter R_F (rate factor, also called front ratio):

$$R_F = d_S/d_{SF} \qquad (2)$$

where d_S denotes the distance covered by the solute on the plate and d_{SF}, the distance covered by the solvent front since the time of deposition (Fig. 15.2). A simple relation can be demonstrated between R_F and the capacity factor (*see* Chapter 13 and eqn. 11 below) $k' = KW_a/V_m$ where W_a denotes the mass of the stationary phase, V_m the volume of the mobile phase on the plate and K the distribution coefficient of the solute. If u is the speed of the solvent front on the plate, the speed of the solute is $u' = u/(1 + k')$. As the distances travelled are proportional to the speeds, $R_F = u/u' = 1/(1 + k') = 1/(1 + KW_a/V_m)$. If the stationary and mobile phases used are the same in a thin layer (TL) as in a column (COL), the coefficients K are the same; the two methods differ only in the values of W_a and V_m, and a transposition coefficient can be defined by

$$K_{tr} = [W_a/V_m]_{COL}/[W_a/V_m]_{TL}$$

The transposition relation is written as

$$k'_{COL} = K_{tr}[(1/R_F) - 1] \qquad (3)$$

The value of K_{tr} can be obtained from eqn. 3 and the relation between k' and the retention time (*see* eqn 9, Chapter 13), by measuring for a given substance the retention time in the column and the R_F on the thin layer.

1.5 Classification of LC Techniques

In LC the exchange phenomena, more complex than in GC, and the wider range of mobile and stationary phases allow a variety of applications from ionic elemental analysis to determination of heavy polar and non-polar compounds. A classification of these applications according to the nature of the exchange phenomena was presented earlier (*see* Fig. 13.1 in Chapter 13):

—adsorption chromatography: separation based on differences in adsorption of molecules of the mixture on a solid adsorbent (LSC),

—partition chromatography: fractionation according to differences in solubility of the molecules to be separated in a liquid impregnating a solid (LLC) or differences in interaction with molecules bonded on the solid (BPC),

—ion exchange chromatography: between the ions of the solution and those of functional groups of a porous solid exchanger,

—ion pair chromatography: between the ions of the mixture to be separated and a suitable counterion,

—ligand-exchange chromatography: formation of complexes between a species fixed on the stationary phase and the solutes to be separated,

—exclusion chromatography (gel permeation or filtration): the stationary phase is a porous solid in which the pores are close in size to the molecules to be separated; molecules too large to enter the pores are eluted first, the others later.

The application of these different techniques is described in §3.

2. Apparatus for Column LC

2.1 Columns

In LC as in GC (*see* Chapter 14), the stationary phase contained in the chromatographic column is the base of the system, though an optional supplementary system for adjusting the properties of the mobile phase is also available. The stationary phase used depends on the nature of the chromatographic separation planned (*see* §1.5 above and §3 below). It is most often composed of silica particles carefully calibrated to 3 μm and 5 μm to reduce the hydraulic pressure gradient (*see* §1.3.2 above), which can be used pure or impregnated or bonded with the desired absorbent. Considering the fine texture, short columns are most often used, of 3 cm to 25 cm length and 4 to 5 mm inner diameter (1/4" externally). Various materials can be used for the body of the column, but the most popular is stainless steel (Fig. 15.3a), with joint fittings that minimize as far as possible the dead volumes at the injector end and the detector end (Fig. 15.3b).

Regulation of column temperature, imperative in GC (*see* Chapter 14) is not necessary in LC. It can be useful in certain cases, however, for ensuring reproducible results when the ambient temperature is subject to fluctuations,

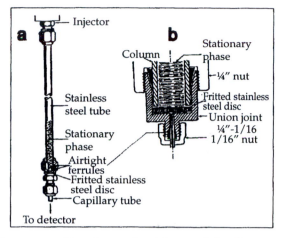

Fig. 15.3 a - chromatographic column of stainless steel; b - lower end of the column minimizing the dead volume (after Rosset *et al.*, 1982).

or for improving the efficiency of some separations. The column can also be introduced into a thermostatic furnace or surrounded with a glass jacket fed by a thermostatic water bath.

It is often recommended that short columns be used, which can be filled by the user, thereby eliminating contamination of the column while ensuring performance. Solvents and liquids injected should likewise be carefully filtered and degassed by bubbling helium through them.

2.2 Systems for Pumping and Regulation of the Mobile Phase

The hydraulic pressure gradients inherent in modern LC necessitate delivery of the mobile phase under high pressure. Most of the time the required pressures do not exceed a hundred bars but with highly viscous mobile phases the hydraulic pressure gradient can reach several hundred bars. Most commercially available equipment deliver the feed at pressures between 300 and 600 bars.

There are pneumatic pumps that function by direct gas pressure on the eluent phase or by increasing pressure by means of a double piston. Electrical pumps can provide a constant volumetric flux. Among them, syringe-type pumps drive the liquid through at constant speed without pulses, but are limited by the volume of the syringe (between 250 and 500 mL), since refilling during a series of analyses is inadvisable.

Electrical pumps with piston or diaphragm are generally preferred; the former operate with a simple two-valve system (Fig. 15.4), the latter similar in principle but using a diaphragm. Both avoid all contact of the eluent with the piston. The disadvantage of these pumps is that they provide a pulsed delivery; however, this drawback can be remedied with an apparatus having two opposed pistons, adopted by many designers (Fig. 15.4).

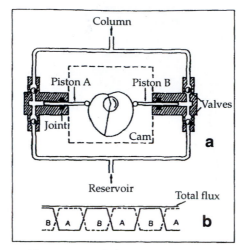

Fig. 15.4   Piston pump with two opposed pistons:
a - experimental appliance; b - profile of flux delivered by each of the two pistons and total flux.

But the fact remains that no pump functions with a strictly constant flux. It has to be provided with flow regulators, which may be based on different types of sensors: one of the best known sensors works on the principle of Darcy's law (eqn 1) by measuring the hydraulic pressure gradient from one part to another in a calibrated capillary tube or a tube packed with microballs of rigorously controlled diameter.

2.3 Elution Gradient

Programming compositional variation of the mobile phase at constant flow during analysis enables optimization of chromatographic separations. Different kinds of programmers are available, which can be grouped into two categories not described here in detail: low-pressure mixers (ahead of the pump) and high-pressure mixers that require two pumping units or a very sophisticated proportional-valve mixing system. A gradient of two solvents allows resolution of most problems (one of the two solvents could be present in the mixture), but some equipment allow programming of ternary mixtures.

2.4 Injectors

As in GC (§2.3 in Chapter 14), this is the interface between the delivery of the mobile phase and the column (high pressure) that allows introduction of the test sample (atmospheric pressure). There are two broad types of injectors: syringe injectors and injectors with sampling ring or valve.

Syringe injectors are of two types: injecting through a septum with the mobile phase in motion as in GC (continuous flow), and with the mobile phase stopped (stop flow). The former type is recommended for moderate pressures (<200 bars); the two techniques are comparable in efficiency. The advantage of this type of injector is that the sample is introduced directly into the column; its chief disadvantage is that it is less reproducible than those with ring or valve. Automated needle-injectors with satisfactory reproducibility are available, however.

In ring-valve injectors the solution to be injected is filled to the calibrated volume of the sampling ring. Rotation of the valve then causes the mobile phase to sweep along the sample to the head of the column (Fig. 15.5). Slide-valves function similarly except that the valve is pushed in instead of being rotated. These injectors are suitable for smaller sample volumes (<5 µL).

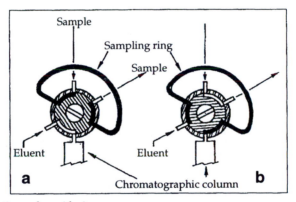

Fig. 15.5   Injection valve with ring:
a -ring filled with sample; b - injection into the column.

2.5 Detectors

2.5.1 Generalities
The detector is an essential component that enables us to continuously follow the separation and identify and quantify the fractionated solutes. However, there is no detector more universally used than in GC. The principal reason is that these solutes are not in a neutral gas but in various solvents that can influence the signal or have an aggressive effect on the instrument. Nevertheless, there are some universal or specific detectors, of which the best known are the differential refractometer and the absorptiometric detector for UV/ visible radiation.

2.5.2 Detection by UV, visible or IR absorptiometry
Detection by absorptiometry is the most commonly used in LC, especially the UV detector because of the relative transparency of many solvents to

UV (Table 15.1). It is, however, a specific detection system as all compounds do not absorb in the UV or visible range. Its range of application has been extended, however, by use of colour reactions in the effluent from the column and of solvents transparent to UV to 200 nm. The signal obtained is proportional to the concentration of the absorbing solute in the effluent from the column according to the Beer-Lambert law (*see* Chapter 9):

$$A = k_{(\lambda)}ec \tag{4}$$

where $k_{(\lambda)}$ is the molar absorptivity (molecular absorption coefficient) of the solute, e the solution thickness traversed and c the concentration of the solute in the effluent.

Table 15.1   Wavelength limit used in the UV and properties of various solvents for LC.

Solvent	UV wavelength limit used, nm	Refractive index at 20°C	Viscosity at 25°C, cP	Boiling point °C
n-Pentane	195	1.358	0.22	36
n-Hexane	190	1.375	0.30	69
Cyclohexane	200	1.421	0.90	81
n-Heptane	195	1.387	0.40	98
Trimethyl 2,2,4-pentane (Isooctane)	197	1.391	0.47	99
Benzene	280	1.498	0.60	80
Toluene	285	1.496	0.55	110
Carbon tetrachloride	265	1.460	0.90	77
Chloroform	245	1.446	0.53	61
Dichloromethane (methylene chloride)	233	1.424	0.41	40
Dichloro 1,2-ethane	228	1.241 (25 °C)	0.78	83
Diethyloxide (ethyl ether)	218	1.353	0.24	35
Di-isopropyloxide (isopropyl ether)	220	1.368	0.38	68
Tetrahydrofuran	212	1.408	0.46	66
Dioxane	215	1.422	1.20	101
Ethyl acetate	256	1.372	0.43	77
Acetonitrile	190	1.344	0.34	82
Acetone	330	1.355	0.30	56
2-propanol (isopropanol)	205	1.378	1.90	82
Ethanol	210	1.361	1.08	78
Methanol	205	1.329	0.54	65
Dimethylformamide	268	1.429	0.80	153
Water		1.333	0.89	100

Thus, e should be as large as possible and the dead volume should concomitantly be as small as possible so that the efficiency of the analysis is not affected. This contradiction is resolved by cell geometry. Appropriate geometry also serves to minimize the background noise caused by refraction and turbulence.

Four types of detectors are distinguished in order of increasing complexity and cost:

—detectors with fixed wavelength, generally 254 nm (mercury vapour lamp),

—detectors with a limited number of wavelengths (interference filters),

—detectors with wavelength continuously variable from 190 to 700 nm,

—multichannel detectors (detectors with diode bars) allowing three-dimensional absorbance-wavelength-time recording.

Absorptiometric detectors for the IR are generally less used than UV-visible detectors, chiefly because of the opacity of many solvents to IR radiation.

2.5.3 Differential-refractometry detector

This detector works by continuously measuring the difference in refractive index between the mobile phase and the effluent from the chromatographic column. The reading R of the refractometer is given by the relation

$$R = Z(n - n_0) = Zp_i(n_i - n_0) \tag{5}$$

where Z is the instrument constant, n the refractive index of the effluent, n_0 the refractive index of the mobile phase, n_i the refractive index of the pure solute and p_i the gravimetric fraction of the solute 'i' in the effluent.

Although of very general use, this detector is much less sensitive than the UV-absorptiometry detector. However, it is very sensitive to variations in solvent caused by temperature and pressure changes. Its use is thus complementary to that of UV-visible absorptiometry when the latter is faulty, even though different techniques might have been developed to improve the performance of the refractometry detector. It is often used in preparative chromatography and exclusion chromatography.

2.5.4 Fluorimetric detector

This is based on measurement of rays emitted by certain organic compounds following absorption of UV radiation. Fluorescence radiation is emitted at a wavelength longer than that absorbed and its intensity is proportional to the intensity absorbed:

$$I_f = k(I_0 - I) \tag{6}$$

where k is a proportionality constant and I_f, I_0 and I denote respectively the intensity of fluorescent radiation, intensity of the incident UV radiation and intensity of transmitted UV radiation. According to the Lambert-Beer law, $I_0/I = 10^{\varepsilon l c}$ and therefore,

$$I_f = kI_0(1 - 10^{-\varepsilon l c}) \tag{7}$$

Response of this kind of detector is not linear with concentration, except at low concentrations at which $10^{-\varepsilon l c} \cong 1 - 2.3 \, \varepsilon \, l c$. Sensitivity at these levels is better than that of absorptiometry and this detector is used for trace analysis. Although its use is limited to fluorescent compounds, its range of application can be extended, as in absorptiometry, by forming fluorescent derivatives in the effluent from the column.

2.5.5 Electrochemical detection

This technique takes advantage of the oxidation-reduction properties of solutes and operates with conductive eluent phases or those that can become so after the addition of an indifferent electrolyte ($LiClO_4$, KCl, etc). An electroactive substance contained in the effluent, when placed in contact with an electrode carrying a potential at least equal to its redox potential, exchanges with the electrode a charge Q given by Faraday's law:

$$Q = n_e FN_m \qquad (8)$$

where n_e denotes the number of electrons taking part in the oxidation or reduction reaction, F the Faraday and N_m the number of moles of the substance transformed. The current i_t passing through the cell at instant 't' is given by

$$i_t = \frac{dQ}{dt} = n_e F \frac{dN_m}{dt} = n_e F(C_i - C_f)D \qquad (9)$$

where C_i and C_f denote respectively the concentrations of the active substance at the cell inlet and outlet (mol L^{-1}) and D the liquid flow through the cell. In principle, the application of equation 9 when the compound has been completely electrolysed ($C_f = 0$) gives absolute detection of the compound without calibration if the number of electrons exchanged (n_e) is known; this is coulometric detection. This mode of detection, however, is difficult to apply to HPLC, chiefly because of the very large flow of the effluent that makes complete electrolysis of the solute impossible.

Thus one prefers most of the time to be satisfied with measurement of the current i_t, which is proportional to the concentration of the solute when its hydrolysed fraction C_f/C_i is constant as shown in the writing following eqn. 9:

$$i_t = n_e FC_i \left(1 - \frac{C_f}{C_i} \right) D \qquad (10)$$

This is the principle of the partial electrolysis amperometric detector, which has several variants (thin layer, polarographic, etc.). This detector can be used for a very large number of families of compounds often with very low minimum detectable limits.

2.5.6 Other detectors

—*Conductometric detector*: this is often based on the detection of inorganic and organic cations and anions by ion chromatography. Certain organic compounds can also be detected by conductometry (photoconductivity).

—*Mass spectrometry (MS)*: the technique is attractive because of its universal application, high sensitivity and the possibility of sequential qualitative identification of the effluents. However, the LC-MS combination is much more complex than the GC-MS combination because of the presence of the solvent and the non-volatility of some of the compounds separated by LC. The apparatus are therefore complicated and reserved for delicate interpretations.

—*Other detection procedures*: radioactivity for labelled molecules, flame ionization by transposition in LC of this excellent detector (difficult because of the presence of a solvent), light scattering, etc.

3. Practice

3.1 Adsorption or Liquid-Solid Chromatography (LSC)

The adsorption properties of the stationary phases, usually silica and alumina gels, are directly put to use. Fractionation depends on the number of free sites on the surface of the material (silanol groups in the case of silica gels- Fig. 15.6).

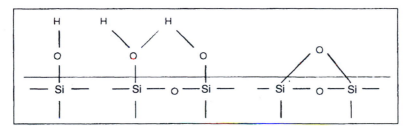

Fig. 15.6   Functional groups on the surface of a silica gel: free silanols, bound silanols and siloxane bridges corresponding to the dehydration of two adjacent silanols.

Water also plays an important role in adsorption chromatography, to the extent that it is preferentially adsorbed on the surface of the gels. The capacity factor k' (eqn. 6 in Chapter 13) is, in this case, given by

$$k'_S = K_S W_a / V_m \tag{11}$$

where K_S denotes the distribution coefficient of the solute between the stationary and mobile phases, W_a the mass adsorbent contained in the column and V_m the total volume of the mobile phase in the column.

Silica is generally more advantageous for solutes with acid character and alumina for solutes with basic character. When selecting a suitable adsorbent, it is necessary to also take into account its specific surface area and the activity of the water it contains. Snyder (1968) defined a relation between the capacity factor K'_S (equation 11) and the characteristics of the adsorbent:

$$\log K'_S = \log V_a + \beta(E_0 - A_S\varepsilon_0) + \log (W_a/V_m) \tag{12}$$

where V_a denotes the volume of the mobile phase adsorbed per gram of adsorbent, β is a dimensionless number between 0 and 1 for measuring the activity of the adsorbent taking into account the number of free (not covered by water molecules) silanol groups, E_0 denotes the free energy of adsorption solute molecules under conditions of standard activity ($\beta = 1$), which is expressed by the relation $E = \Delta G^0_S/2.3RT$ (dimensionless corrected energy, R = the gas constant, $T =$ absolute temperature in K, ΔG^0_S = free energy of adsorption of one molecule of solute), ε_0 denotes ΔG^0_M, the free energy of adsorption of molecules of the mobile phase, under the same conditions of standard activity per unit area A_m occupied on the adsorbent by one molecule of the mobile phase, given by the relation $\varepsilon_0 = \Delta G^0_M/2.3RTA_M$, A_S is the area occupied on the adsorbent by one molecule of the solute and W_a and V_m are as previously defined (eqn 11).

It is advantageous, for reproducible separations, to use solvents that do not alter the water content of the adsorbent. Such solvents are called isoactivants and are characterized by a water content corresponding exactly to the thermodynamic equilibrium with the adsorbent. Several techniques are available for determination of isoactivant water content of various solvents.

The most important parameter for characterizing the mobile phase is its eluting power ε_0 (eqn. 12). The higher the ε_0, the smaller the capacity factors, a change of 0.05 unit in ε_0 causing a nearly twofold or fourfold change in k'. The solvents used in adsorption chromatography should be classified in increasing order of their eluting power; such a classification is called an 'eluotropic series'. Table 15.2 shows an eluotropic series of a few solvents selected from among the most commonly used ones.

The solvent with an appropriate eluting power can be searched for by column or thin-layer chromatography. Rarely is a single solvent the most suitable; it is often necessary to use solvent mixtures, making use if needed of calculators and curves giving the eluting power of the mixtures as a function of concentration. Occasionally it is not possible to continuously vary the eluting power by, for example, mixing isooctane and methanol because they are not miscible in all proportions. It is necessary to add a third solvent that enables solubility of these two (ternary mixture).

Inexpensive HPLC apparatus allow working only with one solvent or one mixture of solvents (isochratic). More advanced models enable varying the eluting power during analysis by programming solvent mixtures as a function of time. They give access to much more sophisticated fractionations

Table 15.2 Eluotropic series of a few solvents from those most generally used in adsorption chromatography.

Solvent	Eluting power ε_0 Silica	Alumina	Viscosity at 20°C cP	Refractive index at 20°C	Wavelength limit, nm	Boiling point °C
Non-polar						
Hexane	0	0	0.31	1.375	190	68.7
Isooctane	0.01	0.01	0.50	1.391	190	99.2
Medium polarity						
Dichloromethane	0.32	0.42	0.44	1.424	233	39.8
Tetrahydrofuran	0.35	0.45	0.55	1.408	212	66.0
Polar						
Acetonitrile	0.50	0.65	0.37	1.344	190	81.6
Methanol	0.73	0.95	0.60	1.329	190	64.7

but rational use of these apparatus is delicate. Analytical conditions should be optimized (Rosset *et al.*, 1982).

3.2 Partition Chromatography on Bonded Stationary Phases

Partition chromatography makes use of a support generally of the same type as those defined in §3.1 (silica gels) except that it is impregnated with a liquid. Separation is based on the partition of solutes between the impregnated phase and the mobile phase. Simple impregnation, earlier used, had the problem of instability, which could be partly corrected by saturating the mobile phase with the impregnating phase in order to avoid redissolution.

Bonded phases, which have progressively replaced impregnated phases enable, without diminishing the partition properties, great improvement in stability, avoidance of all contamination of eluted solutes and working with elution gradients without any problem. Nowadays practically only bonded phases are used (bonded-phase chromatography-BPC), among which are distinguished:

—classic partition chromatography where the molecules bonded to silica gel have a polar group such as —CN or —NH_2; the mobile phase is non-polar or slightly polar;

—partition chromatography with reversed polarity (reversed-phase chromatography) in which the bonded molecules are non-polar (hydrocarbon chains) and the mobile phase is polar.

This terminology is only historical and lacks a scientific criterion. The phase termed 'reversed' made its appearance after the bonding of polar groups; it has very great importance to the extent that most liquid phase chromatographic separations are done nowadays in this manner. The major advantage of the reversed phase in the analysis of natural products such as organic extracts of soils lies in the fact that these extracts can be directly injected *in aqueous medium*, which makes the preliminary operations of sample preparation easy.

Bonding of stationary phases is generally achieved by a silanization reaction consisting of substituting the hydrogen atoms of the silanol groups (Fig. 15.6) by an organosilyl group (Fig. 15.7).

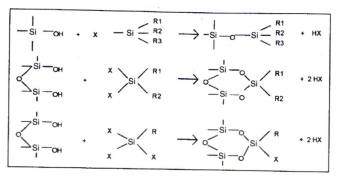

Fig. 15.7   Silanization reactions of silica with a monosilane (bottom), disilane (middle) or trifunctional silane (chlorosilane or alkoxysilane).

Commercial abbreviations occasionally used to designate the bonded phases are derived from the nature of the radicals fixed on the silanol groups ('R' in Fig. 15.7). For example, the Supelcosil LC-18 columns from Supelco have radical chains with 18 carbon atoms (octadecyldimethylsilyl).

3.3 Ion-exchange Chromatography and Related Techniques

Ion-exchange chromatography uses stationary phases on which are fixed ionized functional groups carrying positive or negative charges with mobile ions of opposite sign ensuring electroneutrality. The ions in the neighbourhood of the functional groups are exchangeable with those of the solution that passes through the exchanger. With a stationary phase 'S' carrying exchangeable cations A^+ and a mobile phase containing B^+ ions, the following equilibrium is attained:

$$A^+_S + B^+_M \Leftrightarrow A^+_M + B^+_S \tag{13}$$

The affinity of the exchanger for one constituent or the other is evaluated by the relative proportion of A and B at equilibrium in each of the two phases. In the case of chromatography, the eluting ion of the mobile phase is the same as that originally present on the exchanger. The differences in affinity of the ions of the sample and of the mobile phase result in a different partition coefficient (eqn 1, Chapter 13) for each ion and differential migration of the ions through the column.

The functional groups exchanging cations are negatively charged and are often of sulphonate ($-SO_3^-$) type. The anion exchangers are positively charged and are often of the quaternary ammonium type. These two groups of exchangers are called strong acid ($-SO_3^-$) and strong base ($-NR_3^+$) type respectively, that is, their salts are completely dissociated in acid medium and their exchange capacity is independent of pH.

There are also exchangers in which the functional groups have the characters of weak acid or weak base. The carboxylate group ($-COO^-$), for example, exchanges cations only at rather high pH. In acid medium, the groups exist as non-ionized $-COOH$.

The structures of ion exchangers are of three types:

—ion-exchange resins comprise a matrix of a three-dimensional macromolecular network, often a styrene-divinylbenzene copolymer to which are bonded the functional groups,

—coated exchangers consist of a thin layer of ion exchange resin (coating resin) on a support impermeable to the liquid phase (glass microballs, for example),

—ion-exchange silicas are composed of the type of support silica gel used in §3.1 and §3.2 with specific bonded groups; for example, a silylated derivative with a functional group of the alkylphenyl type can be bonded to the silanol group by a reaction of the type shown in Figure 15.8.

Bonding of functional groups is later achieved by reactions with the phenyl

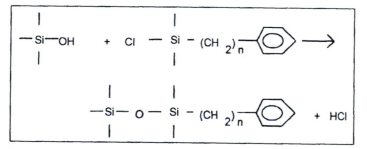

Fig. 15.8 Alkylphenyl silanization of silanol groups, the first step in obtaining certain ion-exchange silicas.

radical, such as sulphonation (sulphonate group, $-SO_3^-$) or even chloromethylation then amination for quaternary ammonium groups.

The separation mechanisms can simultaneously bring in several phenomena. In the case of simple ion exchange the equilibrium in equation 13 is characterized by the exchange coefficient defining a selectivity:

$$\alpha = \frac{[B^+]_S [A^+]_M}{[B^+]_M [A^+]_S} = \frac{K_B}{K_A} \tag{14}$$

where K_A and K_B are the relative distribution coefficients (eqn 1, Chapter 13) of both these ions. A scale of affinity of the ions for each stationary phase can be defined as follows:

—with an anion exchanger of the quaternary ammonium type, the order is: citrate > SO_4^{2-} > oxalate > I^- > NO_3^- > Br^- > SCN^- > Cl^- > formate > acetate > OH^- > F^-;

—with a cation exchanger of the 'strongly acid' sulphonate type, the order is Ba^{2+} > Pb^{2+} > Sr^{2+} > Ca^{2+} > Ni^{2+} > Cd^{2+} > Cu^{2+} > Co^{2+} > Zn^{2+} > Mg^{2+} > Ag^+ > Cs^+ > Rb^+ > K^+ > NH_4^+ > Na^+ > H^+ > Li^+.

This order of affinities partly reflects the eluting power of the mobile phases used. Thus, in the two instances above, a nitrate-based solution gives a more rapid elution of anions than a chloride-based solution, and a potassium-based solution will elute cations more rapidly than a sodium-based solution.

But the phenomenon can be complicated by chemical reactions combined with simple ion exchange. The influence of pH, for example, enters in many equilibria: weak acids in the anionic form, amino acids in cationic form at low pH, etc. Complexation reactions can also participate: cationic complexes of the lanthanides with α-hydroxybutyric acid, anionic complexes of sugars with boric acid. Ion exclusion by an ion of the same type (Donnan equilibrium) and other interactions can take place too. Thus the area of ion exchange chromatography has grown considerably from the classic fractionation of anions and cations to various domains in analytical chemistry and biochemistry.

With certain apparatus specifically designed for such use, ion-exchange chromatography is termed *ion chromatography*. Ion chromatographs generally have a conductometric detector. Then there should be a system for eliminating the actual conductivity of the eluent phase in order to record the peaks corresponding to variations in conductivity due to separated ions; these are suppressors (of conductivity of the eluent phase). Electronic suppressors and various kinds of chemical suppressors (membranes, specific exchange resins) are available for insertion between the column and the detector.

Ion pairs result from associations between two ions of opposite charge and can be due to electrostatic interactions or hydrophobic effects; their fundamental property is their tendency to pass from aqueous solutions to media of low dielectric constant. *Ion-pair chromatography* can be operated by two groups of techniques:

—partition of ion pairs between two immiscible liquid phases (aqueous stationary phase of a stationary silica gel and an organic mobile phase);

—distribution of ion pairs between the mobile phase and an alkyl-bonded silica with a counterion (ion added to form ion pairs with the solutes to be separated) carrying one or several hydrophobic chains to promote partitioning with the alkyl chains of the silica.

Ligand-exchange chromatography is based on the formation of complexes between the solutes to be separated and a metallic ion present only in the stationary phase or in both stationary and mobile phases. Mostly transition metal cations (Cu^{2+}, Zn^{2+}, Cd^{2+}, Ni^{2+}) are used, which give stable complexes with various ligands; two broad modes are distinguished:

—static mode, in which the complex-formation reactions take place on the stationary phase;

—dynamic mode, in which the complex-formation reactions take place in the mobile phase; the separation uses differences in the distribution coefficients of these complexes between the mobile and stationary phases.

3.4 Exclusion Chromatography

The theory of exclusion chromatography is slightly different from that of other chromatographic techniques in that the technique does not rely on a thermodynamic equilibrium of the solute between a mobile phase and a stationary phase (LSC, LLC, BPC), or with other solutes (ion exchange and other ionic techniques). In this technique, the separation is based on the molecular size of the solute and the ability of its molecules to enter the pores of the stationary phase filled with the solvent. Large molecules are totally or partly excluded from the pores and migrate more rapidly than small molecules. The technique has acquired several names, to wit, gel-permeation chromatography (GPC) when the mobile phase is organic and gel-filtration chromatography (GFC) when the mobile phase is aqueous. This latter name is preferentially retained in analysis of biopolymers in aqueous medium derived from natural substrates such as soils.

The equation for retention volume (eqn. 5 of Chapter 13) then takes the form

$$V_R = V_i + KV_p \tag{15}$$

where V_i is the interstitial volume of the column (between the particles of the stationary phase), V_p the pore volume (within the particles of the stationary phase), K the distribution coefficient of the molecules as defined earlier (eqn 1, Chapter 13). There are two limiting values of K:

$K = 0$, then $V_R = V_i$ and all the molecules are excluded from the stationary phase; this condition is otherwise often applied for determining the interstitial volume V_i by injecting molecules large enough to be excluded from the stationary phase;

$K = 1$, all the pores of the stationary phase are accessible to the molecules under consideration; then $V_R = V_i + V_p = V_T$, where V_T is the total permeation volume (V_m in eqn 5, Chapter 13), which can be determined by injection of a solute with low molecular mass.

If $V_R > V_T$, that is, if $K > 1$, it is no longer exclusion chromatography, because another phenomenon is superposed (partition or ion exchange, etc.). Stationary phases are sought in which these secondary phenomena are minimal and for which one may be certain that all the constituents of any mixture whatsoever will be eluted between the retention volumes V_i and

V_T. It also becomes possible to pass the effluent repeatedly through the same column (chromatography with recycling).

Numerous stationary phases are available for exclusion chromatography. The technique has also evolved towards phases with finer particle size (about 10 μm), enabling preparation of very efficient columns. Classic dextran framework gels (Sephadex type), initially used are now rivalled by many other phases belonging to two broad groups:

—organic polymers or gels such as hydroxylated dextran, hydroxylated polyester, polyvinyl, polystyrene-divinylbenzenes;

—rigid phases such as porous silicas, silicas bonded with various types of compounds to reduce parasitic adsorption phenomena or augment hydrophilic character, etc.

Choice of the appropriate stationary phase requires knowledge of its compatibility with the solvent used and also its calibrated range as a function of the molecular weights to be separated. It is necessary to work within the linear portion of the curve log $M = f(V_R)$, where M and V_R respectively denote the molecular mass and retention volume of the solute (Fig. 15.9).

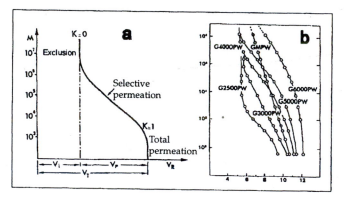

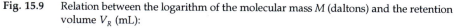

Fig. 15.9   Relation between the logarithm of the molecular mass M (daltons) and the retention volume V_R (mL):
a - theoretical curve; b - application to a range of biopolymers (Supelco; 30 cm $\times$ 7.8 mm columns; mobile phase: water, 1 mL min^{-1}; standard samples: polyethylene glycol and polyethylene oxides).

3.5 Liquid Chromatography and Soil Analysis

All fractionation methods, particularly chromatographic, are valuable tools for embarking on soil analysis because of the complexity of the medium.

Liquid chromatography may be used in two broad ways for studying organic substances:

—fractionation of humic acids: the term humic acids identifies stabilized soil molecules having high molecular weight covering a wide range; their study thus belongs to the domain of exclusion chromatography, in particular gel filtration,

—study of biochemical compounds of soils: these are all the molecules, sometimes called non-humic, essentially originating from living organisms (proteins, lipids, polysaccharides, etc.); in this domain, LC is rivalled by GC for analysis of compounds forming the structural base of these molecules (amino acids, fatty acids, carbohydrates, etc.). Its chief disadvantages are unsatisfactory resolution on occasion and lack of detectors as sensitive as in GC for certain compounds. These disadvantages are largely offset mainly by greater ease in sample preparation [especially with reverse-phase partition (*see* §3.2 above) and ion chromatography (*see* §3.3 above)], in which aqueous solvents are directly injected, and also absence of thermal degradation. Furthermore, LC has the great advantage of allowing study of compounds with higher molecular weight than GC; thus one is not restricted to analysis of compounds forming the structural base of biopolymers but can embark on study of intermediate degradation products (peptides, etc.), and even fractionation of these macromolecules themselves.

Liquid chromatography is also useful for study of organic pollutants of soils, chiefly pesticides, but it is ranked second after GC in this domain because of the selectivity and availability of a number of sensitive and selective detectors for the latter technique.

Liquid chromatography is also a method of choice for soil inorganic analyses, particularly for ionic and derivative techniques (*see* §3.3 above). The progress achieved by ion chromatography in the analysis of anions in the soil solution makes these instruments particularly useful tools in the modern laboratory.

Lastly, it should be emphasized that the phenomena of solute transport in soils recall principles analogous to those in liquid chromatography. Theoretical study of the chromatographic process has occasionally provided the basis for establishment of certain pedological transport equations.

From another viewpoint, the status of our knowledge of materials for chromatographic separation is also worth contemplating. The best that humankind has succeeded in selecting today are alumina and silica, two fundamental constituents of soil minerals.

References

Rosset R., Caude M. and Jardy A. 1982. *Manuel pratique de chromatographie en phase liquide.* Masson, 2nd ed.,374 pp.

Snyder L.R. 1968. *Principles of Adsorption Chromatography.* Marcel Dekker, Inc.

Bibliography

Bièvre C. de and Munier R.L. 1979. Chromatographie de surface en phase liquide. *Techniques de l'ingénieur, Paris,* **P 1475**, 18 pp.

Cunico R.L., Gooding K.M. and Wehr T. 1998. *Basic HPLC and CE of Biomolecules*. Bay Bioanalytical Laboratory, 388 pp.

Hanai T. and Hatano H. (eds.). 1996. *Advances in Liquid Chromatography-35 Years of Column Liquid Chromatography in Japan (Methods in Chromatography, Vol. 1)*. World Scientific Publishing Co.

Lindsay S. 1992. *High Performance Liquid Chromatography (Analytical Chemistry by Open Learning Series)*, John Wiley and Sons, 2nd ed., 360 pp.

Poole C.F. and Schuette S.A. 1984. *Contemporary Practice of Chromatography*. Elsevier, 708 pp.

Rosset R. 1980. Séparations chimiques: chromatographie d'exclusion. *Techniques de l'ingénieur, Paris*, **P 1465**, 7 pp.

Rosset R., Caude M., Foucault A. and Jardy A. 1979. Séparations chimiques: chromatographie en phase liquide sur colonnes. *Techniques de l'ingénieur, Paris*, **P 1455**, 16 pp.

Rosset R., Caude M. and Jardy A. 1991. *Chromatographies en phases liquide et supercritique*. Masson, 936 pp.

Rosset R., Caude M. and Jardy A. 1995. *Chromatographie en phase liquide-Exercices et problèmes*. Masson,, 400 pp.

Rosset R., Caude M. and Jardy A., 1996. *Manuel pratique de chromatographie en phase liquide*. Masson, 3rd ed., 400 pp.

Snyder L.R., Kirkland J.J. and Glajch J. 1997. *Practical HPLC Method Development*. John Wiley and Sons, 765 pp.

Weston A. and Brown P.R. (eds.). 1997. *HPLC and CE: Principles and Practice*. Academic Press, 219 pp.

16

Elemental Analysis for C, H, N, O and S

1. Principle and Evolution

1.1 Introduction

Dry-combustion and wet-digestion techniques of elemental analysis are very ancient. Liebig, Dumas and Kjeldahl in the nineteenth century lent their name to reactions still in use. However, profound changes have been added with respect to the nature of catalysts and use of ultrasensitive detectors to eliminate manual weighing and titration. Automation of operations enables simultaneous rapid analysis of traces of C, H, N, O and S (Fig. 16.1)

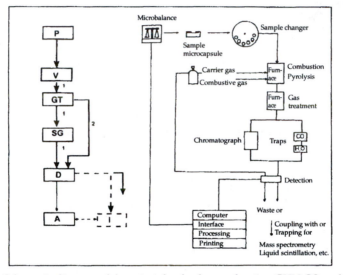

Fig. 16.1 Schematic diagram of the principle of a dry-combustion CHN-OS analyser: P - sample preparation; V - volatilization by combustion or pyrolysis; GT - gas treatment (reduction of oxides, elimination of O_2, elimination of halogens); SG - separation of gases to be determined (not needed in IR-detection CH analysers and electrochemical-detection CS analysers, path 2); D - detection; A - data acquisition; I - optional isotope analysis (coupling with or trapping for mass spectrometry, ^{14}C by liquid scintillation, ^{15}N by spectrometry).

Progress in instrumentation has made possible differentiation of stable isotopes of the elements used as tracers by trapping the effluents and later analysis or by coupling with mass spectrometers. Indeed the isotope natural-abundance ratios $^{15}N/^{14}N$, $^{14}C/^{12}C$, $^{13}C/^{12}C$, $^{18}O/^{16}O$ and D/H are used in numerous research programmes (climatic and palaeotemperature changes, hydrology and hydrothermal or surface circulation of water, pedogenesis and geochemical behaviour of the elements, dynamics of organic matter, nitrogen and carbon cycles, etc.).

1.2 Evolution of Dry-Combustion Analysers

Liebig's (1830) method consisted of igniting a substance containing carbon and hydrogen in the presence of copper oxide (Fig. 16.2):

$$2CuO + C \rightarrow 2Cu + CO_2$$
$$CuO + 2H \rightarrow Cu + H_2O$$

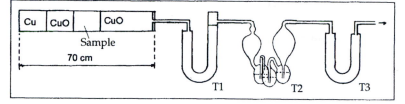

Fig. 16.2   Liebig's apparatus:
T1 - absorption trap for water (anhydrous $CaCl_2$); T2 - absorption trap for CO_2 (Liebig tube with KOH solution); T3 - absorption trap for atmospheric moisture and CO_2 (KOH pellets).

Carbon is converted to carbon dioxide and hydrogen to water. Water is retained in an anhydrous-$CaCl_2$ trap, which can absorb water to $6H_2O$ per mole and CO_2 is retained in an aqueous KOH solution:

$$CO_2 + H_2O \rightarrow H_2CO_3$$
$$H_2CO_3 + 2KOH \rightarrow K_2CO_3 + 2H_2O$$

The increase in weight of the tubes enables calculation of the C and H contents of the sample. To obtain satisfactory precision it is necessary to operate with weighings to the nearest gram.

The Liebig reactions, highly simplified, do not account for the nitrogen sometimes present in the form of oxides and peroxides (or N_2). They can be absorbed by the KOH and vitiate the carbon values.

Dumas (1831), for determining nitrogen, ignited the sample mixed with copper oxide, sweeping the products with a current of CO_2 (Fig. 16.3). The gaseous mixture was led through a copper wire-gauze, which reduced the nitrogen compounds to N_2. Fixation of CO_2 was done in a nitrometer filled with KOH solution. The volume of gas freed was measured at standard pressure to give N_2.

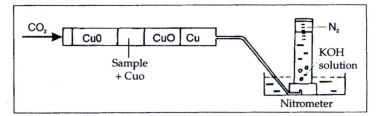

Fig. 16.3   The Dumas method.

Presence of carbonates is incompatible with determination of organic carbon. Carbonates of the soil are slowly destroyed at 500°C liberating CO_2, which is added to that produced by oxidation of organic matter.

$$CaCO_3 \quad \rightarrow \quad CO_2 + CaO$$
$$\text{(calcite, aragonite)}$$

$$FeCO_3 \quad \rightarrow \quad CO_2 + FeO$$
$$\text{(siderite)}$$

It is therefore necessary to eliminate them prior to analysis by a chemical treatment (generally with hydrochloric acid), as the systems used to eliminate the interference are not totally reliable.

Since 1970, automation of equipment has led to preference for vertical furnaces in which the sample can be introduced easily by gravity (Fig. 16.4).

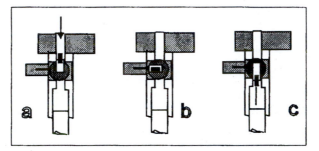

Fig. 16.4   Principle of introduction by gravity (Heraeus):
a - receipt of sample; b - purging by carrier gas; c - introduction into furnace.

The samples, each enclosed in microcupels of tin or silver, are arranged in a rotary dispenser. At programmed intervals, one sample is led to the axis of the furnace and drops by gravity into a chamber purged of air by a current of helium. A temporary addition of oxygen (for C, H, N and S) at 1000°C oxidizes the samples spontaneously, raising the combustion temperature to 1800°C (exothermic reaction: flash ignition). Complete oxidation is thus very rapidly obtained, but the violence of the reaction necessitates arrangement of protection of the quartz furnace by a small nacelle (of nickel, for example) that collects the residual tin dioxide as well as the 'ash' from the sample.

Ignition with CuO is not total if CH_4 is present. In this case Cr_2O_3 should be used for total oxidation.

For determining oxygen, pyrolysis in a carrier current of helium is used, and the gases pass through a column of amorphous carbon or, better still, a graphite-nickel mixture taken to red heat (1100°C):

$$C + O_2 \xrightarrow{\ 600°C\ } CO_2$$

$$2C + O_2 \xrightarrow{\ 1120°C\ } 2CO_2 + 52 \text{ kcal}$$

The sample should not contain chlorides, which interfere by the reaction

$$2C + O_2 + 2Cl_2 \rightarrow 2COCl_2 + 188.2 \text{ kcal}$$

The gas can optionally be led through a column packed with a molecular sieve that traps the impurities but allows N_2 and CO to pass through, the gases being separated in the chromatographic column and determined by a catharometer detector (*see* Chapter 14).

For heterogeneous substrates such as whole soils, systems permitting introduction of samples of the order of 100 mg are preferred. Samples of 50 mg or larger, ground to –0.2 or –0.1 mm, are often needed for giving acceptably uniform and adequately representative samples. On the contrary, for obtaining compounds extracted by chemical dissolution (organic matter), one can use equipment designed for samples of pure organic compounds, injectable at the milligram level.

2. Apparatus and Analytical Products for Dry-Combustion Analysis

2.1 Detectors

The *catharometer detector* (*see* Chapter 14) is used in most CHN analysers. Being non-selective it necessitates separation of compounds by chromatography or by traps. Catharometers are handy, easy to produce and affordable. Their detection limit, of the order of 1 to 10 ng, is adequate for levels found in soils, for sample sizes of the order of 10 to 100 mg.

Infrared detectors (Fig. 16.5) have a detection limit of the order of 10 ng. Their selectivity for water and CO_2 has led some manufacturers since 1970 to use rapid-sweep detection systems with very short response times. The sensitivity can be improved by multiple-reflection systems (reflection from internal coating).

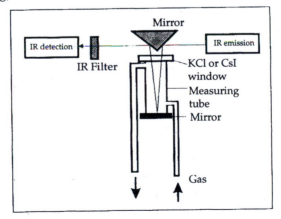

Fig. 16.5   Selective infrared detector.

The electron-capture detector, very sensitive and highly selective, is used in gas chromatography, particularly for analysis of organo-halogen pesticides in traces (*see* Chapter 14). In elemental analysis it is used for sulphur in traces, as SO_2 has affinity for free electrons. The detection limit is of the order of 0.1 pg.

Coupling with a mass spectrometer, not popular nowadays because expensive, was proposed by Fisons Instruments. The gases separated by chromatography and detected with a catharometer emerge unaltered from the apparatus. They can be used for isotope analysis of H, C, N, O and S for determining the isotopic ratios $^{13}C/^{12}C$, D/H, $^{15}N/^{14}N$, etc. The difficulties in coupling are due to differences in pressure between the outlet from the CHN-OS analyser and the secondary vacuum of the mass spectrometer. The measuring time of the mass spectrometer should also be compatible with recording of a spectrum. Volatilization by combustion is advantageous for precise determinations as compounds that can contaminate the mass spectrometer stay in the furnace. Flow times and rates are sufficient to allow recording of spectra. A Ryhage-type jet separator allows separation of the carrier gas helium (Fig. 16.6).

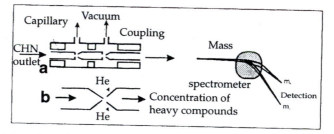

Fig. 16.6   (a) Coupling of CHN-OS elemental analyser with mass spectrometer by a two-stage Ryhage separator; (b) Principle of concentration of heavy constituents ('ion stripping') by helium diffusion.

2.2 Consumables

The precision of CHN-OS apparatus is directly related to the quality of the gases used as combustive gas and carrier gas. A purity of 99.99% is required. Traps can be inserted in the line before connection to the equipment.

Compounds for the oxidation stages should have a large reactive surface area and should be periodically changed or regenerated. Some apparatus have an automatic monitor that sets off an alarm when the performance of the compound becomes unsatisfactory.

Copper oxide (CuO) facilitates oxidation to CO_2, H_2O, N_2 and nitrogen oxides. Mixtures are also used for improving performance in analysis of compounds difficult to oxidize completely:

CuO + Pt + Ag (750°C)

$CuO + PbCrO_4 + PbO_2$

CuO + Ag (this mixture is less susceptible to fritting and retains its porosity for a long time).

Chromium oxide (Cr_2O_3) placed following CuO allows the conversion of CH_4 to CO_2. It does not generate nitrogen oxide and does not retain the gases emitted.

Platinum (Pt) is used for oxidation of CO to CO_2. It is attacked by halogens and at red heat can retain O_2, CO, CO_2, H_2, H, etc.

Hopcalite is a mixture of manganese and copper oxides used to catalyse the oxidation of CO to CO_2. It is cheaper than palladium, or platinum on silica gel.

Vanadium pentoxide (V_2O_5) serves to oxidize SO_2 to SO_3.

Tungstic acid anhydride (WO_3) enables conversion of S to SO_2.

Silver permanganate ($AgMnO_4$) is an oxygen donor.

Vanadate, tungstate and various salts of Ag are used when halogens are present.

Nickelized or platinized charcoal is used for determination of oxygen by its conversion to carbon dioxide, which is detected. The charcoal, containing nickel up to 40%, gives very low blank values. It restricts reduction by the charcoal of the furnace silica at high temperature, up to 1150°C.

The reduction stage uses silver, but more often copper shavings, foil or wire, to fix the excess oxygen after combustion and reduce nitrogen oxides. The trap stage contains sequentially compounds for temporary fixation of water and CO_2.

Anhydrous magnesium perchlorate, $Mg(ClO_4)_2$, is also supplied under the trade names *Dehydrite* or *Anhydrone*. This compound can fix up to $6H_2O$ per mole representing about 35% of its weight (it is effective up to about $3H_2O$); like all perchlorates it is unstable. Silica gel is often used to fix water if it has to be released (after nitrogen estimation) by heating to 400°C. The water can be converted to acetylene by calcium carbide: $CaC_2 + H_2O \rightarrow C_2H_2 + CaO$.

Carbon dioxide is fixed in KOH or ascarite (NaOH-asbestos) tubes.

Elimination of oxides of S and N, and halogens can be done in tubes with MnO_2. Fluorine, which reduces the life of chromatographic columns packed with Porapak, can be fixed by Mg nitride.

Delaying traps enable quick separation of the effluent gases (Porapak 5-Å molecular sieve). Packing of chromatographic columns with Porapak Q.S. enables very efficient separation.

Calibration compounds

The cupels or nacelles of tin used to insert the samples are tested for determining the blank.

Certain apparatus can be standardized directly with the gases. Most of the time, however, standard organic or inorganic compounds are used. It is necessary to use standards with elemental composition close to that of the sample, which is difficult to achieve when C, H, N and S are simultaneously determined (Table 16.1). It is also desirable to always do at least one standardization with a reference soil sample, particularly for mineral soils.

Table 16.1   Standard compounds: classification in order of decreasing C content (theoretical concentration, %).

	C, %	H, %	N, %	O, %	S, %	Precaution
Graphite (C)	100.00	—	—	—	—	
Polyethylene $[(C_2H_4)_n]$	87.70	14.30	—	—	—	
Atropine $(C_{17}H_{23}NO_3)$	70.56	8.01	4.84	16.59	—	(dangerous)
Benzoic acid (C_6H_5COOH)	68.85	4.95	—	26.20	—	
Sulphanilamide $(4H_2NC_6H_4SO_2NH_2)$	41.84	4.68	16.27	18.58	18.62	
Picric acid: 2,4,6 trinitrophenol $[(O_2N)_3C_6H_2OH]$	31.45	1.32	18.34	48.88	—	(inflammable, toxic)
Thiourea (H_2NCSNH_2)	15.78	5.30	36.80	—	42.12	(carcinogenic)
Na carbonate (Na_2CO_3)	11.33	—	—	—	—	
Ammonium sulphate $[(NH_4)_2SO_4]$	—	6.10	21.20	—	24.26	

Some traditional standards are toxic or carcinogenic compounds. It is thus imperative to observe strict precautions when handling them.

3. Dry-Combustion Analysers

The chief solutions currently retained by manufacturers can be grouped according to the modes of fractionation and identification of the gases from combustion or pyrolysis.

3.1 Simultaneous Selective Infrared Detection of C and H

The manufacturer Laboratory Equipment Corporation (Leco[1]) supplies a wide range of apparatus designed for determination of C, H, N, O and S in metals and apparatus for simultaneous determination of N and O, C and S, Si-P-Mn, etc.

In this varied range some apparatus useful for soils can be mentioned: the CR12 analyser, two CHN analysers, of which one is able to handle 200-mg samples (CHN-600 and CHN-800), one CHNS apparatus series 932 and apparatus for determination of sulphur (SC-32 and SC-132).

---

[1]*See* Appendix 6 for addresses.

The Leco CHNS 932 uses samples of the order of 2 to 5 mg. This small nominal sample size allows its use only for organic extracts from soil. The principle of determination is conventional: combustion in a furnace at 1000°C in oxygen, sweeping by a carrier gas (helium), determination by IR detectors for CO_2, H_2O and SO_2. Determination of nitrogen is simultaneously accomplished with a catharometer after reduction of the oxides and separation.

Oxygen can be determined after connection to a high-temperature pyrolysis furnace with carbon for converting O_2 to CO, then by catalysis to CO_2.

The Leco CHN 800 allows sample sizes of 3 to 15 mg according to the elemental composition. Analysis for three elements requires about five minutes. The Leco CHN 600 (Fig. 16.7), based on the same principle, allows use of samples up to 200 mg, such samples being more representative of the whole soil.

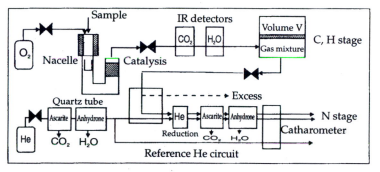

Fig. 16.7   Working principle of the LECO CHN-600.

The sample, weighed in a tin nacelle on a microbalance that can be coupled to the apparatus, is placed in a sample changer. The tin nacelle drops into a furnace heated to 950°C filled with oxygen. Flash combustion of the tin takes the temperature to 1800°C. The gases CO_2, H_2O, N_2, nitrogen oxides and sulphur compounds pass through a catalysis zone in which the combustion is completed and sulphur oxides fixed. The effluent gas is collected and mixed in a syringe with a PVC piston that ensures mixing and takes the combustion products to a standard volume. The CO_2 and H_2O contents are then measured by IR absorption and an aliquot drawn for determination of nitrogen. The aliquot is swept by a current of helium into a furnace containing Cu to reduce nitrogen oxides to N_2, then into an Ascarite tube for eliminating CO_2 and an Anhydrone tube to withdraw H_2O, and finally to a catharometer where it is measured.

This type of apparatus is precise and sensitive for C and N in soils. It should be borne in mind, however, that the aliquot taken for determination of N is satisfactory for the usual C/N ratio of about 10, but can lead to an underestimated value with samples having high C/N ratio. For samples poor in organic matter, like some tropical soils with C content lower than 1%, the determination of carbon is precise (if the sampling has been properly done), but that of nitrogen by this technique is impossible.

Infrared detection is also used in a range of apparatus for determination of carbon in water, soil solution or water extracts. In this case, oxidation is effected not on solid samples but in solution by powerful oxidizing systems (cold persulphate under UV, hot persulphate under UV, O_2 at high temperature, etc.). Infrared detectors are also used in determination of CO_2 in respirometric studies on soils.

3.2 Detection by Catharometer after Chromatographic Separation of Gases

The classic apparatus *Fisons*[1] *1108 CHN-OS (samples from 1 to 100 mg)* (Fig. 16.8) is composed of two furnaces with coupled circuits. The effluent gas is separated by a chromatographic column and analysed with a detector of catharometer type.

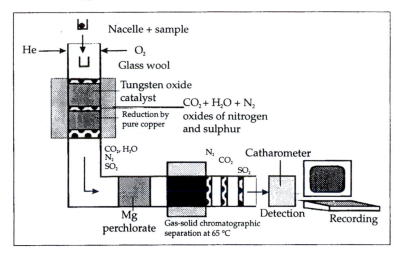

Fig. 16.8   Fisons model 1108 CHN-OS.

For simultaneous determination of C, H and N, the sample is oxidized in a furnace at 1020°C in the presence of oxygen. Flash combustion of the tin capsule takes the temperature to 1800°C. A current of helium sweeps the combustion gases over chromium oxide, then to the reduction stage over copper shavings.

The chromatographic column (Porapak QS) is maintained in a thermostatic enclosure at 100°C and enables separation of N_2, CO_2 and H_2O.

---

[1] *See* Appendix 6 for addresses.

For determination of oxygen, the sample is pyrolysed in the second furnace in the presence of 40% carbon-nickel that instantly liberates the CO formed. The carrier gas, helium, sweeps the combustion products through a filter, then through the second chromatographic column where CO is separated.

For determination of sulphur, the sample is oxidized in the second furnace containing tungstic acid anhydride (WO_3). A current of helium sweeps the SO_2 through a column with copper, then through the chromatographic column and lastly, the thermal-conductivity detector.

Reproducibility is about 0.2%, the detection limit of the order of 10 mg C kg^{-1} soil. A CHN analysis can be done in seven minutes, a CHNS analysis in 12 minutes and an oxygen determination in five minutes. A very sensitive electron-capture detector may be optionally added for determining SO_2, and an interface allows coupling of a mass spectrometer for isotope analysis.

The *Fisons*[1] *NA 1500 (NCS)* apparatus accepts samples up to 100 mg. Nitrogen analysis can be done in three minutes, N-C in six minutes and N-C-S in nine minutes. The principle is identical to that of the 1108 model.

All these equipment are designed to restrict variation in pressure, the precision of the system being dependent on uniformity of gas flow, temperature of the chromatographic column, dead volume in the column and fluctuations in pressure. As the gases are not trapped during their passage through the apparatus, they may be reused for other quantitative analyses.

Perkin-Elmer[1] also has long experience of CHN microanalysis with its ancient 240 appliances and the Coleman 29A nitrogen microanalyser. The *Perkin-Elmer 2400 (CHN)* is fed by a 60-place sample changer. Sample size can go up to 200 mg. The circuit comprises a furnace heated to 950 to 1200°C, a copper reduction tube, a mixing stage with pressure and volume controls, and a chromatographic column for separation of the components. The detection is done with a catharometer sensor. Three options are available: CHN apparatus, CHN-S apparatus and O apparatus with column switch. Times required are 6 minutes for the CHN option, 8 minutes for the CHN-S option and 4 minutes for the O option.

3.3 Detection by Catharometer after Selective Trapping of Gases

The Heraeus[1] company (Foss-Uic) markets specialized apparatus for analysis of metals under the trademark Leybold. This range comprises apparatus designed for C, H, N, O and S as well as C-S and N-O analyses. These equipment have determination ranges not suitable for soils.

The Heraeus CHN-OS apparatus are made in many versions: CHN(OS), CHN, CH, N, O and S. The operating principle is as follows:

Determination of C, H and N: The samples are weighed in tin nacelles (up to 200 mg soil) and calcined under oxygen in a vertical furnace heated to

---

[1] *See* Appendix 6 for addresses.

1050°C (flash combustion at 1800°C). The combustion gases are swept by a current of helium and subjected to a supplementary oxidation over a mixture of copper oxide and cerium oxide. Nitrogen oxides are then reduced to N_2 in a furnace containing copper wire heated to 600°C. Excess oxygen is fixed and sulphur (SO_2) and halogens are reduced and absorbed by silver wool and lead chromate. After this purification the flowing gas comprising the carrier gas He, CO_2, H_2O and N_2 passes to the column-separation unit, where H_2O and CO_2 are fixed.

The nitrogen directly passes into a constant-temperature chamber over a two-channel catharometer (current of helium in the reference channel and the He-N_2 mixture in the measuring channel). After sensing the nitrogen and recording the total signal, the computer starts heating the CO_2 trap to 80°C, releasing that gas and passing it into the catharometer for determination of C. At the end of the signal, the water trap is heated to 400°C, releasing H_2O that passes into the catharometer, enabling determination of H.

Determination of oxygen: A sample in a tin capsule is passed into a cracking furnace taken to 1140°C. The oxygen reacts with the carbon-catalyst mixture to give CO. This gas is swept by a mixture of 95% nitrogen and 5% hydrogen and passed through an NaOH tube, which eliminates the acid halogen fractions. Analysis is effected by a non-dispersive IR detector, as catharometers do not have the required selectivity.

Determination of sulphur: Sulphur can be determined by using a weighed sample with tungsten oxide. After weighing in a tin capsule, the sample (up to 60 mg soil) is calcined in a mixture of 90% helium and 10% oxygen in a furnace taken to 1150°C. Sulphur dioxide results and is dried in a magnesium perchlorate tube with Porapak-N to separate impurities such as methane. The methane occasionally present passes through rapidly while SO_2 is held back. Integration of the SO_2 peak obtained with a selective IR sensor is started only after evacuation of methane.

Determination times are 9 to 14 minutes for CHN, 6 to 9 minutes for CN, 4 to 6 minutes for N and 4 to 6 minutes for O. Precision is of the order of ±0.1%; determination ranges are:

0.03 to 15 mg for C with detection limit of 5×10^{-4} mg
0.03 to 5 mg for N with detection limit of 1×10^{-3} mg
0.03 to 1.5 mg for H with detection limit of 1×10^{-4} mg
0.03 to 2 mg for O_2 with detection limit of 1×10^{-3} mg.

3.4 Detection by Chemiluminescence and Pyrofluorescence

The Antek[1] 700-7000 N (N, S) series apparatus determines total nitrogen by pyro-chemiluminescence in an oxidizing medium and total sulphur by pyrofluorescence.

---

[1] *See* Appendix 6 for addresses.

Determination of total nitrogen

This analysis requires 3 to 10 minutes. The sample (up to 1 g) is pyrolysed in the presence of oxygen in a quartz furnace taken to 1100°C. A current of argon sweeps the nitrogen compounds (Fig. 16.9).

$$\text{N-organic compounds} + O_2 \xrightarrow{1100°C} NO + CO_2 + H_2O + \text{various products}$$
$$(O_2, S, \text{halogens, etc.})$$

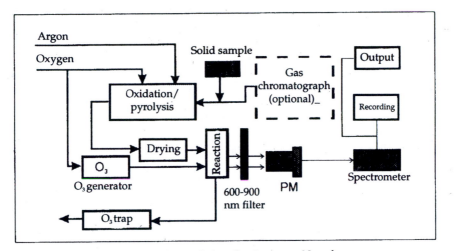

Fig. 16.9 Antek 707°C chemiluminescence N analyser.

This mixture passes through a drying tube, then into the reaction chamber where it is brought into contact with ozone produced in an attached generator:

$$NO + O_3 \rightarrow NO_2^* + O_2$$

The metastable NO_2^* molecule is deactivated and returns to the stable state by emitting a photon that is detected by a photomultiplier:

$$NO_2^* \rightarrow NO_2 + h\nu$$

Photon emission, specific for nitrogen oxide in the 650-900 nm band is free of interferences and is proportional to the measured intensity. The response is linear.

Determination of sulphur

Depending on the sulphur content, 50 mg to 1 g samples can be worked with. Analytical time ranges from 1 min to 10 min.

The sample is introduced into a furnace at 1000°C in oxygen medium:

$$R\text{--}S + O_2 \xrightarrow{1000°C} SO_2 + H_2O + CO_2$$

The gases are dried in a trap, then passed into a fluorescence cell to be exposed to UV radiation:

$$SO_2 + hv \rightarrow SO_2 + hv'.$$

This fluorescence reaction is specific to SO_2 and is free from interference under the above conditions. The fluorescent emission is led to a photomultiplier and the amplified signal is processed in an electronic unit comprising recorder, digital display and computer. The response for SO_2 is linear between contents of a few $\mu g\ kg^{-1}$ and $200\ \mu g\ kg^{-1}$.

3.5 Electrochemical Detection

For the record, the Wösthoff Carmhomat 8 or Carmhographe, Sulmographe and Oxymhograph (Fig. 16.10) are very economical to operate and only require oxygen for C and S determinations. As these apparatus were originally used only for analysis of metals, only the Carmhomat 8 is optimized for soils. They have the disadvantage of requiring constant attention for manual loading and reading of results, which nowadays makes them less attractive despite their high precision in determination of C and S in soils.

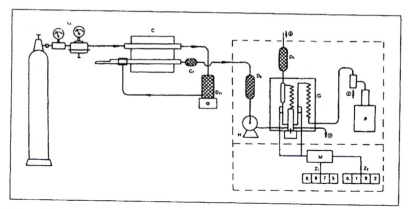

Fig. 16.10   Schematic diagram of the principle of an analytical set up with the Carmhomat 12-G:
L - oxygen cylinder; L_1 - two-stage regulator; C - combustion furnace with two tubes; C_7 - quartz wool filter; D_{11} - soda-lime filter; D_8 - Perhydrite filter; D_9 - soda-lime filter; H - pump; P - dosing pump for the reagent; M - measuring device with amplifier; Z_1 - adjustment of sample weight; Z_2 - digital display of carbon percentage; (1) - inlet for reagent; (2) - outlet for reagent; (3) - outlet for residual gas; G - measuring cell.

4. Wet-Digestion Nitrogen Analysers

The method of wet digestion works by acid mineralization in the presence of a catalyst and distillation after neutralization of the medium (Kjeldahl method). The general eqn. can be written as:

$$R_3\text{—}N + H_2SO_4 \xrightarrow{\text{boiling, catalyst}} (NH_4)_2SO_4 + H_2O + CO_2 + SO_2$$

for an ammonium salt

$$2NH_4^+ + H_2SO_4 \rightarrow (NH_4)_2SO_4 + 2H^+$$

for nitrogen in the amide form, the hydrolysis is

$$RCO(NH_2) + H_2O \xrightarrow{H_2SO_4} RCOONH_4 \rightarrow RCOOH + (NH_4)_2SO_4$$

and with imide nitrogen, the result is

$$R(CO)_2NH + 2H_2O \xrightarrow{H_2SO_4} R(COOH)_2 + 2NH_3$$

Nitro (N—N) and nitroso (N—O) compounds are also taken to the ammonium salt stage with a slightly longer digestion depending on the catalyst used. However, nitrites and nitrates cannot be determined by the classic procedure.

$$3NO_2^- + H_2SO_4 \rightarrow NO_3^- + SO_4^- + H_2O + 2NO$$

With nitrates, even with a mercury catalyst, the reaction is

$$2HNO_3 + 6Hg + 3H_2SO_4 \rightarrow 4H_2O + 3Hg_2SO_4 + 2NO$$

Therefore prior reduction of nitrites and nitrates is necessary if they are abundant. The results can thus differ from those obtained by dry combustion. Bubbling steam into an alkaline medium displaces NH_4^+:

$$(NH_4)_2SO_4 + 2NaOH \xrightarrow{\text{steam}} 2NH_3 + Na_2SO_4 + 2H_2O$$

$$2NH_3 + 2H_2O \rightarrow 2NH_4OH$$

The ammonia thus released is absorbed in a boric acid solution to avoid losses and to permit determination by acidimetry:

$$4H_3BO_3 + 2NH_4OH \rightarrow (NH_4)_2B_4O_7 + 7H_2O$$

$$(NH_4)_2B_4O_7 + 3H_2O \leftrightarrow 2H_3BO_3 + 2NH_4BO_2$$

$$2NH_4BO_2 + 4H_2O \leftrightarrow 2H_3BO_3 + 2NH_4OH$$

Thus it appears as if the released ammonia in the boric acid solution is being directly titrated:

$$H_2SO_4 + 2NH_4OH \rightarrow (NH_4)_2SO_4 + 2H_2O$$

Colorimetry (Berthelot reaction) is also very widely used as an alternative to acidimetry for determination of ammonium and some systems are automated (Skalar).

The Kjeldahl method, despite its disadvantages (handling of dangerous products, long digestion times, uncertainty regarding nitrates and nitrites,

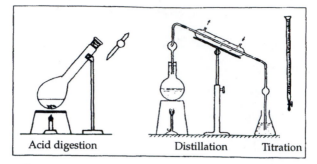

Fig. 16.11 Minimum equipment for the Kjeldahl method.

etc.), has the advantage of requiring only simple, cheap equipment, so that it can be used in places where infrastructure problems are limiting. A Bunsen burner, a glass flask, a rudimentary distillation apparatus and a burette constitute the minimum equipment (Fig. 16.11). However, as the Kjeldahl method is a reference method, several manufacturers have developed complex equipment for automating the determination.

Improvements in this technique can be done at various levels of automation. Manufacturers have sought better-performing and less-contaminating catalysts and also ways to protect operators from problems of handling (boiling H_2SO_4, acid vapours, neutralization with concentrated NaOH, etc.). Thus semi-automated equipment are available with digestion and distillation-determination units at macroscale and microscale. (Bicasa, Büchi, Gerhardt, Skalar, Tecator-Perstorp, Velp[1], etc.). Digestion is done in aluminium blocks heated by electrical coils with programmable temperature-time cycles. Calibrated tubes replace traditional flasks and a reflux cover condenses the acid fumes, eliminating fume hoods.

Prolabo, Cem and Questron[1] suggest microwave digestion and use of hydrogen peroxide instead of catalysts. The distillation unit can be coupled to the estimation unit. It contains a steam generator, an integrated pump for adding NaOH and a safety shield. The determination is done by means of transfer of the distillate by a pump and automatic addition of the titrant. A microprocessor automates the operations and the results are obtained on a printer.

More complete automation has been achieved by several manufacturers. Only the manual transfer of digestion racks is required, no decanting or neutralization.

Tecator-Perstorp[1] markets equipment that enables 180 Kjeldahl analyses per day (Fig. 16.12), the apparatus being able to function unattended for four hours (the racks are loaded all the four hours after digestion).

Gerhardt[1] makes several more or less advanced Vapodest models, with

---

[1] *See* Appendix 6 for addresses.

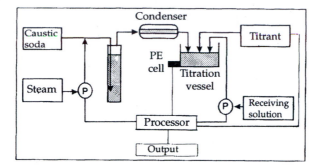

Fig. 16.12 Tecator-auto Kjeldahl distillation-titration unit (P = pumps).

automatic rotary sample changer, serial titration and calculation of results, and optional coupling to an electronic balance. Skalar has a modular digestion block-determination block model, while Uic makes a compact Kjel-auto apparatus with a 20-sample rotary loader and a device for determination by titration

Some manufacturers have sought to eliminate distillation and volumetric titration for doing the final determination directly on the Kjeldahl digest. For example, Mitsubishi[1] uses the diluted Kjeldahl digest for a direct coulometric titration. Similarly, Wescan[1] makes a very sensitive apparatus that can attain a detection limit of the order of mg kg^{-1} (linear response between 10 ppb and 1000 ppm). The analysis requires two minutes and is effected directly on the diluted Kjeldahl digest. This reaction is based on the diffusion of ammonia across a neutral Teflon membrane coupled to a sensor measuring the electrical conductivity. The Kjeldahl digest is mixed with a strongly alkaline solution to convert NH_4^+ to gaseous ammonia:

$$NH_4^+ + OH^- \xrightarrow{\text{heating}} NH_3 + H_2O$$

The membrane is permeable to gases but not to liquids and ions. A neutral solution flowing on the other side of the membrane absorbs the ammonia. Change in the electrical conductivity is measured with a differential conductometer. A reduction device (Zn-Cu) can be added in line for nitrates and nitrites, to ensure that the value for total nitrogen is free from error.

---

[1] *See* Appendix 6 for addresses.

Automation and Robotics in the Laboratory

1. Introduction

The first automatic modular analysis system that underwent considerable development in the domain of soils appeared in the 1960s with the continuous-flow systems segmented by air, comprising a complete range unequalled at the time: preparation of solids, continuous-filtration device, digester, sample changer, peristaltic pump, water bath, spectrocolorimeter, three-channel flame photometer with internal calibration, recorder, etc.

This type of sequence has greatly evolved since then, but still competes with (is complementary to) non-segmented continuous-flow systems.

The boundary between automation and robotics is not well defined and one discipline is often opposed to the other because of the shifting of concepts. Actually, these limits evolve as a function of technical progress. A current definition in the analytical domain can be formulated: 'robotics permits conducting the totality of operations necessary for the execution of one or more analytical tasks, through synergetic combinations of mechanical, electronic and computer assemblages, the process going from sample taking to the useful result. It is the ultimate stage of automation as presently conceived. It enables mastery of the linking of sequences of weighing the sample, extraction, filtration, volume measurement and interactive transfer of analytical parameters, thereby profoundly modifying laboratory strategy'.

Since 1975 the development of microprocessors and microcomputers, a veritable technological mutation, has enabled radical modification of analytical equipment by automation of numerous manual functions (optimization of settings, piloting, etc.) and has pushed old techniques aside. Laboratories are equipped with apparatus from numerous manufacturers, with varied data output, and it is often difficult to interconnect them directly. Computers can act as the link between equipment and become the pillars of an open global architecture through internal networking.

Lastly, around 1981-1982 developments in industrial robotics could find application in the laboratory with the appearance of flexible systems putting to work co-ordinated or anthropomorphic arms, able through electronics and computers to conduct more or less complex analyses. The sample-preparation phase can be computerized, freeing the chemist from repetitive weighings and extractions, the computer and the robot taking charge of the operations up to the obtaining of results. The robot, endowed with a language, becomes an interlocutor whose role will not be limited to automation, but will enable computerization of specific automated operations. Data circulate in the systems, are processed, interpreted and applied, particularly when 'expert' knowledge systems are developed. However, the soil, a solid material, scarcely fluid when powdered, takes poorly to complete automation of analyses, especially at the level of preparation/dissolution.

2. Automated Segmented Continuous-Flow Analysis

Skegg (1956, 1957) lodged the patents for segmentation of a liquid flow by a gas, adapting a process used at a different scale by petroleum engineers to convey fractions of different densities through pipelines by intercalating a segment of immiscible salt water, which works as a self-cleaning piston.

The principle of segmented continuous-flow analysis was developed by Technicon Corp. (Tarrytown NY, USA) and extended to the commercial

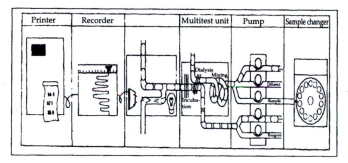

Fig. 17.1   Schematic diagram of an automated sequence (example with sensor for colorimetric measurement).

sector in the form of a modular system (autoanalyser system). This mechanically simple system (Fig. 17.1) has progressively evolved to the Traacs 800 (Technicon Random Access Automated Chemistry System) model, which preceded the purchase of the firm by the Alfa Laval Bran-Luebbe[1] group (1989).

These equipment, in their most popular versions, enable most determinations (ending with spectrocolorimetry, fluorimetry or flame photometry, etc.) to be done on soils, waters and plants, after taking them into solution and continuous dialysis if necessary. The popular applications for soil analysis have long been very varied: ammonium, nitrate, nitrite and total nitrogen, aluminium, iron, silica, titanium, calcium, magnesium, sulphate, chloride, total and extractable phosphorus, sodium, potassium, selenium, organic matter, amino acids after separation in a column, pesticide residues, DCO, etc. [*Technicon Bibliography*, 1968, 1974, 1975 (2), 1975 (3), 1973—1978].

Analytical throughputs of 30 to 100 samples per hour are very suitable for soil-analysis laboratories, which have to process large volumes of repetitive analyses in which reproducibility and precision are essential.

---

[1] *See* Appendix 6 for addresses.

2.1 Sample Preparation

The preparation of solids should first be done manually (Chapter 3), there being no system able to economically computerize preparation of soil samples.

In the Technicon system, outside the industrial establishments for monitoring tablets during manufacture, it is not possible to computerize automatic weighing (Mettler LV-10 automatic feeder + balance). However, assemblies are available for taking elements or compounds to be analysed into solution, starting from complex solid samples (Solid-prep I and II, continuous filtration, continuous digestion, etc.). Although scarcely used in soil analysis, where retention of manual control of these operations is often preferred, such devices deserve to be mentioned for development possibilities.

2.1.1 Extraction/dissolution of solid samples

The solid sample is weighed and placed on the upper plate of a sample feeder turning at a chosen speed. Twenty cups are available (Fig. 17.2a). The cup drops into a homogenization chamber, a quantitative rinsing is effected and rotation of the stirrer at 10,000 rpm enables taking the sample

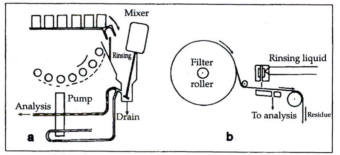

Fig. 17.2   Automated dissolution:
a - systems for extraction or dissolution of solid samples (Technicon); b - continuous filter.

into suspension or solution. At the end of the treatment the suspension is led to a continuous filter. The homogenization chamber is rinsed automatically.

2.1.2 Continuous filter

This filter of defined porosity, seated on a Teflon plate, enables separation of a liquid phase from a solid phase. It is constantly renewed during the filtration and the clear liquid is aspirated below the plate and collected in the sampler tubes or sent directly into the analytical cycle. The precipitate is discarded.

For filtering reagents, 6 mm in-line polythene microfilters are also available, which can be installed at the beginning of the cycle to eliminate precipitates if present before entry into the analytical circuit.

2.1.3 Helicoidal continuous digester

Continuous digestion can be done by the Kjeldahl method using concentrated acid at the end of the dissolution system. A helicoidal Pyrex tube 90 cm long is rotated at 6 rpm (Fig. 17.3a). The soil sample, mixed with acids and catalyst, passes over a resistance taken to 400°C. This is followed by a rinsing with an acid mixture to ensure separation of the samples.

The short digestion time (well suited to organic matter) is insufficient for

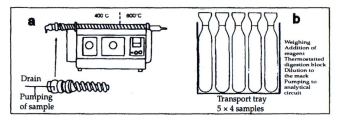

Fig. 17.3 Digestion systems:
a - helicoidal continuous system;
b - discontinuous digestion in digester blocks.

soils, which led Technicon to propose a discontinuous digestion system, with independent aluminium digestion block for 20 or 40 samples (Fig. 17.3b).

A special small distillation column can be attached in series to the outlet for separating NH_4^+. Likewise, reducing columns packed with Cd-Cu for determination of NO_3^- and NO_2^-. Simultaneous determination of ammonium and nitrates in soils has scarcely been considered since the introduction of autoanalysers (Gautheyrou and Gautheyrou, 1965).

A UV microdigester of quartz can also be used for treatment of soil extracts for determination of total nitrogen.

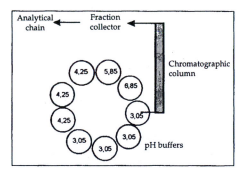

Fig. 17.4 Gradient system for chromatographic analysis of amino acids.

2.1.4 Continuous variable-gradient apparatus for chromatographic columns

Configurations for feeding a chromatographic column (Fig. 17.4) have been proposed for connection to 200-tube fraction collectors.

The equipment consists of a series of interconnected chambers that can contain various buffers or solvents. These systems allow simulation of a gradient (generally of pH) for feeding the chromatographic column.

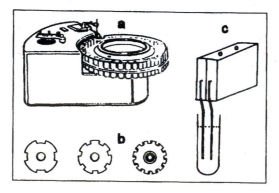

Fig. 17.5   Sample changers for liquids:
a - changer for 40 samples (0.5-8.5 mL); b - cams for regulating the rate of analysis; c - device with vibrating fins to maintain homogeneity.

2.2 Handling of Solutions

2.2.1 Sample changers

Sample changers (Fig. 17.5a) enable the uniform presentation of liquid samples (the volume of the tubes ranges from 0.5 mL to 30 mL) and distribution of 40 to 200 samples.

A device with vibrating foils enables maintenance of homogeneity of the media carrying suspended precipitates (Fig. 17.5c).

A stainless steel tube sucks in the sample solution for a programmed time set by cam systems (Fig. 17.5b) or electronic timer. A reservoir enables rinsing of the aspiration tube after each sample. The pumping time should allow attaining the highest stable signal for a given analysis and the throughput can range from 20 to about 150 samples per hour.

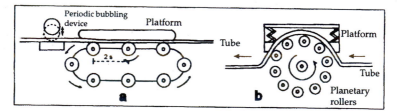

Fig. 17.6   Peristaltic pumps (Technicon):
a - with 'air-bar' bubbling device; b - Traacs 800 model.

2.2.2 Peristaltic proportioning pumps

The peristaltic proportioning pumps (Fig. 17.6) form the heart of the automatic analysis system. They operate by simultaneously pressing on several plastic tubes of different diameters. Selection of suitable tube diameters enables control of the proportions of sample, solvent or reagents that will be mixed to give a reaction for effecting the analysis. One of the tubes sends a small volume of air that segments the flow of liquids upstream to promote mixing, marks the separation between samples and prevents cross-contamination between samples.

The periodic bubbling device (Fig. 17.6a) makes suppression of the pulsation effect possible by synchronizing the injection of air-bubbles with the compression periods of the reagents caused by mechanical crushing of the tubes. The addition of air is very uniform and eliminates a pulsation effect.

2.2.3 Circuits and various accessories

The continuous-flow system is based on crushing calibrated plastic tubes by pump rollers, which on turning drive the liquid and the air segmenting it at constant speed. The tube diameter controls the flow per unit time (Table 17.1). Four qualities of tubes are used for different fluids:

— *Tygon R3603* (polyvinyl chloride-Norton Plastics) for general use;
— *Solvaflex R4000* for pumping certain solvents;
— *Fluran Acidflex R5000* for acids and certain solvents;
— *Silastic*, silicone resistant to acetic acid and organic solvents.

Tygon and *Solvaflex* tubes can be stuck together with cyclohexanone. They are used at room temperature.

The manifolds incorporate all the elements necessary for a given determination. Changing of manifolds is quickly done and enables switching without delay between analyses.

Manifolds can be assembled in the laboratory from various products made available to analysts by manufacturers. A hundred specific components are available.

The T-connectors, H-connectors, etc. (Fig. 17.7) are always used horizontal, but if a precipitate is to be eliminated, the small end will be turned downward. Air is always led in through a small capillary branch of the connector. Bubbles are removed vertically with modified C-type (Fig. 17.7) connectors.

Mixing of reagents in the elementary segments is achieved by means of various coils (Fig. 17.8a) mounted horizontally (convective mixing). In the

Table 17.1   Tubes and couplings with standard colours (Technicon, Tygon, etc.) corresponding to the diameter (flow rate).

Tube type		Internal diameter, mm	PVC 3603	Solvaflex 4000	Acidflex R 5000	Coupling code
			Flow rate, mL min^{-1}			
	Tube colour code		Colourless	Yellow	Black	
Standard microtubes	Orange-black	0.127	0.015	0.015	—	Blue
	Orange-red	0.190	0.030	0.030	—	Blue
	Orange-blue	0.254	0.050	0.050	—	Green
	Orange-green	0.381	0.100	0.160	—	Violet
	Orange-yellow	0.508	0.160	0.160	—	Violet
	Orange-white	0.635	0.230	0.230	—	Violet
Standard macrotubes	Black-black	0.762	0.32	0.32	0.34	Violet-black
	Orange-orange	0.889	0.42	0.42	0.43	Violet-black
	White-white	1.026	0.60	0.56	0.53	Violet-black
	Red-red	1.143	0.80	0.70	0.64	Violet-orange
	Grey-grey	1.293	1.00	—	—	Violet-orange
	Yellow-yellow	1.422	1.20	1.06	1.92	Violet-white
	Blue-yellow	1.524	1.50	—	—	1/8-3/16 coupling
	Blue-blue	1.651	1.60	1.37	1.19	1/8-3/16 coupling
	Green-green	1.854	2.00	1.69	1.44	1/8-3/16 coupling
	Violet-violet	2.057	2.50	2.02	1.71	1/8-3/16 coupling
	Violet-black	2.286	2.90	2.42	2.03	1/8-3/16 coupling
	Violet-orange	2.540	3.40	2.89	2.39	1/8-3/16 coupling
	Violet-white	2.794	3.90	3.39	2.76	1/8-3/16 coupling
Miscella-neous	1/16-1/8 joint	1.580				
	1/8-3/16 coupling	3.170				
	Pyrex tube	1.600				
	BM Pyrex tube	2.000				

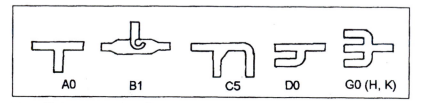

Fig. 17.7   Various connecting tubes:
Code A—T-connectors to mix two reagents (8 models);
Code B—Traps for retaining a gas for determining or eliminating it (6 models);
Code C—Debubbling T (8 models);
Code D—For segmentation of the liquid flow, mixing of two reagents. The current to be segmented is always carried in the straight tube, the reagent or air by the bent branch (5 models);
Code G—Cactus: air is injected in the middle by a capillary branch. Reagents with high flow rates are led through branches 1.6 mm in diameter (4 models);
Code H—Cactus: the reagent with high flow rate is injected through the middle straight arm (6 models);
Code K—Multiple cactus for mixing several reagents (9 models).

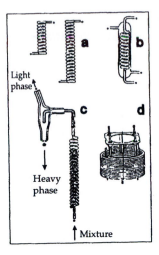

Fig. 17.8   Autoanalyser coils:
a—standard mixing coils: about 25 variants of the 2 models below with 14 and 28
turns (various connectors included); b—constant-temperature coil; c—washing coil;
d—delaying coil for water bath.

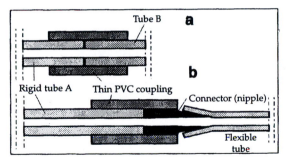

Fig. 17.9   Union joints:
a—direct coupling of two rigid glass or PVC tubes;
b—joining rigid tube and flexible tube by means of a coupling (nipple: 17 models).

water bath, where the temperature reaches 95°C, the internal diameter of
the delaying coils (Fig. 17.8d) is 2.0 mm instead of 1.60 mm to preserve
bubbling without excess pressure despite any increase in volume due to the
temperature. Coils with glass beads (Fig. 17.8c) are used to ensure washing
with solvent. Jacketed coils allow maintenance of constant temperature of
the reagents until they are mixed (Fig. 17.8b)

2.2.4 Manifolds or analytical assemblies

2.2.4.1 GENERALITIES
When a manual operational mode is available, the changeover to automated
mode poses only a slight problem. It suffices to translate to proportional

volumes the volumes of various reagents by adjusting concentrations if need be and regrouping them when possible to reduce the length of circuits and the number of mixing coils.

A well-resolved analytical signal represents the integration of 20 to 30 measurements on a certain number of segments to attain adequate representativeness of the liquid sample. Rinsing times are regulated as a function of the time taken for return to the baseline and the sampling time according to the time required for obtaining an optimum constant analytical signal. The pumping tubes should be taut on the base plate and not twisted.

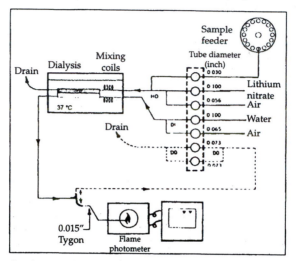

Fig. 17.10 Example of a manifold for segmented continuous flow. Determination of K^+ and Na^+ with dialysis.

A manifold is a dedicated assemblage of various tubes, coils and connectors designed to enable analysis of a given element or compound. Manifolds are mounted on a base-plate or special boxes.

Certain operations of filtration or separation of phases can be done by dialysis, this process leading to dilution. In the case of cycles including a dialysis, the reagents for dialysis and counterdialysis should flow at the same speed and in the same direction (Fig. 17.10).

The speed of reactions is increased by heating. The shape of the signal is an indication of correct development of the reactions as influenced by dispersion of the sample, mixing and stability of flow rate (Fig. 17.11).

Reading of peak height on the calibration curve enables conversion of the signal to concentration (*see* Chapter 18).

2.2.4.2 SPECIAL SYSTEMS

Displacement system

This can be used when the determinations require reagents that attack the

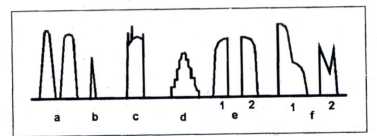

Fig. 17.11 Normal and deformed analytical signals:
  a—well-formed peak;
  b—insufficient pumping;
  c—artefacts, precipitates, etc.;
  d—turbid liquid;
  e—dialysis: 1—counterdialysis flow faster than dialysis flow; 2—counterdialysis
    flow slower than dialysis flow;
  f—contamination between two samples (1) or artefact of pumping (2).

pump tubes. An immiscible liquid with controlled flow is pushed after the
non-pumpable reagent thereby displacing an identical volume of the latter
(Fig. 17.12a).

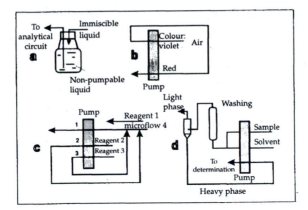

Fig. 17.12 Special systems.
  a—displacement by immiscible liquids (container large enough for a day's work
    load)
  b—segmentation under pressure
  c—differential delivery system:
    flow   1 = 2 + 3 + 4;
    if   1 = red tube 0.80 mL
            2 = black tube 0.32 mL
            3 = black tube 0.32 mL,
    then   4 = 0.16 mL
  d—solvent extraction

Segmentation under pressure
Injection of air under pressure can be used for more regular bubbling, particularly if a delaying coil is operated in the water bath. The air is drawn in by a tube of volume V (mL min^{-1}) and is led into a second pump tube with a flow rate less than about two-thirds of V (Fig. 17.12b). It is then compressed before being injected in the cactus at the level of the liquid to be segmented.

Differential sampling
This system enables avoiding microtubes by working on differences in flow rates of the average tubes (Fig. 17.12c). It can be used for sampling *in vivo* with immediate stabilization of the sample by a reagent.

Solvent extraction
The mixing of a binary water-solvent system is done in coils filled with glass beads (Fig. 17.8c), placed vertically (Fig. 17.12d).

The mixture is fed in from the bottom. The mixing is very efficient at a slow rhythm. The liquid passes over a separator. The light phase or the heavy phase (or both) is determined.

2.2.5 The water bath
The water bath enables speeding up of reactions, which proceed faster above a given temperature and more uniformly in a medium maintained between 37 and 95°C. It contains two delaying glass coils of 2 mm internal diameter, arranged vertically in a heat-carrying liquid with high boiling point, low vapour pressure, low freezing point, high heat capacity and conductivity, medium viscosity, high thermal stability and high resistance to oxidation.

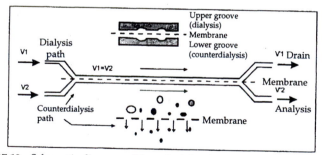

Fig. 17.13 Schematic diagram of the principle of a continuous-flow dialyser.

2.2.6 Dialysers
These enable separation of ions or small organic molecules, of macromolecular and solid species (Fig. 17.13). They are generally made of two flat lucite (acrylic plastic) blocks hollowed in a double spiral and separated by a cellulose acetate membrane. The dialysis path is about two metres long.

The dialysis flow should be about 4 mL min^{-1} and bubbling about 30%.

The sample is led into the upper groove, while the counterdialysis flow is led through the lower groove. The flows should be in the same direction and identical in speed. The substances to be determined pass across the membrane and attain a concentration proportional (dialysis coefficient) to the starting concentration (*see* Chapter 7); the dialysis coefficient depends on the porosity of the membrane, the temperature, the viscosity, the molecular size of the compounds to be dialysed, path length and speed (residence time).

The lengths of the tubes between the pump outlets and the dialyser inlets should be the same. The tautness of the dialysis membrane precludes the breaking of bubbles and limits the accumulation of insoluble substances.

Rectangular microdialysers of lucite, 7.5 to 60 cm long, are available and can be directly installed in the analysis manifolds.

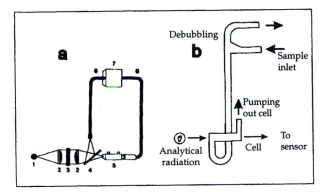

Fig. 17.14 Flow-through colorimeter:
    a—optical arrangement:
        1—tungsten lamp
        2—lenses
        3—filter
        4—splitting of light beam
        5—measuring cell
        6—optical fibres
        7—photocell
    b—spectrophotometric measuring cell.

2.3 Measurements

2.3.1 Flow-through colorimeters

These are installed in most chains (Fig. 17.14a) and operate in accordance with the principle of the Beer-Lambert law (*see* Chapter 9).

Pumping through the colorimeter cuvettes should be done at a rate just lower than that of the liquid reaching the lower horizontal branch of the debubbling 'T' C_5 (Fig. 17.7). Air should not enter the cuvette (Fig. 17.14b) as it might entrain artefacts that would make determination impossible. The volume of 15 mm flow-through cuvettes is about 0.1 mL. The cuvettes can be up to 50 mm long for better sensitivity.

Phototubes are replaced by photoelectric cells; some transmissions by quartz optical fibre enable 5-8 parallel determinations.

Turbidimetric determinations are possible by using suspension stabilizers (polyvinyl alcohol). The Technicon UV-VIS spectrometer with paired quartz cuvettes (15 mm 45 μL; 50 mm 157 μL) allows coverage of the spectral region between 200 and 900 nm.

2.3.2 Flame photometers

A range of instruments is available, essentially for determining sodium (589 nm filter), potassium (768 nm) and lithium (internal standard, 670 nm). A manifold for the simultaneous determination of sodium and potassium with dialysis is shown in Fig. 17.10.

The propane-oxygen burner (2800°C; Fig. 17.15a) is fed through an orange-yellow tube to limit the contamination after debubbling by a connecting

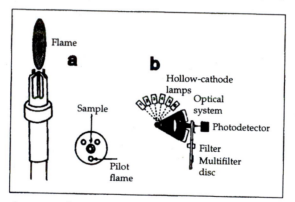

Fig. 17.15 Flame photometer for autoanalysers:
a—Propane-oxygen burner; b—AFS 6 fluorescence apparatus.

tube C_3 (Fig. 17.7). The average flow that can be aspirated is 0.80 mL min^{-1}. According to the concentration and viscosity this may be brought down to 0.60 mL min^{-1} to prevent deposits on the walls of the internal integrating mirror and possible deposition at the bottom of the injection nozzle of the middle capillary tube of the burner.

The AFS 6 analysis system, using atomic fluorescence spectroscopy allows simultaneous analysis of six elements with six light sources and a rotating filter system for selecting the fluorescence radiation of each element (Fig. 17.15b).

2.3.3 Ionometers

The Stat-ion has been designed for simultaneous determination of four elements by ion-selective electrode type ionometry (*see* Chapter 12) for Na$^+$,

K^+, Ca^{++}, F^-, Cl^-, Br^-, NO_3^-, NH_4^+, etc. in soil and plant extracts. The Autoanalyser II ISE enables two simultaneous determinations, generally pH and NH_4^+, etc., at the rate of 30 samples per hour. The measuring chamber is maintained at constant temperature.

2.3.4 *Fluorimeter-nephelometer*

These instruments (*see* Chapter 9) are used essentially for the determination of SO_4^{2-} and organic compounds in soil and plant extracts. They are composed of a double-beam system with interference filters with a narrow bandpass and a flow-through cuvette.

An ultraviolet lamp excites the samples. The presence of interfering fluorescent compounds should be monitored. Plastics and rubber that contain fluorescent compounds, borosilicates with certain solvents, etc., should be eliminated, and the samples protected from sunlight. The temperature should be controlled and the reagents should be brought to the same temperature before analysis.

2.4 Data Recording

The Technicon USA and Bristols Dynamaster (Technicon Europe, Dublin) recorders are simple. The potentiometer circuit is automatically calibrated. The scale-expansion devices (×2, ×4, ×10) give a readable and precise value when the signal is too feeble to allow precise manual reading.

A system for converting the data into tabular and graphic form has been available since 1965. Further development was pursued later with the introduction of a computer for administering the equipment and directly calculating the results.

The bibliography at the end of this chapter cites the work of Furman (1976) and Coakley (1981) and also general references published by Technicon (9000 general references from 1957 to 1978), and those limited to soils, waters and the environment in Gautheyrou and Gautheyrou (1978) (340 references).

3. Automated Non-Segmented Continuous-Flow Analysis

3.1 Principle

Ruzicka and Hansen (1974-1975) and Stewart *et al.* (1976), while reconsidering the idea of continuous-flow analysers, felt that it would be advantageous to omit segmentation by air and inject the test sample directly into a carrier current containing the reagents or diluents. This process was termed flow-injection analysis (FIA).

In segmented continuous-flow analysis (see §2 above) the bubbling ensures cleansing of the walls of the analysis circuits to limit contamination and to minimize losses before measurement; it generates a turbulent flow promoting

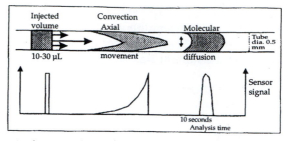

Fig. 17.16 Schematic diagram showing the influence of flow speed and time (viscosity, diffusion coefficient) on the measured signal.

mixing of reagents. The internal diameter of the tubes is generally 1.0 to 1.6 and 2.0 mm for the transfer cycle and analysis throughput is of the order of 60 samples per hour.

When bubbling is omitted, it is necessary to work with very much reduced volumes using injection valves and microcircuits with hydrophobic Teflon tubes to reduce contamination and dead volumes, use short response times, stabilize the inlet-outflow cycles to less than a minute and thus greatly accelerate the analysis throughput (90 to 300 samples per hour). In FIA the response curve does not show a stable plateau, but consists of sharp peaks (Fig. 17.16). As the reaction times are very short and volumes very small, automation is delicate but undistorted for ensuring reproducibility and representativeness of the sample on a level with the sensors. Control of the interfaces is done by a microprocessor. Tube diameters are of the order of 0.2 to 1.0 mm. Consumption of reagents is low and the sample volume required can be just 10 to 100 μL. Samples and reagents are injected in two separate carrier liquids that function as diluents. The two liquid currents are mixed and passed through a coil that enables better reaction and homogenization of the medium before it is sent to the sensor. Washing is done by the carrier-liquid currents. A new sample can then be introduced, this short interval being sufficient for the baseline to be re-established.

Flow-injection analysis handles complex reactions and attains very low detection limits. The circuits allow inclusion of dialysis and effecting distillation, gas diffusion, solvent extraction, dilution, controlled-temperature reactions, etc.

Several sensor-detectors are available: UV-VIS spectrometry, AAS, ICP-AES, ICP-MS, FT-IR spectroscopy, HPLC[1], gas chromatography, ion-selective electrodes, etc., and are cited in numerous bibliographies (*see* Tecator references cited at the end of this chapter). The basic aspect in FIA has been well studied and among the most prominent papers may be mentioned:

---

[1] *See* Appendix 2 for meaning of abbreviations

Ruzicka and Hansen (1974, 1975), Ruzicka *et al.* (1977a, 1977b, 1990), Stewart *et al.* (1976), Betteridge (1978), Poppe (1980), Vanderslice *et al.* (1981), Betteridge *et al.* (1984), Nord and Karlberg (1984), Hansen (1986), Tyson (1987), Valcarcel *et al.* (1987), Brooks *et al.* (1988), Burguera (1989), Clark *et al.* (1989), Fan and Fang (1990), Alves *et al.* (1993), Trojanowicz (1999).

3.2 Theory and Elements of Analytical Chains

A popular FIA chain (Fig. 17.17) may consist of:
—a sample feeder with interchangeable plates (40-100 samples of 2-8.5 mL);
—a peristaltic pump (or syringe or piston pump, etc.; the pumping system should be appropriate for the analysis, piston pump for HPLC for example);
—an analyser that accomplishes mixing (injection valves following specific *chemifolds*);
—a detector to measure absorbance between 400 and 700 nm (or 880 nm)

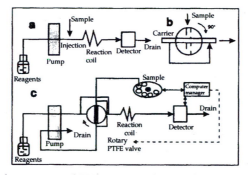

Fig. 17.17　Example of an automated FIA-system analyser (Riley *et al.*, 1984):
a—principle of flow-injection analysis (FIA); b—rotary PTFE valve; c—FIA system (bypass when the valve is in filling position).

of the compounds to be analysed;
—a controller to administer the equipment by means of a microprocessor, execute the calculations and present the results (digital display, computer-graphics printing), and optionally a computer with software for calibration, display of the peaks in real time on a screen, processing, calculation in chosen concentration units, recording and storing the results on the disk.

This equipment allows control of the dispersion of the sample after it has been injected into the system, flow rate and time for development of the reaction before sensing.

The injection valves should dispense very precise volumes, at strictly identical time intervals. When the sample is injected, it forms in the carrier current a segment that is deformed under the influence of the walls affecting the flow of the incompressible liquid (Bernouilli's law). Convection and then molecular diffusion alter the shape of the segment and its concentration until the detector is reached (Betteridge, 1978).

The turbulence during injection can cause random displacement of the equilibrium level measured by the detectors.

Ruzicka *et al.* (1977) proposed a 'dilution factor, D' defined as the ratio of the concentrations before and after the process of dispersion of the components in the flow; it corresponds to the maximum of the dispersion curve. It is also called the 'dispersion number', 'dispersion coefficient', 'dispersion value' or 'dilution ratio' and has expanded the notions of 'standard dispersion', 'total dispersion', 'segment dispersion', 'separate phase dispersion', etc., which are applied to liquid-liquid extractions.

The system should tend to a zero dead volume outside of the indispensable

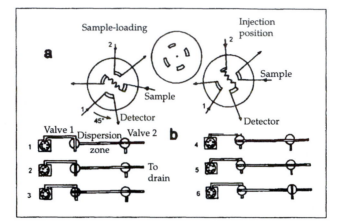

Fig. 17.18   Two commercial injection systems for FIA:
  a—Constant linear flow feeder.
  b—Two-valve system:
     1—sample aspiration; 2—loading of injection valve 1; 3—injection of sample;
     4—dispersion of sample; 5—loading of injection valve 2; 6—injection of the
     dispersed-zone system.

components (rings and connections). The injection valve allows introduction of a reduced volume that scarcely interferes with the total flow.

The ring system (Fig. 17.18a) allows injection of reproducible volumes that can be selected easily. The rings may have the same volume or different volumes. The ideal basic unit for FIA will be a compact module incorporating injection valve, reaction manifold and detector.

The carrier current is generally sent at a constant speed but attempts have been made to modify the technique by using sinusoidal-flow pumps that are replacing peristaltic pumps because of several advantages, though concomitantly with lower speed of operation (Ruzicka *et al.*, 1990).

Dissolution of solid materials (plants, soils, etc.) has been achieved in microcups with ultrasonics and appropriate filters (Lazaro *et al.*, 1991).

The injection module (Fig. 17.18b) allows reproducible introduction of an aliquot into a non-segmented continuous flow. Two rotary valves, driven

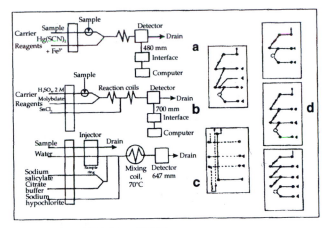

Fig. 17.19   Three examples of FIA 'chemifolds':
   a—determination of chloride, b—determination of phosphate (Fan and Fang, 1990),
   c—determination of NH_4^+ in KCl extracts (Alves *et al.*, 1993), d—standard
   'chemifold' configurations.

by a single motor at high angular velocity and large couple of rotation, and
an electronically controlled fixed-volume ring enable alignment of the analysis
flow and sample flow and vice versa. Filling and emptying of the rings
respectively facilitate sampling and injection.

Figure 17.19 shows three examples of FIA 'chemifolds' built with five
standard configurations of reagent-mixing.

4. Robotics in the Laboratory

4.1 Introduction

Although the objectives of robotics have been well defined for industry,
namely, increasing productivity and quality, and reducing costs (with
consequent effect on economic and social factors), the purposes differ when
applied to a laboratory. The factors of choice will be centred on:

   —improvement in the overall quality of the results (reliability,
repeatability, precision, validation of data, elimination of random factors,
chemometrics, quality assurance, etc.);

   —transfer of repetitive tasks that lead to lowering the attention level of
chemists;

   —safety of remote-controlled operations in a 'hostile' environment
(hazardous, explosive, toxic, radioactive, biological, etc. substances, or high
noise level, odours, etc.);

   —the possibility of following in real time without interference dynamic
processes of long duration.

A robot generally works more slowly than an efficient chemist, but it can work continuously or conduct analyses at site with very great regularity for several days.

A consideration of analytical concepts and associated computer systems will enable modernization and improvement of existing analytical procedures through a chemometric approach. Manual techniques, whose validity has been established and whose management sequence is reproducible, can serve as the bases for installation of robotized systems.

The fabrication of flexible systems (that is, not using a single final assemblage of instruments but one that is modifiable for other operational modes) becomes an unavoidable necessity if the complexity and volume of analyses are sufficient, particularly when programmes requiring repetitive determinations have to be run.

Robotization appears more or less complex at the mechanical and electronic levels, the computer always being the heart of the systems (piloting various functions, apparatus-sensor linkage, interactive data transfer, enhancement of signals, etc.).

Programming and efficiency of the languages of robotics and also the memory level of the commercial apparatus are very variable. Some robots are built around a central arm, others move laterally, enabling use of many peripheral stations with an appropriate flow chart. The speed of execution and precision of the X, Y and Z movements, the mechanical power of vertical and lateral pressures and the variety of specific gear render the systems more or less efficient.

Robotization is distinguishable from automation in its flexibility, each station permitting a modulable sequence of functions to take the place of manual tasks (Table 17.2), while dedicated equipment managed by microprocessors are much more rapid but can only take into account specific exact operations. These special units can be integrated with the robotic chain through an adequate interface. It is thus possible to find a common denominator between the high throughput of an analytical system and the slower preparation of samples.

One of the constraints in the methods putting to use automated or robotic appliances is the strict calibration of the manipulated elements (tubes, flasks, types of closure, configuration and linear and volumetric stability). The thickness of the containers should be uniform and sufficiently low to allow monitoring tests (presence of precipitates using photodiode sensors, for example), yet give the mechanical strength to resist the pressure of robotized fingers (the pressure controlled by the total weight of the tube + sample + reagent).

Another constraint is the need for the robotic assembly to be frequently used, as a long period of disuse with disconnection from the sector will disturb the programming and electronics, leading to hazards arising from uncontrolled movements of such systems.

When the conditions for automation and robotization are reconciled, the changeover from manual methods to automated methods does not take place instantly. Simple copying of manual tasks is not always possible or desirable.

Actually, with repetitive manual methods, one is led to prefer to do the analysis in batches by effecting the steps of dissolution, filtration and pipetting before the actual determination. These steps require large

Table 17.2   Analytical operations possible by robotics with sequential chains.

—Sample identification	—'bar code'
—Grinding (homogenization)	
—Weighing	—balances with RS 232C interface
	—proportioning balances with volume dispenser
	—balance with humidimeter
—Handling of solids	—pouring from transfer flasks
	—individual weighings
—Handling of liquids	—volume measurements (aliquot, dilutions)
	—Vortex mixers, homogenization, etc.
	—volume dispensers
	—proportioning pump
	—piston burettes, syringes, etc.
	—volume transfer
—Working under controlled atmosphere	—inert
	—oxidizing
	—reducing
	—pressure
	—vacuum
	—manipulations in the dark, under UV, etc.
—Dissolution	—dry (fusions)
	—wet digestions
	—microwave digestion
	—oxygenated combustion
	—low temperature (LTA)
	—solvation/concentration
—Measurements with various sensors	—potentiometry, pH, ionometry
	—conductivity
	—spectrocolorimetry, AAS, ICP, etc.
—Separation of various phases	—filtration
	—solvent extraction
	—chromatography
	—centrifugation
	—distillation
	—electrophoresis
	—supercritical fluid

workspaces for preparing the batches. With robotics the operation can be designed in two ways:

—serial analysis, each sample entering the analysis circuit at the rate permitted by the preparative step, in order to limit degradation, evaporation, possible neoformations; storing and recovery are avoided (gain of space and speed);

—batch analysis, several samples being processed simultaneously if certain delays do not allow adaptation of the sample to the planned cycle (microwave digestion, development of a reaction, etc.), the overall speed being controlled by the speed of the limiting step.

To make a choice, it is necessary to take into account:

—changes of the robotic tool in the sequence, reducing the time required (for example, a double finger-pressure pipette arm can be chosen if a reagent is to be added and the tube shaken immediately);

—the spatial environment of the instruments, each displacement of the arm between two given points giving rise to random movements. It is necessary then to fix the intermediate control points (which greatly slows down the process, particularly if the computer guides it very slowly) and to follow the indications of the proximity sensors (sonar, IR beam) that permit avoiding collision with obstacles if any.

This X-Y mapping of operations should be linked coherently to the temporal (T) and spatial (Z) constraints by limiting the amplitude of the movements.

The first step consists of a careful analysis of the operating conditions and translating them from the natural language to unitary operations following a definite syntax.

For this, it is necessary to have thorough knowledge of the chemical processes taking place (Table 17.3), reaction constraints, stability of reactions, reference bases for planning the actions, executing them in sequence and managing the various interferences in the system.

The actions pertaining to experimental conditions should refer to reagents and standards (concentration, volume, pH, indicators, etc.), types of apparatus (wavelengths, sensing methods, etc.), the technique of analysis (peak height, area, etc.) and interferences (masking agents, suppressors, etc.).

Instrument commands include transmission of the data and also reception of results and data from the apparatus to make analytical corrections if necessary. The linking of elementary steps can then be done.

Detailed programming of the robot, even for doing a simple titrimetric determination, may require, according to the interfacing language, several tens of programme steps, which requires the presence of specialists. This has led manufacturers to deliver appliances comprising preprogrammed sections with specific software, simplifying final programming. Artificial intelligence (expert systems) is called upon to rationalize the overall

Table 17.3   Example of soil analysis methods that can be done in a flexible robotic laboratory with a multiprogram computer through simple switching of reagents, time, sampling and sensors.

Analysis	Operation	Determination	
Free Fe (Deb)	weighing-extraction	colorimetry or AAS	
Free Fe (Tamm)	weighing-extraction	colorimetry or AAS	other
(Mehra-Jackson extraction)	weighing-extraction	colorimetry or AAS	elements
(tetraborate extraction)	weighing-extraction	colorimetry or AAS	by AAS or
P fixation	weighing-extraction	colorimetry	ICP
Exchangeable cations	weighing-extraction	AAS, flame emission	
CEC	weighing-extraction	AAS	
pH (water)	weighing-extraction	pH-metry	
pH (KCl)	weighing-extraction	pH-metry	
pH (CaCl$_2$)	weighing-extraction	pH-metry	
pH (NaF)	weighing-extraction	pH-metry	
Conductivity	weighing-extraction	Conductometry	
Extractable P$_2$O$_5$	weighing-extraction	Colorimetry	
Truog	weighing-extraction	Colorimetry	
Ayres		Colorimetry	
Olsen		Colorimetry	
Soluble salts in water extract	Volume	AAS	
Total nitrogen	weighing + digestion	Colorimetry-titrimetry	
Total P	weighing + digestion	Colorimetry	
Total C	weighing-extraction	Colorimetry	

architecture (interpretation, data and sequencing of steps, problem resolution, etc.). Expert systems allow development of quality-assurance protocols required for GLP (Good Laboratory Practices) and for automating them to monitor so that all analytical data not conforming to specifications are detected. These software are now popular in ICP-AES[1] and ICP-MS[1] for evaluating the basic data by comparison with the acceptable limits defined by users and Good Laboratory Practice. They can automatically modify the analytical conditions in order to restore the quality of data. Small expert systems are also used in robotics and managing equipment, reaction chemistry, selection of wavelength for atomic absorption and interpretation of spectra, etc.

A well-designed system should be able to take decisions during the course of analysis by monitoring the results and assigning confidence limits to them. The feedback, in between programmed steps (sensor signal and data, for example), can generate optional operations and modify the procedure: repetition of a highly concentrated sample exceeding the norms and dilution (or concentration if the signal is not significant), determination of 'doubtful' results after calculation of levels of significance, etc.

The programming and software design adapted for robotic systems can

[1] *See* Appendix 2 for meanings of abbreviations.

be ascertained by referring to Eckert and Isenhour (1991), Schliepfer and Isenhour (1988), Verillon *et al.* (1988) and the relations with chemometrics and statistics by referring to Sharaf *et al.* (1986) and Deming (1984).

When the system is changed, the accuracy and reproducibility obtained manually and by a robotized procedure should be cross-checked. All robotic operations can be interfaced with management systems of the Lims (Perkin Elmer-Nelson) or Easy Lims (Beckman-Harley Systems), etc. type.

4.2 Robotic Arms

Robotic arms designed to reproduce manual operations are often constructed with anthropomorphic shape consisting of a shoulder, an elbow and a wrist. Complicated movements are executed by rotation, lateral elongation and vertical movements according to a number of axes generally limited in laboratory models to 5 to 6, rarely 8, as increase in the number of axes leads to greater software complexity. Cycle times and spatial movements are managed by the central computer.

4.2.1 Robotic arm with axial movement

Axial-movement robots such as the Anatech 'Lab Robot' (Fig. 17.20) move on a rail towards the sensors installed on conventional laboratory surfaces. They can be well adapted to the geometry of a laboratory and possibly to envisaging interactive double-robot systems with cumbersome measuring satellites, more numerous than the rotating system. These assemblies are expensive and their programming can only be done by specialists. However, the flexibility provided by this system can be a solution when sensors have to be used independently. The robotic arm with its standard satellites can be moved to whichever work site it is required at.

—The 'PRI-2000 Robot' arm from Precision Robots Inc. is a 5-axis system that can be used in a large room. The rail is placed before the platform is fixed. Lifting power is 6.8 kg, precision ±1 mm.

The 'Isra' (*Intelligent System Roboter und Automatisierung*, 1989) from Systemtechnik (Germany) is a complex robot with 8 axes moving on a central rail in the middle of a double platform carrying the equipment.

—The 'Orca' (Optimized Robot for Chemical Analysis) from Hewlett-Packard (1992) can move on precise rail sections of one to two metres. The model has six axes; it can serve complex peripherals and automate preparation operations.

4.2.2 Robotic arm on circular platform

Robots turning around a central axis trace a cylindrical or complex volume to give a fixed station administered by a central computer and encompassing

---

[1] *See* Appendix 6 for addresses.

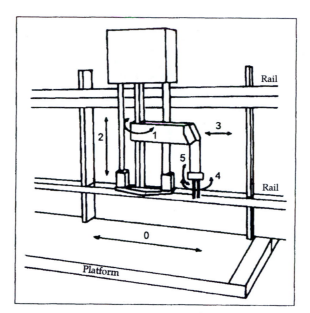

Fig. 17.20   Anatech-Itech 'Lab Robot' axial-movement robotic arm:
 0—linear axial movement 6 m and more
 1—arm rotation 190°
 2—vertical movement 500 mm
 3—horizontal arm movement
 4—rotation of grip ±90°
 5—wrist rotation 90°
          lifting power   range 0 – 10 kg
                                           precision ±500 g
          IBM-compatible computer and ARC robotic language

Modules: hand with pipette   centrifugal
           holders   Vortex mixer
           pipetting station   stirrer
           solvent distributor   balance
           rinsing station   instrument interface
           evaporation station   closing/opening station
           membrane-filtration station
           bar-code reader
Applications: Dissolution test
               Solution preparation, etc.

predefined locations for the installed instruments. The administration of these assemblies is assisted by specific software, which can be easily modified. Command languages can be learnt by laboratory personnel.

Mitsubishi-Perkin Elmer 'Masterlab' robotic arm[1]
Multifunctional programmable 5-axis manipulator arm with command unit controlling the arm movements (Fig. 17.21) provided with optical sensors, supervising computer for all the manipulations and data recording, a

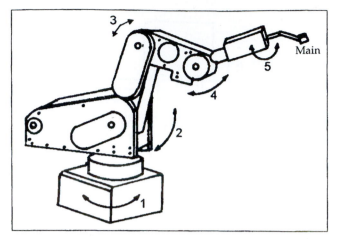

Fig. 17.21 Example of robotic arm (5-axis Mitsubishi Perkin-Elmer MASTERLAB):
1. Base rotation 300°
2. Shoulder swing 130°
2. Elbow swing 90°
4. Wrist swing ±90°
5. Wrist rotation ±180°

program-tutorial system and software.

Attached modules: balance, distributor-diluter, mixer, bar-code reader, closing-opening system, etc.

Maximum height: 696 mm
Lifting with hand: 660 mm
Precision: ±0.5 mm
Lifting power: 1.2 kg
IBM-compatible controller

Applications:
—gas chromatography
—UV-VIS spectrophotometry
—emission spectrophotometry
—sample preparation
—thermal analysis

Microbot[1] 'Minimover-5' robotic arm
Vertical arm movement: 440 mm
Precision: ±0.1 mm
Lifting power: 240g
APPLE II controller
and Applesoft BASIC software

Modules:
—syringe pipette
—pH meter
—diluter, solvent dispenser,
—electronic balance

Prolabo-Aid[1] robotic arm
Robo-Aid V5 (5-6 axes)
Rotation of base: 270°
Shoulder swing: 180°
Elbow swing: 270°
Wrist rotation: 180°
Wrist swing: 180°

vertical arm movement: 900 mm
precision: ±0.2 mm
lifting power: 1 kg
PC-compatible controller
with Expert software

---

[1] *See* Appendix 6 for addresses.

Modules:
–balance, tube rack, etc.
–microwave furnace.

4.3 The Zymark System[1]

This system comprises preparation units enabling the resolution of problems of automation at three levels:

—relatively simple, reasonably priced single-function automated appliances that execute one specific technical operation; they work independently (or can occasionally be integrated with assemblies: Labmate off-line equipment;

—standardized multifunctional appliances enabling execution of an ensemble of linked tasks, for example a cycle consisting of dilution, filtration, injection and transfer to the sensing instrument; they can continuously transfer liquids to an assembly: Benchmate on-line equipment;

—preconfigured robotic assemblies that enable automation of complex repetitive procedures or, with a flexible and programmable configuration, guided by a PC that integrates the different necessary elements on demand: Zymate on-line robot.

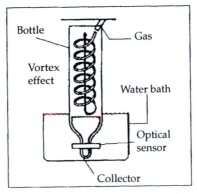

Fig. 17.22   Turbovap-type evaporators.

The installation of preconfigured assemblies is quite simple, though the programmable flexible configuration, even assisted, requires personnel trained in these techniques.

The level of investment necessary is very variable, going from a value Y for the first level to $Y \times 10$ (or more) for the third level, if operational costs are taken into account.

The total cost of a robotic assembly can then be equal to that of an average measuring instrument.

(1) The Labmate equipment brings within the reach of laboratories the repetitive execution of certain operations such as evaporation (Fig. 17.22).

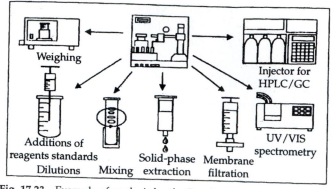

Fig. 17.23 Example of analysis by the Benchmate automated system.

The Turbovap Labmates enable evaporation of samples by the Vortex effect under the pressure of an inert gas (nitrogen) with or without heating (90°C). These appliances can handle from two 500-mL samples to 50 samples of 25 mL.

(2) Standardized automated equipment consist of various assemblies (Fig. 17.23):

—stations for determination of traces by means of solid-liquid cartridges and elution by an organic solvent; it is possible to simultaneously treat three to six samples of 200 to 1000 mL; drying, elution, rinsing and evaporation in line are automatic;

—Benchmate dedicated assemblies, allowing preparation of samples and their transfer to specific sensor-analysers: HPLC, GC, ICP, AAS[1], etc. Only the supply of samples and consumables (reagents, filters, etc.) is manually done.

These appliances are expected to be located on platforms 60 cm wide. Liquid samples are filtered, withdrawn, mixed (Vortex, internal standard, standard additions, etc.) and transferred for analysis through a special software.

Membrane filtration, centrifuge extraction, injection for chromatography, confirming transfer of liquids by weighing, etc. can be done. The operations can be effected at constant temperature (chambers with low temperature maintained, etc.).

(3) Robotized assemblies can be configured to simplify operations by means of a standard program in order to execute an ensemble of repetitive tasks such as titrations, acid digestions, DBO/DCO[1] analyses, analysis of pesticides in soils, dissolution tests and transfer to measuring instruments.

The automatic assembly (Fig. 17.23) for soil analysis (pesticides, herbicides, polyaromatic compounds, etc.) enables extraction of samples by Vortex and

---

[1] *See* Appendix 2 for meaning of abbreviations.

ultrasonic mixing and by multiple solvents, and standard additions. After purification by centrifugation and/or filtration, or by passing through cartridges, the sample is placed in a stoppered bottle. Each sample can be identified by a bar-code. Determination is then done with the appropriate instrument (GC, UV spectrometry).

The dissolution test allows filling, verification of the environmental conditions (temperature, pH), weighing and introduction of samples, analysis at determined time by HPLC or by spectroscopy with choice of wavelengths, and establishment of a cumulative curve of dissolution rate against time. Curve-fitting is automatic.

Acid digestion of samples is done in corrosion-resistant containers. Beakers of PTFE are used, weighing of samples is automated and wet digestion

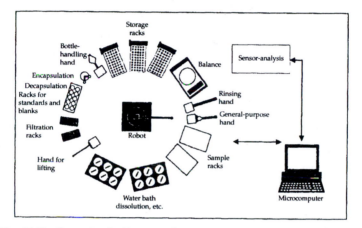

Fig. 17.24   Example of a Zymate robotic station built around a central arm.

(HNO_3-HCl-HF) in an open system can be done with a programmed heating cycle on a hot plate or sand bath. The sample is transferred after dissolution to a measuring instrument. Digestion may also be accomplished in an open system or in a sealed bomb in a microwave furnace, greatly accelerating the dissolution of organic substances, etc.

Zymate flexible robotized assemblies (Fig. 17.24) are more complex in operation. They are composed of a variable number of mechanical and electronic units run by a central computer that integrates the sequences, controls the instruments, acquires the data and ensures their processing (computation, recording, etc.).

The manipulation software can be supplemented by small expert systems enabling detection of handling errors, monitoring of management tools and elements, detection of absence or breakage of a receiver, avoidance of weighing exceeding the capacity of balances, verification of compatibility of capacity of receivers and volume of reagents added, repetition of doubtful samples or those falling beyond the chosen calibration range, correction of baseline, inclusion of quality-control tests, etc.

The third-generation XP arm is provided with a positioning device. It turns at the centre of a circular modular station with 25 PY-section sites with 48 indexing possibilities. Location of the apparatus necessary at each handling step around the robotic arm allows the controller command over the programming software of the PY-sections.

The robotic arm, according to the assigned program, moves from one post to another. It can be equipped with:

—a general-purpose hand that can grip and carry tubes and bottles; the pressure of detachable fingers of various shapes is controlled and programmed, and a tactile device is provided;

—a syringe hand for dispensing volumes of 10 μL, withdrawing 200-μL, 1-mL and 5-mL portions for single use, filters and needles with automatic ejection;

—a dual-function hand, rotation of the fingers can pipette from one position or they can be used as gripping tools after returning to the original position for moving a tube, which immediately follows pipetting;

—a vibrating hand for manipulation of powders and pulverulent solids; the vibration can be regulated to enable weighing and so can the angle for adding the desired weight; the balance gives continuous readout of the weight;

—an IPS (integrated product strategy), preprogrammed robotic-hand assembly with open architecture relying on instructions in macros; the usage parameters are stored in memory;

—an Elisa hand for simultaneous distribution of eight samples on microplates used in biochemistry and biology;

—an unequipped hand for mounting special accessories as required, which can receive two independent analogue signals.

The analytical chain can comprise:

—a management station for bar-coded samples;

—a station for autonomous interception, setting, bolting and unbolting to save robotic arm time;

—a weighing station: balance with interactive control, for weighing solid or liquid and weight data;

—a grinding station for homogenization of samples (mixer);

—a liquid-handling station: volume addition, aspiration, pumping, dilution, injection, transferring, etc.;

—a drying station allowing application and control for a defined time of:
  • temperature (heating, cooling, thermostatting, etc.),
  • atmosphere (controlled atmosphere, inert gas, etc.),
  • stirring (mixing, stirring, Vortex, etc.);

—a station for transfer of samples to instruments for various measurements: pH, conductivity, absorption, fluorescence, AAS, ICP, etc.;

—a station for phase separation and chemical modification: extraction/dissolution, partition, digestion (liquid-solid or liquid-liquid), precipitation, filtration, distillation and electrophoresis.

5. Special Equipment

5.1 Generalities, Various Appliances

The development of microprocessors and microcomputers has enabled design of automated instruments that are easier to handle, more reliable and capable of functioning for long periods without supervision, and all this with an acceptable quality : price ratio.

Certain operations of soil-sample preparation (Chapter 3) remain manual, however, because of the difficulty in and exorbitant cost of robotizing these operations. Research laboratories in particular cannot manage these operations economically except for very specific programmes. So far as extraction of soils by the several existing methods is concerned, some semi-manual improvements have been added such as, for example, the Centurion International[1] syringe extractors that allow adjustment of the cycle between 30 minutes and 24 hours. The Custom[1] range comprises cheap, small appliances that do considerable work (dilutions by syringe and triple-function dispensers that are operated by simple movement of a lever, perpendicular-transport platforms with beaker racks allowing rationalizing and speeding up of handling, and multiple stirrers in series).

The most elementary level at which mechanization and computerization have enabled great progress is that of automatic titration devices with potentiometric sensor. These instruments, generally modular, have a sample distributor with magnetic stirrer and one or more motorized pipettes (Metrohm, Mettler, Radiometer, Schottgerate, Tacussel[1], etc.). They allow printing of results and drawing the titration curve. Metrohm in particular has developed potentiometric (or ionometric) titroprocessors equipped with an alphanumeric keyboard, monitor, printer and RS 232C interface for linking to computers and management systems. The beaker trays (from 10 to 96 beakers) allow samples to be fed for determinations in repetitive analyses.

Lisabio[1] has developed a 120-place sample changer for soils (50-mL beakers) for measurement of pH and five parameters by ionometry. A computer enables programming, management of the parameters and control of instruments (self-calibration, self-monitoring with comparison with standards and repeating the measurements, etc.).

The same approach is used for measurements by voltamperometry and polarimetry.

This approach of automating certain specific operations is less expensive than robotization, but is not totally flexible because there is a separate instrument for each function, whereby interfacing is difficult. Also, a

---

[1] *See* Appendix 6 for addresses.
[2] *See* Appendix 2 for meaning of abbreviations.

relatively large volume of analytical work is necessary for an appliance to be fully utilized.

In this way, certain 'bottlenecks' in laboratories can be removed one by one, but it is preferable to establish an equipment strategy that will allow automation of the different steps in the medium or long term, and to finally end in a robotic system with multiple complementary procedures.

Robots can be integrated with more sophisticated systems with other analytical satellites. They are often dedicated to HPLC, GC, AAS[2], biology, etc. (Gilson, Spectra-physics France AS3000, Euro/DPC Mark V, Malvern X/Y autosampler, Tecan RSP5000 and RSP8000, GBC System 1000, Hamilton Microlab 2100[1], etc.). Analytical balances (Mettler, Sartorius[1], etc.) nowadays have RS 232 interfaces allowing their incorporation in automated systems.

Separation and concentration of elements in liquids are effected by ion exchange in the Jobin-Yvon[1] Labrob JY308 robot. The software ensures ion exchange, elution, rinsing and regeneration of columns; all operations are controlled by microcomputer.

5.2 The Skalar System

The particular domain of soil analysis has seen a few initiatives in mechanization or automation of specialized precise tasks. With the Skalar[1] system the soil is weighed manually, then placed in the digestion medium. Digestion is automatically effected in a standard 20-sample carrier with 20 tubes of 250 mL or 42 tubes of 75 mL.

The carrier, capped by a system carrying 20 reflux condensers, is lowered into an aluminium digestion block programmed to reach a temperature up to 460°C with selectable heating steps. After digestion at the chosen temperature, the assembly is automatically raised and cooled. The condenser rack is separated and the rack with digested samples moves on rails to the processor for the chosen measurement after addition of the diluent. For extracts that do not require digestion, the analyses can be effected on solutions passed through racks with filtration columns or directly on the soil suspensions (conductometry, pH, etc.). A segmented continuous-flow chain (up to 16 channels) can then be used for other determinations and to lead each treated sample to the appropriate sensor.

5.3 Mechanization of Particle-Size Analysis

5.3.1 Mechanized classic pipette method
The TDF 3F automatic particle-size analyser from Technologie Diffusion France-Lisabio[1] enables silt and clay determinations on 24 samples by the classic pipette method in a constant-temperature environment maintained

---

[1] *See* Appendix 6 for addresses.

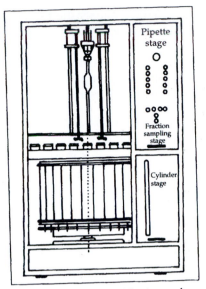

Fig. 17.25   TDF 3F particle-size analyser.

at 20°C as well as recovery of the sands for later separation.
The cylinders are automatically filled to the reference level by means of

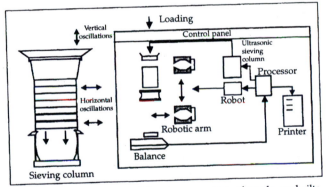

Fig. 17.26   Seishin Robot Sifter for separation of sands and silts.

pumps and sensing probes. The samples are homogenized and brought into
suspension by a helical stirrer. The various fractions are withdrawn by
pipette (Fig. 17.25) at the appropriate time and depth, and are transferred to
cups: one fraction of coarse silt + fine silt + clay, one of fine silt + clay and
one of clay.

The contents of the 12 cups are then evaporated to dryness and weighed
manually. The sands are washed by successive decantation and refilling

---

[1] *See* Appendix 6 for addresses.

according to a factory-set program. The washed sands are dried and weighed, then manually separated by sieving or with the Seishin[1] Robot Sifter RPS-85 apparatus (Fig. 17.26).

The mechanized system does not have an RS 232C interface nor a device for evaporation and automatic weighing of the cups. It has the advantage, however, of enabling isolation of the extracted fractions.

5.3.2 Automatic dry-atmosphere weighing bench

The Isabert weighing bench comprises a balance with 1/10 mg precision associated with a 25-place carousel with automatic commands. The assembly is protected by an airtight enclosure with dry, moisture-free atmosphere. Each cupel is placed on the balance pan; the balance is connected through an RS 232C interface to a microcomputer that enables automation of recording of the initial tare and the weight; each place in the carousel is identified by a serial number.

Clays, sands and silts taken during particle-size analysis may be weighed automatically after drying in a forced-draft air oven.

5.3.3 Automatic apparatus for quantitative sieving of sands and silts

The Seishin[1] Robot Sifter RPS-85 is an automatic apparatus for separation of eight particle-size fractions between 2000 and 20 μm. The analysis is effected by sieving in a sealed column of electroformed sieves, subjected to ultrasonic vertical and horizontal vibrations (Fig. 17.26).

All the operations, apart from the initial loading, are done automatically. The apparatus tares each sieve, ensures sieving in the ultrasonic field for the programmed time, weighs each fraction and prints the particle-size distribution in per cent. An RS 232C interface enables linking to other systems.

The determinations, without loss or contamination, are very uniform even for the very fine 20-μm fractions. The total weight treated is about five to ten grams. Six to ten samples can be fractionated per hour depending on the vibration times and intervals selected.

5.3.4 Automatic particle-size distribution analyser

The *Micromeritics Sedigraph 5100* is based on absorption of an x-ray beam in a homogenized dispersed medium before sedimentation. The rate of fall of particles under the effect of gravity (Stokes law) is proportional to their mean size and the difference in density from the dispersing liquid. These irregular particles are classified according to the equivalent spherical diameter. The attenuation of the x-ray beam is independent of the thickness of the particle; as the wavelength of the radiation is much shorter than the diameter of the finest soil particles, the precision is satisfactory. The *Sedigraph* comprises a measuring device (or two) guided by a microcomputer. The x-

---

[1] *See* Appendix 6 for addresses.

ray source emits a beam that passes through a narrow slit, then through the mobile constant-temperature measuring cuvette containing the suspension. This cuvette is placed in the path of the x-ray beam. The cuvette is automatically filled by a flow-through system, a reservoir of the dispersing liquid, a mechanical stirrer and an ultrasonic dispersing device.

The samples can be fed by a sample changer. The software enables the determinations and storing of up to 50 individual operating protocols. At the end of the cycle, the cuvette is emptied and washed automatically before the next sample is taken in. Depending on the determination and the minimum content looked for, one to six samples can be analysed per hour, with no manual intervention. The results are displayed and printed as continuous quantitative distribution curves.

References

Automated segmented continuous-flow analysis
Coakley W.A. 1981. *Handbook of Automated Analysis: Continuous Flow Technique*. Marcel Dekker.
Furman W.B. 1976. *Continuous Flow Analysis*. Marcel Dekker.
Gautheyrou J. and Gautheyrou M. 1965. Contribution à l'étude de la dynamique de l'azote dans les sols. Dosages simultanés de l'azote nitrique et ammoniacal dans les sols. *Fourth Int. Symp. Technicon.*
Gautheyrou J. and Gautheyrou M. 1978. Méthodologies mécanisées. Introduction à l'automatisation des opérations analytiques concernant les sols, les végétaux et les eaux d'irrigation. *Notes laboratoires ORSTOM Guadeloupe*, 113 pp.
Skegg L.T. 1956. An automatic method for colorimetric analysis. *Clin. Chem. Abstr.* **2**: 241.
Skegg L.T. 1957. An automatic method for colorimetric analysis. *Am. J. Clin. Pathol.* **28**: 311-322.
Technicon. 1968. *Technicon Autoanalyser Bibliography 1957-1967*. Technicon Corporation, ref. 1-1825.
Technicon Corporation. 1974. Ref. 2001-7270, 100 pp.
Technicon. 1974. *Technicon Bibliography Supplement (1)*, Technicon Corporation, ref. 7271-7590.
Technicon. 1975. *Technicon Bibliography Supplement (2)*, Technicon Corporation, ref. 7591-7870.
Technicon. 1976. *Technicon Bibliography Supplement (3)*, Technicon Corporation, ref. 7871-8025.
Technicon. 1973-1978. *Technicon Bibliography Papers* 8026-8965.

Automated non-segmented continuous-flow analysis
Alves B.R.J., Boddey R.M. and Urquiaga S.S. 1993. A rapid and sensitive flow-injection technique for the analysis of ammonium in soil extracts. *Comm. Soil Sci. Pl. Anal.* **24**: 277-284.
Betteridge D. 1978. Flow injection analysis. *Anal. Chem.* **50**: 832-846.
Betteridge D., Marczewski C.Z. and Wade A.P. 1984. A random walk simulation in flow injection analysis. *Anal. Chim. Acta*, **165**: 227.
Brooks S.H., Leff D.V., Torres M.A.H. and Dorsey J.G. 1988. Dispersion coefficient and moment analysis of flow injection analysis peaks. *Anal. Chem.* **60**: 2737-2744.
Burguera J.L. 1989. *Flow Injection Atomic Spectroscopy*. Marcel Dekker.
Clark G.D., Ruzicka J. and Christian G.D. 1989. Split zone flow injection analysis: an approach to automated dilutions. *Anal. Chem.* **61**: 1773-1778.
Fan S. and Fang Z. 1990. Compensation of calibration graph curvature and interference in flow-injection spectrophotometry using gradient ratio calibration. *Anal. Chim. Acta*, **241**: 15-22.

Hansen E.H. 1986. *Flow Injection Analysis*. Dissertation, Technical University of Denmark. ISBN 87-502-0636-2.

Lazaro F., Luque de Castro M.D. and Valcarcel M. 1991. Direct introduction of solid samples into continuous-flow systems by use of ultrasonic irradiation. *Anal. Chim. Acta*, **242**: 283-289.

Nord L. and Karlberg B. 1984. Extraction based on the flow injection principle. Part 6. Film formation and dispersion in liquid-liquid segmented flow extraction systems. *Anal. Chim. Acta*, **164**: 233.

Poppe H. 1980. Characterization and design of liquid phase flow through detector systems. *Anal. Chim. Acta*, **114**: 59.

Ruzicka J. and Hansen E.H. 1974. *Danish patent* No. 4846/84; *U.S. patent* No. 4022575.

Ruzicka J. and Hansen E.H. 1975. Flow injection analysis. Part I. A new concept of fast continuous flow analysis. *Anal. Chim. Acta*, **78**: 145.

Ruzicka J., Hansen, E.H. and Mosbaek H. 1977. Flow injection analysis. Part IX. A new approach to continuous flow titrations. *Anal. Chim. Acta*, **92**: 235.

Ruzicka J., Hansen E.H., Mosbaek H. and Krug F.J. 1977. Pumping pressure and reagent consumption in flow injection analysis. *Anal. Chem.* **49**: 1858.

Ruzicka J., Marshall G.D. and Christian G.D. 1990. Variable flow rates and a sinusoidal flow pump for flow injection analysis. *Anal. Chem.* **62**: 1861-1866.

Stewart K.K., Beecher G.R. and Hare P.E. 1976. Rapid analysis of discrete samples: The use of non-segmented continuous flow. *Anal. Biochem.* **70**: 167.

Trojanowicz, M. 1999. *Flow Injection Analysis: Instrumentation and Applications*. World Scientific Publ. Co.

Tyson J.F. 1987. Analytical information from doublet peaks in flow injection analysis. Part I. Basic equation and applications to flow injection titrations. *Analyst*, **112**: 523-526.

Valcarcel M. and Luque de Castro M.D. 1987. *Flow-injection Analysis: Principles and Applications*. Horwood.

Vanderslice J.T., Stewart K.K., Rosenfeld A.G. and Higgs D.J. 1981. Laminar dispersion in flow injection analysis. *Talanta*, **28**: 11.

Automation and robotics

Deming S.N. 1984. In B.R. Kowalski (ed.). *Chemometrics, Mathematics and Statistics in Chemistry, Reidel-NATO ASI Series C, Mathematical and Physical Sciences*, **138**: 267-304.

Eckert-Tilotta S.E. and Isenhour T.L. 1991. Development of a robotic standard addition method. *Anal. Chim. Acta*, **254**: 215-221.

Schliepfer W.A. and Isenhour T.L. 1988. Complexometric analysis using an artificial intelligence driven robotic system. *Anal. Chem.* **60**: 1142-1145.

Sharaf M.A. Illman D.L. and Kowalski B.R. 1986. *Chemometrics*. Wiley-Interscience.

Verillon F., Pichon B. and Quan F. 1988. Un nouvel automate de préparation d'échantillons liquides pour des analyses complexes en grandes séries. *Analusis*, **16**: 60-63.

Bibliography

Automated non-segmented continuous-flow analysis (FIA)

Burguera J.L. (ed.). 1989. *Flow Injection Atomic Spectroscopy*. Marcel Dekker.

Fang Z. 1993. *Flow Injection Separation and Preconcentration*. John Wiley and Sons.

Fang Z. 1995. *Flow Injection Atomic Absorption Spectrometry*. John Wiley and Sons, 320 pp.

MacDonald A.M.G., Pardue H.L., Townshend A. and Clerc J.T. 1986. Flow analysis III. *Proc. Int. Conf.* (Birmingham, 1985), *Anal. Chim. Acta*, **179** (special publication).

Ruzicka J. and Hansen C.H. 1988. *Flow Injection Analysis*. Wiley-Interscience, 2nd ed.

Tecator. 1985. *FIA Star-Flow Injection Analysis Bibliography, 1974-1984*, Ref. 1-727. Tecator, 154 pp.

Tecator. 1986. *FIA Star-Flow Injection Analysis Bibliography, Supplement 1985*, Ref. 7281061. Tecator, 81 pp.

Tecator. 1987. *FIA Star-Flow Injection Analysis Bibliography, Supplement 1986*, Ref. 1000-1400. Tecator, 98 pp.

Zagatto E.A.G., Reis B.F. and Bergamin H. 1989. The concept of volumetric fraction in flow-injection analysis. *Anal. Chim. Acta*, **226**: 129-136.

Automation and robotics

Beebe K.R., Pell R.J. and Seasholtz M.B. 1998. *Chemometrics: A Practical Guide (Wiley-Interscience Series on Laboratory Automation)*. John Wiley and Sons.

Haralick R. and Shapiro L. 1993. *Computer and Robot Vision*. Lavoisier, 630 pp.

Hegarty C. 1992. *An Introduction to the Management of Laboratory Data. A Tutorial Approach Using Borland's Paradox Relational Data Base*. Lavoisier, 588 pp.

Hurst W.J. (ed.). 1995. *Automation in the Laboratory*. VCH Pub.

Liscouski J. 1994. *Laboratory and Scientific Computing: A Strategic Approach (Wiley-Interscience Series on Laboratory Automation)*. John Wiley and Sons, 224 pp.

Russo M.F. and Echols M.M. 1999. *Automating Science and Engineering Laboratories with Visual Basic (Wiley-Interscience Series on Laboratory Automation)*. John Wiley and Sons.

Warwick K. 1993. *Robotics: Applied Mathematics and Computational Aspects*. Lavoisier, 614 pp.

Zymark Corporation. 1997. *Laboratory Automation and Robotics*. Zymark Corp.

Quality Control of
Analytical Results

1. Introduction

The chapters in Part One pertained to field sampling, then the subsampling in the laboratory. Analysis proper will comprise a certain number of preliminary operations, then a qualitative or quantitative determination by means of an instrumental technique (*see* Chapters 8 to 17).

Qualitative measurements are very variable in complexity. It may be a matter of tests of relative estimation of colour, texture, etc., of a soil (*see* Chapter 2), evaluation of its acidity with a coloured indicator, evaluation of the presence of chlorides, etc. (*see* Chapter 5). It could also be a question of identifying more or less complex compounds by interpretation of chromatograms (*see* Chapters 13 to 15) or of various spectra, as in molecular absorption spectrometry (*see* Chapter 9). It is then necessary to acquire a signal, which poses a new sampling problem, and to effect various operations for improving this acquisition (transformation, filtering, smoothening, straightening, etc.), then exploit the information (comparison of spectra, etc.).

Quantitative data can also emerge from elementary (bulk density, pH, etc.) or complex operations. The latter include, apart from the methods for acquisition and treatment of the signal from qualitative analysis, various measuring and calibration procedures.

In all these cases, each step, from initial sampling to obtaining the final result, can be a possible source of error. The errors will further depend closely on the protocol chosen for each step, the dexterity and diligence of the operator and the settings of the measuring instrument. It is thus necessary to have methods for (1) optimizing the various analysis protocols and instrument settings and (2) studying and evaluating the various causes of error. On the whole, it is a matter of providing for each analysis not just a result, but also verification of its repeatability by a confidence interval, the amplitude of which will indicate reliability or otherwise of the value.

Computation methods pertaining to physicochemical analysis are thus varied, from treatment of the signal to analysis of the data, including methods of optimization and those related to quality control. They have been considerably developed over the past few years with increasing automation of analysis protocols and introduction of computers in laboratories. Though the chief difficulty earlier for the chemist lay in obtaining the result, it is today becoming more and more necessary to ensure control of the quality of the information received from different steps. It is in this sense that the term *chemometrics* has appeared during the past decades, to group together all the most appropriate computation methods for physicochemical analysis. Chemometrics has already become the subject of several books cited in the reference section, and it would be presumptuous to claim to describe it in a single chapter.

This chapter presents to the chief statistical tests related to quality of the results in quantitative analysis of soil. The different types of errors are defined before more fully describing the distribution of random errors. The propagation of errors and detection of 'aberrant' results are described next, followed by the field of comparison of results among themselves or with reference values and the study of causes of error. Lastly the following four particular types of errors are discussed: (1) field sampling, (2) laboratory subsampling, (3) calibration and (4) detection limit.

2. Types of Errors

Quantitative physicochemical analysis is always marred by an error and a result is interpretable only if it is accompanied by the associated error. For example, if the organic activity of a sediment is characterized as a function of depth by a determination of organic carbon, the interpretation can be very different depending on whether the data are considered error-free or taken to be intervals having a high probability of containing the true value (Fig. 18.1).

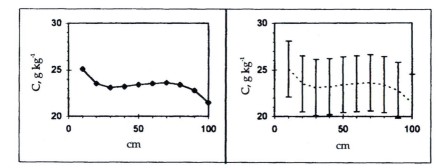

Fig. 18.1   Carbon analysis of a sediment:
Left, determinations presumed error-free; right, intervals containing the true value for a given confidence level (fictitious data).

In the first case, one can surrender to stirring geochemical interpretations of biological activity at different historical periods related to the depth of sediment. In the second case, the interpretation is much more delicate. Figure 18.1 shows noticeable difference between the first and last samples, but all the other determinations appear equal. Tests become necessary for checking whether the results can be considered equal or different.

But what are these errors represented by an interval (Fig. 18.1)? Before they are quantified, it is appropriate to characterize their various types, by examining the simulated data (Table 18.1) from 5 laboratories each of which did 7 carbon analyses on the same standard soil sample with reference content determined to be 21.5 g (C) kg^{-1}. These results (Table 18.1) are reported in graphic form in Figure 18.2.

Table 18.1 Results of carbon analyses obtained by 5 laboratories for a reference standard with content estimated at 21.5 g C kg^{-1} (fictitious data).

	Laboratory 1	Laboratory 2	Laboratory 3	Laboratory 4	Laboratory 5
Reference	**21.5**	**21.5**	**21.5**	**21.5**	**21.5**
1	23.5	22.0	19.5	21.6	21.5
2	23.7	20.1	18.3	21.3	21.3
3	23.2	21.3	20.2	21.8	21.8
4	24.0	19.4	21.8	21.4	24.8
5	23.5	23.1	20.0	21.6	21.6
6	23.4	20.0	18.4	21.7	21.4
7	23.8	21.6	21.0	21.5	21.7

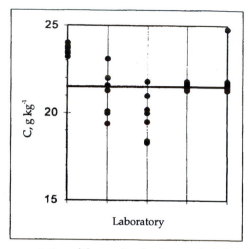

Fig. 18.2 Graphic representation of the results of carbon analysis from Table 18.1. The thick line indicates the reference value.

Data from Laboratory 1 vary little. It could be said that they are *precise*. But they are far from the reference value: they are *erroneous*. They are also said to be marred by a *systematic error* or *bias*. The precision, on the contrary, is related to *random error* or characterizes the *repeatability* of a laboratory.

The results from Laboratory 2 are accurate, but are not repeatable. *Systematic error* is not revealed here but a large *random error*.

The data from Laboratory 3 are imprecise and at the same time marred by systematic errors. Contrarily the results from Laboratory 4 are accurate and repeatable.

Six of the results from Laboratory 5 are accurate as well as precise. However, determination No. 4 is very different from the others. We can suspect a *large error* in this determination leading to an *aberrant value*. Should this value that vitiates the accuracy as well as the repeatability of Laboratory 5 be retained?

Let us now suppose that analysis is required of an unknown sample for a kind of soil close to this reference sample from one or the other of these laboratories.

From Laboratory 4 an accurate result with a narrow confidence interval can be expected. From Laboratory 1 too we can expect a result with a narrow confidence interval but with a value higher than could be expected to be corrected if the systematic error is confirmed. The only way in which reliability can be expected from Laboratory 2 is by demanding a   larger number of repetitions and finding an estimate of the central value. The same holds true for Laboratory 3 with an additional correction for bias of this central value.

Let us now suppose that the same analysis is sought without knowing which laboratory would do it. To know the expected accuracy and repeatability, all the data in Table 18.1 have to be regrouped. The accuracy can be improved but the random error generally becomes greater. Instead of the word repeatability the term *reproducibility* is used to denote this error.

The above reasoning would also be valid if, instead of analysing the results from several laboratories, we concern ourselves with those from the same laboratory obtained by several operators or, better still, according to several analysis protocols, etc. It should be pointed out that the incidence of type of error differs considerably according to the type of determination and the associated interpretation.

Reverting to the example (Fig. 18.1), let us now suppose that the analyses have been done by two laboratories such as 1 and 4 (Fig. 18.2), having high repeatability but with a systematic difference between them. This systematic error will not much impede the interpretations concerning the observed variation in biological activity over geological time in relation to depth (Fig. 18.3). It will, in this case, be much less troublesome than a large random error, which would preclude interpretations (Fig. 18.1). Also, it is necessary to avoid a reflex action that often occurs in analysis, namely, taking the mean of results from laboratories for each depth. The true values will perhaps be approached but the random error is considerably augmented and the results will probably become impossible to interpret.

The effect of types of error would probably differ if, instead of carbon analysis, we had to interpret ^{14}C-dating data, for example. In this case, the effect of a systematic error might be much greater. It is more satisfying to provide an accurate age, for example, 2000 ± 100 y than a more precise but erroneous age of 1800 ± 50 y. In this latter type of example, there is a systematic error in consequence of the atmospheric nuclear tests in the 1960s that changed the composition of the atmosphere and the biosphere by noticeably raising the ^{14}C content, which we can correct for.

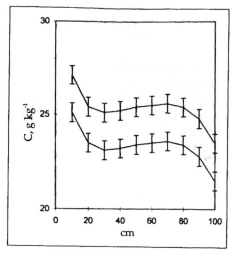

Fig. 18.3 Carbon determinations done on a sediment by two laboratories having high repeatability but with a systematic difference between them.

3. Random Error and its Distribution

3.1 Parameters of a Statistical Series

The data from Laboratory 2 (Table 18.1) are graphically represented in the order they were obtained (Fig. 18.4) and sorted in increasing order (Table 18.2). It is known that the true value (which is practically almost impossible) for the sample analysed is $\mu = 21.5$ g kg^{-1}. It is now a question of finding an estimator of this true value. The most commonly used estimator is the arithmetic mean:

$$\bar{x} = \frac{1}{n} \sum_{i=1}^{n} x_i \tag{1}$$

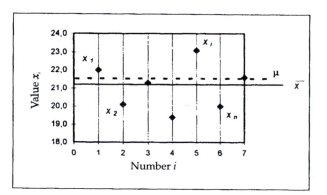

Fig. 18.4 Distribution of the results from Laboratory No. 2 (Tables 18.1 and 18.2).

Computation of the mean in our example gives $\bar{x}$ = 21.1 g kg^{-1}. This is not the true value; underestimation is seen. It should be noted that, apart from the expectation of the true value, the estimator used permits evaluation of the systematic error. How do we characterize the random error, that is, find an estimator of the dispersion of the data? A possible estimator is the range of the series, which is the difference between the highest value and the lowest. This estimator is equal to 3.7 g kg^{-1}. It is easy to compute but has the defect of being highly dependent on the extreme values of the series. For this reason, the dispersion of the results is characterized in relation to the preceding estimator, the mean. One way to proceed consists of computing the differences $x_i - \bar{x}$ (Table 18.2). However, as these differences are positive or negative, their sum and their mean are both equal to zero and we have no estimate of the variability. Another method sometimes used consists of taking the absolute values of these differences. But the most popular and most satisfactory estimation, so far as the properties of the estimator are concerned, is based on consideration of the squares of the differences and taking their mean. It is not exactly their mean that is selected but their sum divided by $(n - 1)$, which is an unbiased estimation of the *variance* of the population from which the statistical series is extracted:

$$s^2 = \frac{1}{n-1} \sum_{i=1}^{n} (x_i - \bar{x})^2 \qquad (2)$$

The division by $(n - 1)$ in formula 2 instead of by n is based on the fact that $\bar{x}$ has already been obtained by equation 1. The value $(n - 1)$ is called the number of degrees of freedom (df). Generally df = total number – number of relations between values used. Computation according to formula 2 can be very drawn out as it is necessary to first calculate $\bar{x}$ before calculating s^2 and it is often preferable to transform this formula. According to Newton's binomial series:

$$s^2 = \frac{1}{n-1} \sum_{i=1}^{n} (x_i^2 - 2x_i\bar{x} + \bar{x}^2) = \frac{1}{n-1} \left[\sum_i x_i^2 - 2\bar{x} \sum_i x_i + n\bar{x}^2 \right]$$

and, taking relation 1 into account:

$$s^2 = \frac{1}{n-1} \left(\sum_i x_i^2 - \frac{\left(\sum_i x_i \right)^2}{n} \right) \qquad (2')$$

Table 18.2 gives details of the computation of variance by formulae 2 and 2'. Actually the parameter most often used is the *standard deviation* estimated by s, the positive square root of the variance. The *coefficient of variation* or *relative standard deviation* RSD is also used. The use of RSD permits comparison of the precision of analytical results without reference to the unit of measure selected:

$$\text{RSD, \%} = 100 \, s/\bar{x} \qquad (3)$$

Table 18.2 Parameters of the distribution of the results of Laboratory 2 (Table 18.1)

Determination No., i	Value x_i g kg^{-1}	$x_i - \bar{x}$	$(x_i - \bar{x})^2$	x_i^2
4	19.4	−1.7	2.79	376.36
6	20.0	−1.1	1.15	400.00
2	20.1	−1.0	0.94	404.01
3	21.3	0.2	0.05	453.69
7	21.6	0.5	0.28	466.56
1	22.0	0.9	0.86	484.00
5	23.1	2.0	4.12	533.61
Number of values n	7			
Sum	147.5	−0.2	10.19	3118.23
Mean, g kg^{-1}	$\bar{x} = 21.1$	−0.03	1.46	
Median	$M = 21.3$	0.2		
Range	3.7	3.7		
Variance (eqn. 2)			1.7	
Variance (eqn. 2′)				1.7
Standard deviation g kg^{-1}			1.3	1.3
RSD, %			6.2	6.2

Another parameter is often recommended for the true value m. This is the *median, M*, the central value of the series of selected values (the observation than which half the observations in the series are lower). This parameter is sounder than the mean because it allows better elimination of the extreme values of the series. In the example chosen, $M = 21.3$ g kg^{-1} (Table 18.2) is actually closer than $\bar{x} = 21.1$ g kg^{-1} to the true value $\mu = 21.5$ g kg^{-1}. This is more glaring in the results from Laboratory 5 where one aberrant value is suspected: the median $M = 21.6$, the mean $\bar{x} = 22.0$ g kg^{-1}.

3.2 Study of a Distribution

Taking the example of Table 18.1, let us now assume that $5 \times 7 = 35$ analytical results were available without distinction, without knowledge of their source. We shall now study the distribution of these values. The extreme values observed are 18.3 and 24.8 g kg^{-1}. The interval between these two values is divided into ten classes and the number belonging to each class is counted. By choosing intervals centred around 18, 19, 20, etc. with respective class limits of 17.5, 18.5, 19.5, etc., the results presented in Table 18.3 are obtained, constituting an observed frequency distribution.

When the data in Table 18.3 are presented as a graph with the classes values on the *x*-axis and the corresponding population on the *y*-axis, a histogram is obtained. If, instead of numbers, the relative frequencies (number in class/total number) are considered, a similar histogram is obtained in which the cumulative frequency of all the classes is equal to 1 (Fig. 18.5).

Table 18.3 Study of the distribution of the 35 carbon determinations of Table 18.1 (g kg^{-1}) taken without distinction and knowledge of source.

Class	Lower limit	Upper limit	Midpoint	Total number	Frequency	Cumulative frequency
1		17.5		0	0	0
2	17.5	18.5	18	2	0.058	0.057
3	18.5	19.5	19	2	0.057	0.114
4	19.5	20.5	20	4	0.114	0.229
5	20.5	21.5	21	8	0.229	0.457
6	21.5	22.5	22	10	0.286	0.743
7	22.5	23.5	23	5	0.143	0.886
8	23.5	24.5	24	3	0.086	0.971
9	24.5	25.5	25	1	0.029	1
10	25.5			0	0	1

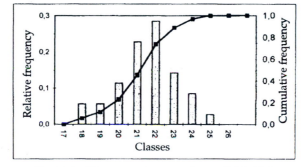

Fig. 18.5 Histogram corresponding to Table 18.3.

Replacement of numbers by the relative frequencies thus permits elimination of the scale factor and enables progress to the search for a unique law governing the distributions. The frequency polygon is obtained by joining the distribution frequencies *f(x)* corresponding to each class midpoint (*x*). By reducing the class interval, if the number of determinations with the same precision is augmented, the frequency polygon tends to a smooth curve such as $\int_{-\infty}^{+\infty} f(x)dx = 1$. The function *f(x)* represents the probability density function; *F(x)* is the cumulative distribution function defined by

$$F(x) = \int_{-\infty}^{x} f(x)dx \qquad (4)$$

To specify the probability density function $f(x)$, a mathematical model should be found that most closely approximates the observed values in the frequency histogram. In physicochemical analysis the theoretical distribution model most often (but not always) verified is the Gaussian distribution, also called normal distribution:

$$f(x) = \frac{1}{\sigma\sqrt{2\pi}} e^{\frac{-(x-\mu)^2}{2\sigma^2}} \qquad (5)$$

The symbols μ and σ respectively denote the mean and the standard deviation of the population (*see* § 3.1). Figure 18.6 presents the adjustment of the observed distribution to a normal distribution. This distribution passes through a maximum for $x = \mu$ and tends to zero when x tends to ∞. To eliminate the scale the change is often effected to the variable $u = (x - \mu)/\sigma$. The reduced centred normal distribution is then obtained:

$$f(u) = \frac{1}{\sqrt{2\pi}} e^{-\frac{u^2}{2}} \qquad (5')$$

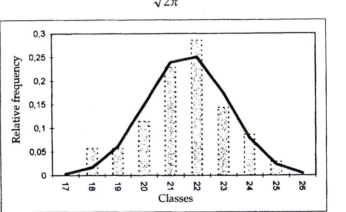

Fig. 18.6 Adjustment of the observed distribution (Table 18.3) to the normal distribution.

In a normal distribution, 68% of the values are located in the interval $\mu \pm \sigma$, 95% in the interval $\mu \pm 2\sigma$ (actually $\mu \pm 1.96\sigma$) and 99.7% in the interval $\mu \pm 3\sigma$). Expressed with eqn 4H: $[F(x)]_{\mu-\sigma}^{\mu+\sigma} = 0.68$, $[F(x)]_{\mu-2\sigma}^{\mu+2\sigma} = 0.95$, $[F(x)]_{\mu-3\sigma}^{\mu+3\sigma} = 0.997$. For other values of distribution the normal-distribution table in Appendix 4 can be referred to.

Most of the tests developed in this chapter have normality of distribution as a necessary condition. Although laboratory measurements often follow normal distribution, it is important to know how to test for normality of distribution of common determinations. For these tests to be applied, it is necessary to have at least 30 repetitions. They are not detailed here, but are listed in the appended bibliography, particularly in the publication of the CEA (1986). The best known are: use of the Pearson coefficient (symmetry, levelling), comparison of the numbers observed and predicted in the classes by the χ^2 test, and tracing the Henry line.

3.3 Confidence Interval

It was shown above that the reduced centred normal distribution (eqn. 5′) is deduced from the normal distribution by changing the variable $u = (x - \mu)/\sigma$. The reduced centred normal distribution allows definition of the limits $u1$ and $-u1$ such that a given probability or confidence level is obtained containing the expected value. In general, the probability level (eqn. 4) is fixed at 0.95. For this value the tables of normal distribution give $u1 = 1.96$ (Fig. 18.7).

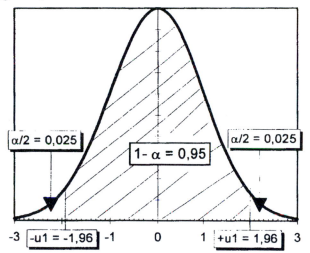

Fig. 18.7　Using the reduced normal distribution for determination of the confidence interval.

If several estimations of the mean of the population are done with several series, each of n repetitions, the mean $\bar{x}$ of the observed series follows a normal distribution of the mean μ and the variance σ^2/n: $\bar{x} \sim N(\mu, \sigma^2/n)$. Then $-1.96 < (\bar{x} - \mu)/\sqrt{(\sigma^2/n)} < 1.96$. The confidence interval of the mean will be:

$$\bar{x} - 1.96\frac{\sigma}{\sqrt{n}} < \mu < \bar{x} + 1.96\frac{\sigma}{\sqrt{n}} \tag{6}$$

Replacement of σ by its estimate s, if the total number n is not very large, necessitates a modification in the computation of the confidence interval of the mean. The actual formula becomes

$$\mu = \bar{x} \pm t_{1-\alpha/2}^{n-1} \frac{s}{\sqrt{n}}$$

(7)

in which t is the value of the Student variable given by the probability level $(1 - \alpha/2)$ table chosen, and the number of degrees of freedom used $(n - 1)$. The distribution of t tends to u of the normal distribution when the df rises (Table 18.4 and the Student table in Appendix 4).

Table 18.4   Some values of t for calculating the confidence interval at 95% and 99%. For the normal distribution the corresponding values are $u_{0.975} = 1.96$ and $u_{0.995} = 2.58$.

df	$t_{0.975}$	$t_{0.995}$
1	12.71	63.66
2	4.30	9.92
3	3.18	5.84
4	2.78	4.60
5	2.57	4.03
10	2.23	3.17
20	2.09	2.85
30	2.04	2.75
50	2.01	2.68
100	1.98	2.63
∞	1.96	2.58

If a new series of analyses is now done according to the same protocol, with n repetitions, the expected result will be:

$$\mu = \bar{x} \pm t_{1-\alpha/2}^{n-1} \frac{s}{\sqrt{n'}}$$

(8)

for which t is given, as before, by the Student table for the probability level $(1 - \alpha/2)$ chosen and the number of degrees of freedom used for the determination of the standard deviation $(n - 1)$.

Example 1. Give the confidence interval of the mean of the 35 results in Table 18.1, here considered as constituting a homogeneous population (source unknown, or one single laboratory).

Formula 1 gives an estimate of the mean: $\bar{x} = 21.62$ g kg^{-1}.

Formula 2 or 2' gives the estimate of the standard deviation: $s = 1.55$ g kg^{-1}.

For 95% probability level ($\alpha/2 = 0.025$), the Student table (Appendix 4) gives $t = 2.03$. Formula 7 gives the confidence interval of the mean

$$\mu = 21.62 \pm 1.55 \times 2.03/\sqrt{35} \text{ or}$$
$$\mu = 21.62 \pm 0.53 \text{ g kg}^{-1}.$$

It is encouraging that the true value (21.5 g kg^{-1}) is located within the confidence interval. If not, a systematic error should be suspected. With the normal distribution, it will be found that

$$\mu = 21.62 \pm 1.96 \times 1.55/\sqrt{35}$$
$$= 21.62 \pm 0.52 \text{ g kg}^{-1}$$

For a total number greater than 30, the estimates given for the normal distribution and by the Student distribution are practically the same.

Example 2. A carbon analysis is done according to the same protocol on a soil sample with texture close to that in Example 1, with 2 repetitions that give the values 14.5 and 13.9 g kg^{-1}. Give the content with its confidence interval at 95% for this sample.

Tables give $t^{34}_{0.975} = 2.03$. Formula 8 gives

$$\mu = (14.5 + 13.9)/2 \pm 1.55 \times 2.03/\sqrt{2} \text{ or}$$
$$\mu = 14.2 \pm 2.3 \text{ g kg}^{-1}$$

Example 3. The same carbon analysis is done in Laboratory No. 4 (Table 18.1), giving the same values as in Example 2: 14.5 and 13.9 g kg^{-1}. Give the new confidence interval.

Laboratory No. 4 has been identified to be the most precise and the most accurate. For the 7 repetitions in Table 18.1, it gives the mean value of 21.56 with standard deviation of 0.17 g kg^{-1}. For 6 degrees of freedom the Student tables (Appendix 4) give at the confidence level of 0.95, $t = 2.45$. Thus for the present sample,

$$\mu = 14.2 \pm 0.17 \times 2.45/\sqrt{2}$$
$$= 14.2 \pm 0.3 \text{ g kg}^{-1}$$

The value of the confidence interval is thus much higher if we confine ourselves to the analysis without identifying the laboratory that has done it (Example 2), or if we address ourselves to a laboratory identified as precise (Example 3).

4. Propagation of Errors

4.1 Propagation of Random Errors

If E and var denote the expected value and variance respectively and k is a constant, for two distributions of X and Y:

E $(X + Y) =$ E $(X) +$ E (Y) and E $(kX) = k$E (X),
var $(X + Y) =$ var $(X) +$ var $(Y) + 2$cov (X,Y) with cov $(X, Y) = 0$ if X and Y are independent,
var $(kX) = k^2$var (X)

4.1.1 Linear combinations

Let it be assumed that the final result y of an analysis was obtained by linear combination of intermediate results x_1, x_2, x_3, etc. If k_1, k_2, k_3, etc. are constants, we have

$$y = k + k_1x_1 + k_2x_2 + k_3x_3 + \ldots$$

If the standard deviation pertaining to the different results of y, x_1, x_2, etc. are denoted by σ_y, σx_1, σx_2, etc., we have

$$\sigma_y = \sqrt{(k_1\sigma_{x_1})^2 + (k_2\sigma_{x_2})^2 + (k_3\sigma_{x_3})^2 + \ldots} \tag{9}$$

Example 1. Independent determinations of four humic-carbon fractions are summed to determine the total carbon of a soil. If the standard deviation associated with the determination of each fraction was estimated at 0.5 g kg^{-1}, that of the sum will be

$$s = \sqrt{(4 \times 0.5^2)} = 1 \text{ g kg}^{-1}.$$

Example 2. If the carbon is assumed to be uniformly distributed in the first ten centimetres of the soil, express the above standard deviation in Mg ha^{-1}. For a bulk density of 1.3, the weight of soil for one hectare = $10,000 \times 0.1 \times 1.3 = 1300$ Mg = 1.3×10^6 kg. Thus 1.3 is the factor for converting g kg^{-1} to t ha^{-1}. The standard deviation will be given by

$$s = [(1.3 \times 1)^2]^{1/2} = 1.3 \text{ Mg ha}^{-1}.$$

The relative standard deviation is the same in both systems of units.

4.1.2 Multiplicative expressions

If the final result is now obtained from intermediate results by a formula of the type $y = kx_1x_2/x_3$, the relation of transmission of errors is expressed using the relative standard deviation:

$$\frac{\sigma_y}{y} = \sqrt{\left(\frac{\sigma x_1}{x_1}\right)^2 + \left(\frac{\sigma x_2}{x_2}\right)^2 + \left(\frac{\sigma x_3}{x_3}\right)^2} \tag{10}$$

Example 1. To find the relative standard deviation pertaining to the determination of C/N ratio in soils, given that the relative standard deviation of carbon (C) determination is 4% and that of nitrogen (N) determination is 3%, we have:

$$\text{RSD}_{C/N} = \sqrt{(4^2 + 3^2)} = 5\%.$$

Example 2. Let us see what the coefficient of variation for Example 1 becomes when the error in C and N is reduced by 1%.

On C: $\quad \text{RSD}_{C/N} = 4.3\%$
On N: $\quad \text{RSD}_{C/N} = 4.5\%$.

Thus improvement in the final precision is greater if the greater error is reduced.

4.1.3 Other functions

If the final result is obtained with the aid of any relation $y = f(x)$, transmission of the error is obtained by differentiation to get the expression:

$$\sigma_y = \left| \sigma_x \frac{dy}{dx} \right| \tag{11}$$

Example 1. The pH of a solution is obtained by the relation pH = $-\log a_H$, where a_H is the activity of the H^+ ions, which is assumed to be comparable to concentration. For a solution containing 10^{-3} mol (H^+) L^{-1}, what is the standard deviation for pH if the standard deviation for concentration is 10^{-4}? We have

$$\frac{dpH}{da_H} = \frac{-\log e}{a_H} = \frac{0.434}{0.001} = 434,$$

from which the standard deviation $\sigma_{pH} = 434 \times 10^{-4} \cong 0.05$ for a pH of 3.

Example 2. What is the effect of error in concentration in Example 1 for pH = 1 and pH = 5?

For pH = 1, $\sigma_{pH} = 0.0001 \times 0.434/0.1 \cong 0.0005$;

For pH = 5, $\sigma_{pH} = 10^{-4} \times 0.434/10^{-5} \cong 5$.

At acid pH the error in concentration has little effect on the pH. When neutrality is approached, error in concentration renders measurement of pH impossible.

4.2 Propagation of Systematic Errors

4.2.1 Linear combinations

For an expression of the same type as that calculated in § 4.1.1, the systematic error Δy can be calculated by:

$$\Delta y = k_1 \Delta_{x_1} + k_2 \Delta_{x_2} + k_3 \Delta_{x_3} + \dots \tag{12}$$

It should be remembered that, unlike random errors, systematic errors have signs. So they can be added or deducted.

Example. A soil sample of 20.5 g is weighed and oven-dried to give a dry weight of 18.2 g. Knowing that the balance used gives weights systematically higher by 0.5 g, what is the effect on determination of the water of the soil?

With $k_1 = 1$ and $k_2 = -1$, equation 12 gives:

$$\Delta y = \Delta x_1 - \Delta x_2 = 0.5 - 0.5 = 0.$$

No systematic error is added by the lack of accuracy of the balance.

4.2.2 Multiplicative expressions

The relative systematic error is obtained by addition of the relative systematic errors of the intermediate errors:

$$\frac{\Delta y}{y} = \frac{\Delta x_1}{x_1} + \frac{\Delta x_2}{x_2} + \dots \tag{13}$$

Example. The bulk density of a soil is determined by withdrawing a 100-mL core. The soil core weighs 130 g. Knowing that determination of the volume is subject to an error estimated to be 10 mL in excess, and that of weight to a shortage of 10 g, what is the systematic error in the result?

$$\Delta \rho / \rho = 10/130 - 10/100 = 0.177.$$

The volumic weight is thus 1.30 kg L^{-1} with a negative error of 1.3×0.177 = 0.23. The systematic error can therefore be corrected and:

$$\rho = 1.30 + 0.23 = 1.53 \text{ kg } L^{-1}.$$

5. Verification of Analytical Results

5.1 Tests of Accuracy

5.1.1 Comparison of determinations on a standard sample to its reference value

It is desired to test the hypothesis of zero bias in the estimation when μ is estimated by $\bar{x}$, or the hypothesis $H_0 : \bar{x} - \mu = 0$. In this case, assuming normal distribution, eqn. 7 shows that μ should be within the interval $\mu = \bar{x} \pm t^{n-1} s / \sqrt{n}$.

The Student variable is calculated thus:

$$t_{obs} = \frac{|\mu - \bar{x}| \sqrt{n}}{s} \tag{14}$$

The hypothesis of equality will be rejected if the value found for t exceeds the theoretical value for the chosen probability ($t_{1-\alpha/2}^{n-1}$, risk α). The standard deviation can be computed directly for the tested population or independently. In the latter case, the above test is valid only if the variances of the two populations are equal.

Example 1. The determination by atomic absorption (*see* Chapter 10) of lead in a reference soil containing 4.5 mg (Pb) kg^{-1} gave the following results (fictitious data): 3.5, 4 and 4.5 mg kg^{-1}. Can a systematic error be proved?

The mean of the results is 4.0 and the standard deviation is 0.5 mg kg^{-1} (eqns 1, 2 and 2'). Equation 14 gives:

$$t = [(4.5 - 4)\sqrt{3}]/0.5 = 1.73.$$

For df of 2, the critical value of t for a probability of 0.05 is 4.3 (Table 18.4 and the t table in Appendix 4). Thus we cannot reject the hypothesis of equality between the true value and the measured value. No systematic error is detected.

Example 2. What happens if the same mean analysis result is obtained in Example 1 but with a more precise protocol giving a standard deviation of 0.1 mg kg^{-1} over 3 repetitions? We have:

$$t = [(4.5 - 4)\sqrt{3}]/0.1 = 8.66.$$

This time the hypothesis of equality is rejected with risk of error smaller than 5%. A systematic error is thus found, which can be evaluated with $\mu - \bar{x} = 0.5$ and corrected for. It is noted that the decision depends as much on the difference (value obtained − actual value) as on the precision with which the data have been obtained. It should also be noted that with a risk of 1%, the critical value of t becomes 9.9 and the hypothesis of equality cannot be rejected.

5.1.2 Comparison of means of two sets of results

As in the preceding case, there are several tests. Only the t test in the following validity domain will be described here:

—the populations should have normal distribution or the sample size should be large (≥ 30),

—their variances should be equal and the determinations should be independent.

Denoting by $\bar{x}_1$ and $\bar{x}_2$ the means of the two populations with respective numbers of samples n_1 and n_2, the test used is written:

$$H_0 : \bar{x}_1 - \bar{x}_2 = 0$$

$$t_{obs} = \frac{\bar{x}_1 - \bar{x}_2}{s\sqrt{\dfrac{1}{n_1} + \dfrac{1}{n_2}}} \tag{15}$$

If $t_{obs} > t_{1-\alpha/2}^{n_1 + n_2 - 2}$: H_0 is rejected with risk α.

where t is expressed for df of $n_1 + n_2 - 2$ and s denotes the estimate of the standard deviation common to both populations:

$$s^2 = \frac{\sum_{i=1}^{n_1}(x_{1i}-\bar{x}_1)^2 + \sum_{j=1}^{n_2}(x_{2i}-\bar{x}_2)^2}{n_1+n_2-1} = \frac{\{(n_1-1)s_1^2 + (n_2-1)s_2^2\}}{n_1+n_2-1} \tag{15'}$$

Example. An analysis of nitrogen in a soil by the Kjeldahl method done on equivalent samples by two variants of the operational protocol gave the following results in g kg^{-1}:

| Series 1: | 1.19 | 1.16 | 1.18 | 1.18 | 1.20 | 1.15 | 1.27 |
| Series 2: | 1.28 | 1.26 | 1.30 | 1.21 | 1.21 | | |

Is there any systematic difference between the two protocols, assuming the conditions for valid application of the test in formula 15 are met?

Series	$\bar{x}$	s	s^2	n	$(n-1)s^2$
1	1.19	0.036	0.0013	7	0.079
2	1.25	0.041	0.0017	5	0.067
$\bar{x}_1-\bar{x}_2 = -0.6$				$s^2 = 0.015$	
				$s = 0.038$	

where $t = -0.06/[0.038\sqrt{(1/7+1/5)}] = -2.8$.

For $7 + 5 - 2 = 10$ df, the Student table (Appendix 4) gives for risk $\alpha = 0.05$: $t = 2.23$.

At a confidence level of 95%, the H_0 hypothesis of equality of the means is rejected. A bias between the two experimental protocols is detected.

5.1.3 Comparison of results from different samples

Each sample is analysed according to procedure 1 and procedure 2, and the difference between each of the n pairs of results is estimated by (procedure 1 – by procedure 2). If the differences follow a centred normal distribution and the mean of the differences is 0, let:

$$H_0 : \bar{x}_d = 0$$

$$t_{obs} = \frac{\bar{x}_d \sqrt{n}}{s_d} \tag{16}$$

If $|t_{obs}| > t_{1-\alpha/2}^{n-1}$: rejection of H_0 with α risk,

where n = number of pairs of results,

$\bar{x}_d$ and s_d = estimates of the mean and standard deviation of the differences.

The absolute value of t observed is compared with the theoretical value in the tables.

Example. It is desired to compare two techniques of determination of nitrogen in soils: the Kjeldahl method and the CHN combustion method (*see*

Chapter 16). The comparison is done on six samples of varying nitrogen content. Is there a systematic difference between the two methods?
The results are given in the following table:

Sample No.	Kjeldahl N	CHN N	Difference
1	0.40	0.39	0.01
2	4.39	4.37	0.02
3	3.31	3.60	−0.29
4	1.41	1.39	0.02
5	3.11	3.20	−0.09
6	1.29	1.36	−0.07
		Mean	−0.067
		s	0.119

Formula 16 gives $t = (-0.067\sqrt{6})/0.119 = -1.38$.
For 5 df, Table 18.4 and the table in Appendix 4 give $t_{0.05} = 2.57$. The null hypothesis cannot be rejected. The results by both methods seem to be equal.

5.1.4 Relative bias and absolute bias

The Student test on the pairs described in § 5.1.3 enables detection of the presence or otherwise of a total difference between the series of values obtained from the same samples. The results can also be reported graphically with the x-axis giving the values by the reference method and the y-axis those by the method tested. The intersection gives the experimental points that have to be close to the bisector if the two methods are equivalent (Fig. 18.8).

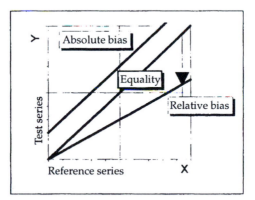

Fig. 18.8 Relative bias and absolute bias.

An independent difference in concentration indicates absolute bias, otherwise a relative bias. The regression line $y = a_0 + a_1x + \varepsilon$ is fitted (*see* § 6.5). For the two methods to be identical, we should have $a_0 = 0$ and $a_1 = 1$. Let s_{a0} and s_{a1} be the standard deviations in the determination of a_0 and a_1 respectively. If a bias is suspected,

$$-s_{a0} < a_0 < s_{a0} \Rightarrow \text{no absolute bias;}$$

$$-s_{a1} < (a_1 - 1) < s_{a1} \Rightarrow \text{no relative bias.}$$

If a bias is detected it should be tested for significance. Computation of the confidence intervals for a_0 and a_1 will be presented in § 6.5.

Example. Using again the example in § 5.1.3 and taking the Kjeldahl nitrogen values on the x-axis and the CHN nitrogen values on the y-axis, the curve fitting gives:

$$a_0 = 0.0095 \qquad\qquad s_{a0} = 0.0101$$
$$a_1 = 1.025 \qquad\qquad s_{a1} = 0.038.$$

Thus, $\quad -0.101 < 0.0095 < 0.101 \Rightarrow \text{no absolute bias,}$

$$-0.038 < (1.025 - 1) < 0.038 \Rightarrow \text{no relative bias.}$$

The results were predictable in this case given the results of the Student test for the pairs in § 5.1.3.

5.2 Tests of Repeatability

In statistical language, tests of comparison of variance are undertaken. Several such tests are available: Fisher—Snedecor test, Cochran test, Hartley test, test of ranges, etc. The tests are necessary in analytical chemistry for comparing the repeatability (and not the accuracy as in § 5.1) between laboratories, operators, series of determinations, etc. They also have to be done before the tests of accuracy, study of the causes of error, etc., when the validity conditions of the last-mentioned demand equal variances. Only the f test for comparison of variance is described here.

Let there be two populations of size n_1 and n_2 with estimated variances of S_1^2 and S_2^2. Putting the larger of these variances in the numerator, the following variable is used:

$$F_{(n_1 - 1, n_2 - 1)} = \frac{s_1^2}{s_2^2} \tag{17}$$

The null hypothesis tested is that the two populations have equal variances. If the hypothesis is true, the ratio of the variances will be close to 1. The observed variable F will be compared with that seen in the Fisher-Snedecor tables for $(n_1 - 1)$ and $(n_2 - 1)$ degrees of freedom according to the confidence level chosen. If one of the methods is a reference method, a unilateral test is used; this signifies that there is no possibility that the variance tested will be more precise than that of the reference method; the test seeks only to ascertain whether the variances studied can be of the same order as that of the reference method. In the opposite case, where no hypothesis can be applied to the variances, a bilateral test is used.

Example 1. Again taking the example of Table 18.1, it is desired to test the repeatability of laboratory 1 compared to laboratory 4.

For 7 determinations in each case, the following variances are obtained (formulae 2 and 2′):

    laboratory 1: 0.0714,
    laboratory 4: 0.0295,

whence $F = 714/295 = 2.42$.

At a risk level of 5% for a bilateral test with $(n_1 - 1) = (n_2 - 1) = 6$, the F table (Appendix 4) gives a critical value of 5.82 for F. Thus the hypothesis of equality of variances cannot be rejected. The repeatability of the two laboratories is the same.

Example 2. What happens to the test in Example 1 when laboratory 1 is considered a reference laboratory?

This time a unilateral test is considered for a risk level of 5%. The F table (Appendix 4) gives a critical value of 4.28 for F. As the value found is again smaller than the critical value, the equality hypothesis cannot be rejected. The repeatability of laboratory 1 is also as good as that of the reference laboratory.

5.3 Detection of Large Errors

The detection of systematic errors (§ 5.1) and random errors (§ 5.2) does not take into account the third type of error mentioned in § 2, which often has a great effect on the results of the determination; these are the large errors represented by data of unexpected values that appear aberrant in the distribution (outliers). These errors can be entirely due to the operator (wrong transcription of a result, for example) or to a failure in the measuring system. Know-how for detection and elimination of such errors is essential. Furthermore, they may not be errors at all and one should always be cautious in eliminating them. For example, if gold is being determined in soils of auriferous regions by taking small samples, the distribution of this element in the ground soil (*see* § 6.3 below) may be such that 30 determinations have small random error and a low mean of results. Then the 31st determination may give a value 50 times larger. Tests will show that this latter value is aberrant. Still, it is the elimination of this error that will constitute an aberration. Actually, gold often appears in the form of micronuggets in sediments and rocks. As long as the sample taken does not contain a micronugget, the results of determination are quite low; presence of one micronugget in a sample is going to suddenly indicate the presence of a large quantity of gold. The same reasoning is valid for other determinations too. In carbon determination, if the samples are not well prepared, the result can be greatly influenced (by a factor of 2, for example) by whether or not a plant fragment has been included.

In the search for aberrant values, whether the values deviate significantly from the distribution must be checked, assuming that the distribution is generally normal. If this hypothesis is not verified, results that appear to be aberrant may not actually be so.

One of the most commonly used tests is Dixon's, for which the CEA (1986) formulation is given here. Let $x_1, x_2, ..., x_{n-2}, x_{n-1}, x_n$ be the results selected in increasing order from the population X, numbering n values. According to whether the doubtful result is the lowest x_1 or the highest x_n, and whether n is greater than 10 or not, the discriminant function is expressed by r_1 or r_2, as follows:

	Doubtful x_i	Doubtful x_n	
$n \le 10$	$r_1 = (x_2 - x_1)/(x_n - x_1)$	$r_1 = (x_n - x_{n-1})/x_n - x_1)$	
$n > 10$	$r_2 = (x_3 - x_1)/(x_{n-2} - x_1)$	$r_2 = (x_n - x_{n-2})/(x_n - x_3)$	(18)

Should the value of r_1 or r_2 exceed the critical value given by the Dixon tables, the suspect value can be considered aberrant, and rejected.

Example. Can the suspect value for carbon of 24.8 g kg^{-1} given by laboratory 5 (Table 18.1) be considered aberrant? We have $n \le 10$ and x_n doubtful. The corresponding eqn. 18 gives:

$$Q = (24.8 - 21.8)/(24.8 - 21.3) = 0.86.$$

For a sample size of 7 the Dixon table (Appendix 4) for a probability of 0.05 gives the critical value for r_1 as 0.507. The suspect value can then be considered aberrant with risk of error smaller than 5%.

5.4 Quality-Control Diagrams

These diagrams are important for controlling the quality of determinations over time. The laboratory will have a reference sample on hand that can be periodically analysed with each large set of analyses. It is advisable to test in advance for normality of the distribution of the measurements on this reference sample, and to know precisely the expected value μ and the standard deviation σ associated with this value. If n is the number of repetitions of the reference sample, the formula for the confidence interval for a normal distribution (eqn. 7) indicates that approximately 95% of the means of the reference sample should be located within the interval $\mu \pm 2\sigma/\sqrt{n}$, and approximately 99.7% within the interval $\mu \pm 3\sigma/\sqrt{n}$.

The aim of the quality-control diagrams is to show graphically the possible evolution of the values of the reference sample as a function of time in such a way that the need for a correction can be detected as quickly as possible. Figure 18.9 represents one type of quality-control diagram, the Shewhart diagram.

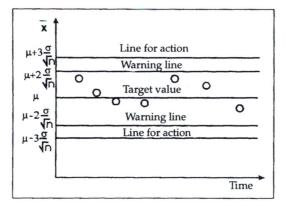

Fig. 18.9 The Shewhart diagram. The line $\bar{x} = \mu$ is the target value; the lines $\bar{x} = \mu \pm 2\sigma/\sqrt{n}$ represent the upper and lower warning lines, and the lines $\bar{x} = \mu \pm 3\sigma/\sqrt{n}$ are the lines where action has to be taken.

If the values of the reference sample go beyond one of the lines for action, there are only three chances in a thousand that the process is still under control. It is then necessary to stop the analyses and check from where the error originated. If a value is between a line for action and the corresponding warning line, there is a 5% chance that the process is still under control. The analyses can be continued with a new reference sample; if the value still crosses a warning line, it is again necessary to stop the series, even if the line for action has not been crossed.

These diagrams are useful for monitoring not only change with time of the accuracy of the analyses, but also their precision. In this case, instead of following the changes in mean value of the reference sample, the change in standard deviation is monitored.

There are more efficient methods than the Shewhart diagram for detecting changes in a process very quickly, before the warning line is reached. The reader is referred to publications on techniques of quality control in the bibliography at the end of the chapter.

6. Study Of Causes Of Errors

6.1 General Case

Chemical operations are complex. In each step, from the original sampling to the final result, a preponderant error can be imposed on the quality of this result. Investigation of the causes of error consists of breaking down the general analysis protocol into different operations with a number of repetitions for each operation. Thus experimental designs should be drawn up.

It is then necessary to proceed to the acquisition of results corresponding to the design and thereafter to analysis of these results to ascertain which component operation has the greatest effect on the quality of the analysis. The technique generally applied is variance analysis. The techniques of experimental designs and variance analysis in analytical physical chemistry are not restricted to the study of the causes of error, although they are unavoidable tools in this case.

Only the case with a single monitored factor is described below. Does this factor represent a preponderant source of error or not for the final result? Let $i = 1, 2, ..., p$ be the modalities associated with repetitions of the monitored factor; let $j = 1, 2, ..., q$ be the modalities of other factors (unmonitored). For simplification, it is assumed here that q is constant for each level of i, but this condition is not indispensable. For physicochemical analyses generally $q < p$, the monitored factor being assumed to give the preponderant error. After judicious choice of p and q, the work plan is drawn up as in Table 18.5.

Table 18.5 Experimental design for study of a cause of error.

Order number	i	j	Value of x
1	1	1	1_{1l}
2	1	j	x_{1j}
—	1	q	x_{1q}
—	i	1	x_{i1}
—	i	j	x_{ij}
—	i	q	x_{iq}
—	p	1	x_{p1}
—	p	j	x_{pj}
pq	p	q	x_{pq}

The total number of determinations to be done is pq. In order to mitigate the effect of variation with time of certain unmonitored factors, it is recommended that the determinations be done in random sequence of their order number (Table 18.5). After collection of the data corresponding to the work plan, it is necessary to proceed to their treatment by an analysis of variance, corresponding to the graphic representation in Figure 18.10.

The variance (eqn. 2) is defined by the sum of squares of the difference between each value and the mean of the values. Variance analysis is defined in the same way. Total variance is expressed by the total sum of squares (TSS) of each determination compared to the general mean:

$$TSS = -\sum_{i=1}^{p}\sum_{j=1}^{q}(x_{ij} - \bar{x})^2 \qquad (19)$$

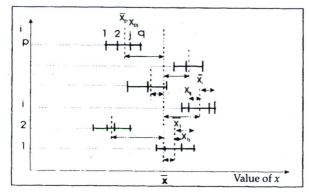

Fig. 18.10 Graphic representation of the data for investigation of a cause of error by variance analysis.

Let $\bar{x}_i$ be the means of all modalities of the monitored factor i. The difference TSS can now be broken down into two terms. The first term (factor sum of squares, FSS) expresses the variability of the monitored factor compared to the general mean:

$$FSS = \sum_{i=1}^{p} q(\bar{x}_i - \bar{x})^2 \tag{20}$$

The second term (residual sum of squares, RSS) expresses the remaining variability or residual variability, by the sum of squares of the differences between each value and the mean at each level of the monitored factor:

$$RSS = \sum_{i=1}^{p} \sum_{j=1}^{q} (x_{ij} - \bar{x}_i)^2 \tag{21}$$

The RSS can also be found out by the relation RSS = TSS − FSS. By analogy with the two expressions for the variance (eqns. 2 and 2′), eqns. 19 to 21 can also be expressed differently to simplify computation; the other expressions, which do not facilitate understanding of the reasoning, are not given here. As for the variance (eqn. 2) it now remains to divide these sums of squares by the number of degrees of freedom (df) associated with each determination. The TSS uses qp values with one relation among themselves, the general mean; thus $df_t = qp - 1$. The FSS uses p values with one relation, whence $df_f = p - 1$. The residual term uses qp values with p relations among themselves, so that $df_r = qp - p$. We thus have, as for the sums of squares, $df_t = df_f + df_r$. Division of each sum of squares by the number of degrees of freedom gives FSSm, RSSm and TSSm (Table 18.6), called mean squares.

Table 18.6 Variance analysis for study of a cause of error.

	Sum of squares	df	Mean squares	F
Model	FSS (eqn. 20)	$p - 1$	$FSSm = FSS/(p - 1)$	$F = \dfrac{FSSm}{RSSm}$
Residue	RSS (eqn. 21)	$qp - p$	$RSSm = RSS/(qp - p)$	
Total	TSS (eqn. 19)	$qp - 1$	Variance	

The hypothesis to be tested is 'the means $\overline{x}_i$ are equal', which is equivalent to 'the monitored factor has no preponderant influence compared to other unmonitored factors on the accuracy of the analytical result'. The greater the influence of the monitored factor, the larger the ratio $F = FSSm/RSSm$. The Fisher-Snedecor F table (Appendix 4) gives the critical value of F for the confidence level chosen. If the F obtained (Table 18.6) is greater than this critical value, the equality hypothesis has to be rejected.

The result of the test allows directing the operator to priority efforts at improving quality. The test also leads to the computation to be done for a better estimation of the general mean and its confidence interval.

(1) If the effect of the factor has not been proved

The general mean is estimated as in equation 1 by:

$$\overline{x} = \frac{\sum_i \sum_j x_{ij}}{qp} = \frac{\sum_i q_i \overline{x}_i}{qp} \tag{22}$$

Then the confidence limits at probability level $(1 - \alpha)$ are:

$$\overline{x} \pm \sqrt{\frac{TSSm}{qp}} t_{1 - \alpha/2} \tag{23}$$

where t is defined for df $qp-1$.

(2) If the effect of the factor has been proved (case of the same number of repetitions, q_i)

The true value is again estimated by the general mean (eqn. 22). The confidence limits at the probability level $1 - \alpha$ are defined by

$$\overline{x} \pm \sqrt{\frac{FSSm}{qp}} t_{1 - \alpha/2} \tag{24}$$

with t defined for df $p-1$.

If the effect of the factor has been proved, in the case of different q_i the precise computation of the best estimates of the mean and its confidence interval is more complex. The weighted coefficients inversely proportional to σ_{xi}^2 (*see* bibliography) should be brought in.

Validity of the tests

The populations X_i should have normal distribution with equal variances (*see* § 5.2) and the determinations should be independent.

Example. Determination of polysaccharides in soils necessitates the following operations:

—acid hydrolysis to release the sugars,

—determination of released sugars by colorimetry.

We wish to test whether, under fixed conditions, hydrolysis introduces a preponderant error compared to other unmonitored factors, essentially colorimetric measurement. It is assumed that the determination is sufficiently well known to admit to normality of distributions and uniformity of their variances.

Six (6) hydrolyses ($p = 6$) are done and two colorimetric determinations on each hydrolysate ($q = 2$). The work plan is given in Table 18.7. The results of determination obtained after randomization of the analyses followed by rearrangement in the work plan are also indicated (spectrometer readings in arbitrary units).

Table 18.7 Example of the work plan for determination of soil sugars and results of corresponding determinations in arbitrary units.

	Work plan		Measured value
Analysis No.	i	j	
1	1	1	120
2	1	2	112
3	2	1	118
4	2	2	109
5	3	1	109
6	3	2	105
7	4	1	121
8	4	2	97
9	5	1	118
10	5	2	101
11	6	1	107
12	6	2	122

Eqns. 19, 20 and 21 give

TSS = 752.9

FSS = 127.4

RSS = 625.5

The relation $TSS = FSS + RSS$ is verified. We have $p = 6$ and $q = 2$, whence $df_t = qp - 1 = 11$; $df_f = 5$, $df_r = 6$. The relation $df_t = df_f + df_r$ is also checked. Lastly, the table of variance analysis (Table 18.6) gives

Cause of error	SS	df	SSm	F
Hydrolysis	127.4	5	25.5	0.24
Other	625.5	6	104.3	
Total	752.9	11	68.5	

At the probability level of 0.95, the Fisher-Snedecor table (Appendix 4) gives the critical value

$$F_{(5,6)} = 4.39.$$

The hypothesis of equality of means cannot be rejected. Hydrolysis does not induce a preponderant error compared to other unmonitored factors, essentially the spectrometric determination. As the residual error is larger than that originating from hydrolysis, it is necessary to ameliorate the spectrometric determination to improve the quality of this analysis.

Eqns. 22 and 23 give the estimates of the mean and their confidence interval. At the confidence level of 95%, the Student table (Appendix 4) gives $t_{0.975}^{11} = 2.20$, whence

$$\bar{x} = 111.6 \pm 2.20 \times \sqrt{(68.5/12)} = 111.6 \pm 5.3.$$

6.2 Sampling Error

Sampling comprises two broad phases that can themselves be broken down into elementary operations: taking of the sample from the field (*see* Chapter 1) and sample preparation in the laboratory (*see* Chapter 3). Most of the time the greatest variations in relation to phenomena are seen to originate primarily from the quality of field sampling, then from the sample preparation and lastly from the different analytical operations. If s_s^2 and s_a^2 respectively denote the variance from sampling and that from analysis, the variance of the result can be estimated by $s_s^2 + s_a^2$. If the analytical method is known, knowledge of s_a^2 allows calculation of the variance of sampling from the total variance. For acceptable precision the following relation can be adopted:

$$s_s \leq 3s_a \qquad (25)$$

Actually, if relation 25 holds, we have $s_a = s_s/3$, whence $s^2_{total} = s_s^2/9 + s_s^2$, or $s_{total} = 1.05s_s$. The standard deviation of the analytical method comes to less than 5% of the total standard deviation. Thus it is of no use to improve the analytical method before sampling has been perfected.

6.3 Error in Field Sampling

The quality of field samples is closely related to the more or less learned or intuitive knowledge of the soil cover to define, for example, homogeneous two-dimensional or three-dimensional zones (*see* Chapter 1). This process can be aided by conducting various field tests (*see* Chapter 2). The array of conventional statistics, comprising the methods described above and other techniques, will also be of great use in interpretation of the preliminary tests or determinations.

However, in the case where the measured variables are spatially dependent, classic statistics are no longer a sufficient tool. Without taking into consideration spatial organization, they may tend to overestimate the error. The content of a sample can depend on the geographic location it occupies in relation to its surroundings. The method for defining spatial dependence is taken from the theory of regionalized variables developed by Matheron in the early 1960s. A regionalized variable is actually a function $Z(x_i)$ defined at all points x_i with coordinates (x_u, x_v, x_w) in three-dimensional space.

It can be very irregular and have two characteristics: (1) a random local characteristic and (2) a structured spatial characteristic.

In the vast domain of geostatistics, only the principle of drawing a semi-variogram is touched upon below because it is a useful tool for improving field sampling and directly concerns the quality of analyses.250

Let a sampling be done according to a grid or a fixed protocol in the field (rectangular spatial zigzagging, triangular spatial zigzagging, etc.) in such a manner that m pairs of points are obtained with distance h. The intrinsic hypothesis of second-order stability of the increments shows (1) the expected increments do not depend on x, $E[Z(x + h) - Z(x)] = 0$, regardless of x and h, and (2) the increments have finite variance related to h and independent of x, $\text{var}[Z(x) - Z(x + h)] = 2\gamma(h)$. The function $2\gamma(h)$ is termed the semi-variogram. It is estimated by:

$$s^2(h) = \frac{1}{2m_j} \sum_{i=1}^{m_j} [z(x_i) - z(x_i + h)]^2 \tag{26}$$

where m_j is the number of pairs of points associated with different values of h. The estimates by formula 26 can be graphically represented as in Fig. 18.11. When the semi-variance at $h = 0$ is not negligible, it represents the local variability of the property and thus of the sampling. We sometimes say there is a nugget effect. The distance at which the samples are no longer correlated among themselves and become independent is termed the range.

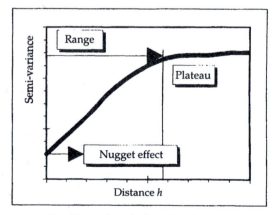

Fig. 18.11   A typical semi-variogram.

The semi-variograms computed by formula 26 can be fitted to different shapes (spherical, exponential, infinite) not having a plateau at all (*see* Bibliography).

When anisotropy exists, the semi-variogram is no longer the same in all directions. The semi-variograms will be valid only in one direction of the transect.

Example. Determinations (fictitious, size too low) of bulk density were done at regular intervals of 10 m in a 100-m transect. The following results were obtained:

$$1.4, 1.3, 1.4, 1.4, 1.4, 1.5, 1.6, 1.5, 1.6, 1.5, 1.5.$$

Compute the semi-variogram.

Formula 26 gives for Class 1 (10-m spacing of samples):

$$[(1.4-1.3)^2 + (1.3-1.4)^2 + (1.4-1.4)^2 + ... + (1.5-1.5)^2]/2 * 10 = 0.0035$$

and for Class 2 (20-m spacing):

$$[(1.4-1.4)^2 + (1.3-1.4)^2 + (1.4-1.4)^2 + ... + (1.6-1.5)^2]/2 * 9 = 0.0039.$$

The results are presented in Table 18.8. Without modelling the semi-variogram it is seen that the range or the distance at which the bulk densities are no longer correlated among themselves is between 60 and 70 mm.

Table 18.8 Calculation method of the semi-variogram of bulk density.

Class	Number of pairs of observations	h(m)	$s^2(h)$ (eqn. 26)
1	10	10	0.0035
2	9	20	0.0039
3	8	30	0.0056
4	7	40	0.0100
5	6	50	0.0133
6	5	60	0.0140
7	4	70	0.0150
8	3	80	0.0150

6.4 Error in Laboratory Sampling

After it is drawn from the field, the sample is received in the laboratory and subjected to the operations of drying, grinding, sieving and partitioning prior to the actual analyses. These various operations represent sources of error in accuracy because of the physical and chemical changes suffered by the sample. These errors can, in certain cases, gravely disturb the analytical results (*see* Chapter 3).

There is another random error inherent in sampling of solid materials, which can be modelled. Solids are always heterogeneous; it can be understood that the more finely a solid is ground, the less the heterogeneity in bulk (heterogeneity between particles increases, but so does the number of particles); this is the reason why samples are always ground before subsampling for analysis. Figure 18.12 represents schematically a sample taken from ground solid material. Most of the elements or compounds are not uniformly distributed in the grains of the powder. In some cases, certain particles (black in Fig. 18.12) will contain the element to be determined, others might not. In other cases, the concentration will simply be higher in the black grains than in the white grains. It can then be imagined that the sampling error will be smaller the larger the number of grains taken from the total. For a given ground particle size the sampling error will be smaller the greater the weight of the sample taken (assuming as well that the sampling is random, *see* Chapter 3).

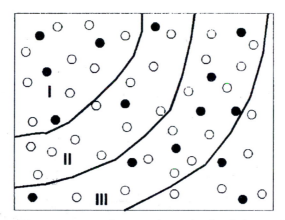

Fig. 18.12 Schematic representation of a ground sample of soil: (•—particle containing the element or compound to be determined; o – ancther particle of the matrix; I, II, III—subsamples of increasing weight.

It can be understood that the variance associated with 10 g samples s_{10}^2 will be ten times smaller than that associated with 1 g samples (s_1^2); that of 100 g samples (s_{100}^2) will be 10 times smaller than that for 10 g samples or 100 times smaller than for 1 g samples. In other words, if s_M^2 is the variance due to sampling for a sample weight M and K a constant for the material studied, the following can be written:

$s_1^2 = 10s_{10}^2 = 100s_{100}^2 = Ms_M^2 = K$. By multiplying the last two terms by $10^4/(\bar{x})^2$, we end with

$$M(RSD)^2 = k_s \qquad (27)$$

where RSD represents the relative standard deviation in per cent (formula 3) and k_s the sampling constant for the given material such that $k_s = K\,10^4/(\bar{x})^2$.

Formula 27 shows that k_s can be defined for each determination as the mass of solid material to be taken for obtaining a relative standard deviation RSD = 1%. The model (formula 27) is the first approximation for sampling of solid materials. It presumes that a single heterogeneity level will be obtained if the grinding, sieving, subsampling, etc. are uniform. It may be thought that an extremely fine grinding would minimize k_s. Actually, electrostatic forces in very fine powders sometimes lead to secondary heterogeneities. At any rate, it is easy to verify whether the model in formula 27 is followed by a given material by doing several determinations of RSD for different subsample weights M.

Example 1. The data in examples 1 and 2 have been taken from determinations of carbon in a reference sample of soil at the IRD laboratory[1].

Ten carbon analyses were done with the Leco 600 CHN Analyser (*see* Chapter 16) on ten 250 mg subsamples from the same lot ground to 200 μm. Found $\bar{x}$ = 36.8 g kg^{-1}, s = 0.4 and RSD = 1.09. To calculate k_s.

Formula 27 gives k_s = 250 × 1.09^2 = 297 mg.

For a precision of 1% in the first approximation, a 300 mg subsample of the test material should be taken.

Example 2. For testing the model on this material, the experiment of example 1 is done with a series of 10 repetitions for different subsample weights. The data are assembled in Table 18.9. Model the sampling error.

The aim is to express RSD = $f(M)$. Equation 27 becomes

$$\log (RSD) = 0.5 \log k_s - 0.5 \times \log M.$$

Table 18.9 Error in precision as a function of the sample weight for carbon determination in a reference sample.

Sample, mg	$\bar{x}$, g kg^{-1}	s, g kg^{-1}	RSD, %
20	23.5	9.3	39.57
45	34.8	6.4	18.39
100	35.3	1.5	4.25
150	34.2	0.9	2.63
200	36.7	0.5	1.36
250	36.8	0.4	1.09

It is then proposed to study the fit of log RSD = $f(\log M)$, which should be a straight line of slope −0.5 if eqn. 27 holds. The fit obtained (Fig. 18.13) is very close to a straight line (coefficient of determination R^2 = 0.99). However, the model is not exactly that in eqn. 27 because the slope is −1.47 instead of −0.5. The equation obtained enables definition of the value of k_s as 274 mg. The subsample weight for 1% error can then be slightly smaller than 300 mg (value obtained in example 1).

---

[1] Institut de recherche pour le développement, BP5045, F34032, Montpellier Cedex 1, France.

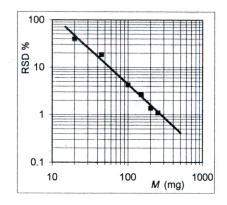

Fig. 18.13 Modelling the error in precision as a function of subsample weight: determination of carbon in a reference sample of soil.

6.5 Calibration and Associated Error

6.5.1 The calibration curve

Calibration is an unavoidable operation in quantitative analytical physical chemistry. There are several methods of calibration. The most common technique consists of doing the determinations on samples of known content (or other measured property). The measured signal is then graphically reported as a function of the content (or other measured property) for each standard sample. This enables tracing of a calibration curve with content (or property) on the x-axis and the measured signal on the y-axis. The curve is then used for finding the content corresponding to the signal measured for the unknown sample. Most of the time the analyst seeks to work under conditions where the observed curve is a straight line, so that the error in the content or measured property would be least dependent on its magnitude. For example, a flattening of the curves at high contents augments the error in the corresponding determinations. We describe here the classic straight-line model obtained by the method of linear regression of least squares.

Let **X** be the matrix containing the known value of each concentration (or property) for the n standards of the calibration range. Let **Y** be the vector containing the corresponding measurement for each standard. Let $\hat{\mathbf{Y}}$ be the vector containing the estimates by the model of these measurements. In the case where the measurements are independent of **X**, calibration is not possible and only an estimator of **Y** of the following type can be obtained

$$\mathbf{Y} = \mu + \varepsilon \tag{28}$$

An estimation of μ is done such that $E(\varepsilon) = 0$. The most popular is the mean $\bar{y}$. In the case where the measurements are dependent on **X**, we expect to find an estimator of **Y** of the following type:

$$\mathbf{Y} = \mathbf{X}\alpha + \varepsilon \tag{29}$$

where α represents the matrix containing the parameters of the estimated model. The estimator $\mathbf{A}$ of α should be found for fitting of the values of $\mathbf{Y}$ such that $E(\varepsilon) = 0$. The fitting is accomplished by minimizing the sum of squares of the residues (method of least squares) represented by $\varepsilon^t\varepsilon$ where ε^t is the transposition of the vector ε. If $\mathbf{X}^t$ is the transposed matrix of $\mathbf{X}$, it can be shown that the solution is given by

$$\mathbf{X}^t\mathbf{X}^{-1}\,\mathbf{X}^t\mathbf{Y} = \mathbf{A} \tag{30}$$

where $\mathbf{X}^t\mathbf{X}^{-1}$ is the variance-covariance matrix. The terms in the principal diagonal permit estimation of the variance associated with each parameter. The residues of the estimate are represented by the vector

$$\varepsilon = \mathbf{Y} - \hat{\mathbf{Y}} = \mathbf{Y} - \mathbf{X}\mathbf{A} \tag{31}$$

The fit obtained is depicted in Fig. 18.14.

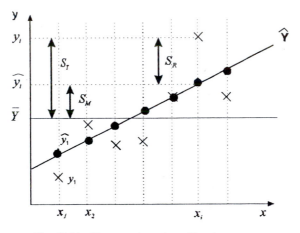

Fig. 18.14   Construction of a calibration curve.

The sum of squares to be minimized is $\Sigma(y_i - \hat{y}_i)^2 = S_R$. The total variance is computed with the sum of squares $S_T = \Sigma(y_i - \bar{y})^2$. The variance due to the model is computed with $S_M = \Sigma(\hat{y}_i - \bar{y})^2$ (Fig. 18.14).

As the computations always give a fit, it is indispensable that its validity be tested: does the model chosen (eqn. 29) explain $\mathbf{Y}$ better than when $\mathbf{Y}$ is considered independent of $\mathbf{X}$ as in eqn. 28? The coefficient of determination is defined by

$$R^2 = \frac{S_M}{S_T} = \frac{\Sigma(\hat{y}_i - \bar{y})^2}{\Sigma(y_i - \bar{y})^2} \tag{32}$$

It is understood that when all the experimental values fall on the line of predicted values, that is, for a perfect fit, $R^2 = 1$. Consequently, the closer R^2

is to 1, the more significant the test of adequacy of the model with the experimental values. The R^2 value (eqn. 32) is often given after multiplication by 100. It then expresses the percentage of variation in **Y** explained by the model.

For p estimated parameters (without the constant) of the n values in **Y**, the table of variance analysis of the fitting (*see* Table 18.10) is also drawn up.

Table 18.10   Variance analysis of the fitting.

Origin of variation	Sum of squares	df	Mean squares	F test
Model	$S_M = \sum (\hat{y}_i - \bar{y})^2$	p	$V_M = S_M/p$	
Residue	$S_R = \sum (y_i - \hat{y}_i)^2$	$n - p - 1$	$V_R = S_R/(n - p - 1)$	$\dfrac{V_M}{V_R}$
Total	$S_T = \sum (y_i - \bar{y})^2$	$n - 1$		

The higher the F, the greater the influence of the model compared to the residue of fit (Fig. 18.14). For a chosen risk α, the Fisher-Snedecor tables (Appendix 4) give the critical value of F $(p, n-p-1)$.

Generally the models include estimation of two parameters a_0 and a_1 of the equation $y = a_0 + a_1 x$. However, in calibration a value of $a_0 > 0$ indicates that the measured signal cannot be considered zero even when the content (or property to be measured) is zero. A value of $a_0 < 0$ indicates that the signal may be zero even when the content is not zero. Thus there will be risk of systematic error at low contents, positive in the former case and negative in the latter. Estimations of a_0 and a_1 are accompanied by estimations of their standard deviation, and their confidence interval can be computed. In the case where a_0 can be considered zero, recalculation of the model should be recomputed according to $y = ax$.

Example. Returning to the example in Chapter 11, § 2.4.2, on calibration for potassium by flame emission (range of 4 standard solutions with 1, 2, 3 and 4 mg L^{-1} potassium, and corresponding emission intensities of 0.27, 0.42, 0.55 and 0.67).

To fit the two parameters the matrix **X** is written as

$$\begin{bmatrix} 1 & 1 \\ 1 & 2 \\ 1 & 3 \\ 1 & 4 \end{bmatrix}$$

The result vector **Y** is written

$$\begin{bmatrix} 0.27 \\ 0.42 \\ 0.55 \\ 0.67 \end{bmatrix}$$

The matrix X^t is

$$\begin{bmatrix} 1 & 1 & 1 & 1 \\ 1 & 2 & 3 & 4 \end{bmatrix}$$

The matrix X^tX is

$$\begin{bmatrix} 4 & 10 \\ 10 & 30 \end{bmatrix}$$

The matrix X^tY is

$$\begin{bmatrix} 1.91 \\ 5.44 \end{bmatrix}$$

The matrix X^tX^{-1} is

$$\begin{bmatrix} 1.5 & -0.5 \\ -0.5 & 0.2 \end{bmatrix}$$

The matrix A of the parameters is given by eqn. 30:

$$\begin{bmatrix} 0.145 \\ 0.133 \end{bmatrix}$$

The predicted values are given by $\hat{Y} = XA$:

$$\begin{bmatrix} 0.278 \\ 0.411 \\ 0.544 \\ 0.677 \end{bmatrix}$$

The residues are given by ε (eqn. 31):

$$\begin{bmatrix} -0.008 \\ 0.009 \\ 0.006 \\ -0.007 \end{bmatrix}$$

It is seen that the mean of the residues is zero.

The mean is $\bar{y} = 0.4775$. The sums of squares associated with the variance of the model, the residual variance and the total variance are:

$$S_M = \Sigma(\hat{y}_i - \bar{y})^2 = 0.08845$$

$$S_R = \Sigma(y_i - \hat{y}_i)^2 = 0.00023$$

$$S_T = \Sigma(y_i - \bar{y})^2 = 0.08867.$$

It is verified that $S_T = S_M + S_R$.

The coefficient of determination is given by equation 32:

$R^2 = 0.997 = 99.7\%$.

Table 18.10 becomes:

Origin of variation	Sum of squares	df	Mean squares	F test
Model	0.08845	1	0.08845	769
Residue	0.00023	2	0.000115	
Total	0.08867	3		

At a confidence level of 99%, the Fisher-Snedecor table (Appendix 4) gives the critical value $F(1.2) = 98.5$. As the value found is higher than this limit, the calibration model can be accepted.

The residual error and the matrix of X^tX^{-1} enable computation of the standard deviations S_A associated with each coefficient, that is

$$\left| \begin{matrix} \sqrt{0.00015 \times 1.5} \\ \sqrt{0.00015 \times 0.2} \end{matrix} \right| = \left| \begin{matrix} 0.01313 \\ 0.004796 \end{matrix} \right|$$

These values permit computation of a confidence interval (CI) for each estimated parameter of the model $y = a_0 + a_1x$ by the relation

$$CI_A = s_A t_{n-2}^{0.975}.$$

At the confidence level of 95% for $n-2 = 2$ df, the Student table in Appendix 4 gives $t = 4.3$. Lastly, the estimation gives the parameters

$$a_0 = 0.145 \pm 0.056$$
$$a_1 = 0.133 \pm 0.021.$$

The null hypothesis for the constant a_0 should be rejected for this calibration. The reading is not zero for a zero K content and determinations at very low contents risk being overestimated. This disadvantage (*see* Chapter 11) can be remedied by choosing a more complex model, of the polynomial type in this case.

6.5.2 Error associated with the calibration curve

After it has been drawn, the calibration curve is used to find out the concentration of unknown samples by marking the measured reading on the y-axis for reading the result on the x-axis. In the case where the equation has been computed as in § 6.4.1, it can possibly be transformed to a graphic representation. If y_s is the reading corresponding to the unknown sample in the calibration eqn. $y = a_0 + a_1x$, the corresponding concentration is

$$x_s = (y_s - a_0)/a_1 \tag{33}$$

Equation 33 is not sufficient, however, and does not give an understanding of the error that depends on many factors: analysis of the unknown sample, transmission of the error during computation (eqn. 33) and error committed in the calibration. The concentration of the unknown sample and its confidence interval are computed by

$$x_s = \frac{y_s - a_0}{a_1} \pm \frac{t_{0.975}^{n-2} \sqrt{V_R}}{a_1} \sqrt{1 + \frac{1}{n} + \frac{\left(\dfrac{y_s - \bar{y}}{a_1}\right)^2}{\sum(x_i - \bar{x})^2}} \qquad (34)$$

If eqn. 34 is used for half-unknown samples corresponding to the calibration readings, the absolute error in the known concentrations as well as the confidence interval for each concentration is computed (Table 11.4 in Chapter 11). By using these latter values and the calibration curve, two curves of confidence are drawn for determinations on the unknown sample (Fig. 11.2 in Chapter 11). To avoid variations in the error with concentration, the analyst should work in the linear range of calibration (*see* § 6.5.1). However, even in this case, it should be noted that the error is not constant at all concentrations. The confidence curves are hyperbola above and below the calibration line. The most precise determinations will be those corresponding to the concentrations in the middle of the range. The error is larger at higher and lower concentrations.

Example. Continuing with the example in § 6.5.1, let us estimate the concentration and confidence interval for the readings of emission intensity 0.27, 0.42, 0.55, 0.67.

The results of the example in § 6.4.1 give $a_0 = 0.145$, $a_1 = 0.133$, $\bar{x} = 2.5$, $\bar{y} = 0.4775$, $V_R = 0.000115$, $\sqrt{V_R} = 0.010724$, $n = 4$, $\Sigma(x_1 - \bar{x})^2 = 5$. The Student table (Appendix 4) gives for 95% confidence level, $t(2) = 4.303$. It is found that $t\sqrt{V_R}/a_1 = 4.303 \times 0.010724/0.133 = 0.346958$. The results from application of formula 34 are presented in Table 18.11.

Table 18.11 Details of calculation of K content (eqn. 33) and confidence interval (eqn. 34) for 4 unknown samples corresponding to the calibration signals calculated in § 6.5.1

Value	Calculated content, mg L^{-1}	Absolute error	$(y_s - \bar{y})/a_1$	$\sqrt{}$	Confidence interval
0.27	0.94	−0.06	−1.5601	1.31788	0.46
0.42	2.07	0.07	−0.4323	1.134629	0.39
0.55	3.05	0.05	0.54511	1.144303	0.40
0.67	3.95	−0.05	1.44737	1.291888	0.45

These results are those indicated in Chapter 11 (Table 11.4 and Fig. 11.2).

6.5.3 *Method of standard additions*

Although the calibration method described in § 6.5.1 is by far the most widely used in analytical chemistry, it is not free from criticism in some cases. In addition to the error in precision inherent in this technique (*see* § 6.5.2) reduction in accuracy can result from introduction of a systematic error when the sample is found in a complex matrix liable to interfere with determination of the test element (*see* Chapters 9 to 11, 'interferences in spectrometry', for example). In this case, many solutions are available to the analyst: addition of a modifier that eliminates the interference, attempts at duplication of the composition of the matrix in the standard solutions and practising the method of standard additions described here.

The method of standard additions is generally more time-consuming that that of external calibration. The sample has to be divided into several equal portions. To each portion is added an exactly known quantity of the element to be determined, and all the solutions are then made up to the same volume. The quantity added is expressed in concentration relative to the original unknown sample. Each solution is analysed. The reading, graphically reported in Fig. 18.15, can be modelled as a function of the addition. If the relation is linear, the equation obtained is of the same type as before, that is, $y = a_0 + a_1 x$ with a translation of the line along the y-axis equal to the value of the reading of the sample. The original content then corresponds to a reading of zero:

$$x = a_0/a_1 \tag{35}$$

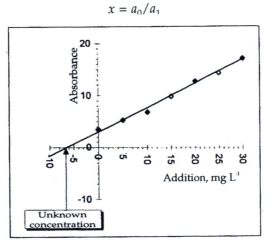

Fig. 18.15 Method of standard additions.

Example. Copper extracted from a polluted soil is determined by atomic absorption spectrometry using the method of additions. The following results are recorded (fictitious data):

Cu added to original solution, $\mu g \ mL^{-1}$	Absorbance reading
0	3.5
10	6.8
20	12.7
30	17.1

Find the Cu content of the sample solution.

The line $y = a_0 + a_1 x$ (x = addition, y = absorbance reading) is obtained as in § 6.5.1. The constants are $a_0 = 3.0200$ and $a_1 = 0.4670$ with $R_2 = 0.99$. The experimental and predicted values are presented in Fig. 18.15. Thus, the content of the original solution is

$$[Cu] = 3.02/0.467 = 6.47 \ \mu g \ mL^{-1}$$

6.5.4 Error associated with the method of standard additions

Although the method of additions is sometimes attractive for eliminating matrix effects, it has the defect of being a method of extrapolation, less precise than interpolative methods. The confidence interval for the calculated concentration of the sample can be computed by the formula:

$$x = \frac{a_0}{a_1} \pm \frac{t_{n-2}^{0,975} \sqrt{V_R}}{a_1} \sqrt{\frac{1}{n} + \frac{(\bar{y}/a_1)^2}{\Sigma(x_i - \bar{x})^2}}$$

where a_0 and a_1 are as defined in § 6.5.3, n is the number of additions, x_i the added amounts, $\bar{x}$ and $\bar{y}$ the mean content and the mean reading respectively of the additions, and V_R the residual variance of the fit of the values (Table 18.10).

Example 1. Give the confidence interval of the determination computed in the example in § 6.5.3.

We have $\bar{x} = 15$, $\bar{y} = 10.025$, $n = 4$. At a confidence level of 95%, the Student table (Appendix 4) gives $t(2) = 4.303$. The analysis of variance for the regression gives $V_R = 0.5715$. Formula 36 gives

$$x = 3.02/0.467 \pm (4.303 \times 0.756/0.467) \sqrt{[1/4 + (10.025/0.467)^2/500]}$$

or $$x = 6.5 \pm 7.5 \ \mu g \ L^{-1}.$$

As the confidence interval is larger than the concentration found, the determination is not possible under these conditions. To remedy this it is necessary to make the additions more precise (reduction in V_R) or to increase the number of additions; generally at least 6 additions are recommended. Augmentation of n permits diminution of the confidence interval through its influence on several parameters: decrease in t, reduction in $1/n$ and augmentation of $\Sigma(x_i - \bar{x})^2$.

Example 2. For improving the determination in example 1, 3 further additions corresponding to 5, 10 and 15 $\mu g \ mL^{-1}$ are done, giving respective absorbances of 5.3, 9.8 and 14.3. Find out the new value of the content and the corresponding confidence interval.

The fitting gives $a_0 = 2.996$ and $a_1 = 0.4621$. The residual variance is 0.2542. We have $n = 7$, $\bar{x} = 15$, $\bar{y} = 9.9286$. For 95% confidence level, the Student table (Appendix 4) gives $t(5) = 2.571$. Formula 36 gives:

$$x = 2.996/0.4621 \pm (2.571 \times 0.5042/0.4621) \sqrt{[1/7 + (9.9286/0.4621)^2/700]}$$

or $x = 6.5 \pm 2.5 \ \mu g \ mL^{-1}$.

The random error is reduced to one-third but is still high. It is possible to improve the determination by a greater precision in the additions, whereby the residual V_R can be greatly reduced.

6.6 Detection Limit

Why is the concept of detection limit grouped along with causes of error? It is important at the outset to recall certain definitions:

—*detection limit of an analytical method*: the lowest value of the content to be measured that the operational technique can confirm is not zero;

—*blank determination*: determination done under the same conditions as the sample but without it; if the determination comprises chemical operations, the blank will include the same operations with the same reagents; in the other case the blank will only be the background noise of the instrument;

—*determination limit*: this should not be confused with the detection limit; it is the lowest value of the content to be measured that may be determined with a still satisfactory margin of error;

—*sensitivity*: this concept, also often mistakenly confused with detection limit, represents the ratio of increase in value measured to the increase in content, within the determination zone; in the case of a calibration line (*see* §6.5.1) it is the slope a_1.

The detection limit is a concept that is becoming more and more important today with expansion of the requirement of environmental analyses (determination of traces of pollutants) and earth sciences (geochronology, isotope ratios, etc.). The perfecting of new instruments and methodological research are often aimed at lowering the detection limit. It is then important to group this concept along with the causes of error. The errors caused are not random but systematic. Their amplitude is 100%: the test element may be detected or not at all. The risk of error is of two kinds (Fig. 18.16): (1) the test element is absent though its presence has been detected (risk α) and (2) the element is present though it has not been detected (risk β).

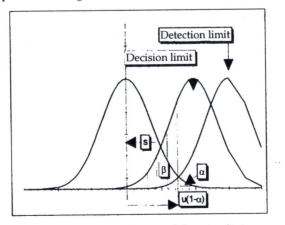

Fig. 18.16 Decision limit and detection limit.

Actual determination of the detection limit requires a certain number n_0 of repetitions of the blank determination. Let s_b be the standard deviation determined for the blank according to equations 2 and 2', and y_b the blank reading estimated by the mean of the determinations. The problem is now to determine the threshold of concentration beyond which the reading y is significantly greater than that of the blank. Considering only the risk α, the problem reverts to that in § 5.1.2, but only a detection limit or decision limit (Fig. 18.16) is determined. There are many definitions of the detection limit. It used to be generally said that the detection limit is the analyte concentration that gives a reading equal to $y_b + 2s_b$. More recently, especially in the US, a still higher limit of $y_b + 3s_b$ has been recommended. The disadvantage of these definitions is introduction of the standard deviation rather than the confidence interval. In France CEA (1986) recommends:

$$y = y_b + u_{1-\alpha} s_b \sqrt{2} \tag{37}$$

for the detection threshold and $y_b + 2u_{1-\alpha} s_b \sqrt{2}$ for the detection limit. The value obtained for y should then be expressed in concentration according to the calibration method chosen.

It should be recalled that it is useful to properly define the medium in which the detection limit is determined. In a blank determination, it can be shifted relative to the limit of determination in a different medium. The determination limit is by definition (*see* above) higher than the detection limit, the error becoming larger the closer the concentration is to these limits.

Example. It is desired to calculate the detection limit for determination of potassium by flame emission, the calibration of which is described in § 6.5.2. The readings of 25 blank determinations are 5 readings of 0.15, 10 of 0.16, 5 of 0.17 and 5 of 0.18.

The calculations give:

$$\Sigma(y_i) = 5 \times 0.15 + 10 \times 0.16 + 5 \times 0.17 + 5 \times 0.18 = 4.1$$
$$\Sigma(y_i^2) = 5 \times 0.15^2 + 10 \times 0.16^2 + 5 \times 0.17^2 + 5 \times 0.18^2$$
$$= 5 \times 0.0225 + 10 \times 0.0256 + 5 \times 0.0289 + 5 \times 0.0324 = 0.675.$$

Formula 2' gives $s_y^2 = (0.675 - 4.1^2/25)/24 = 0.00011$, whence $s_y = 0.0104$. The estimate of the mean of blanks = 0.164.

The normal distribution (Appendix 4) at 95% confidence level gives $u_{1-\alpha} = 1.96$. The detection threshold thus corresponds to the reading

$$y = 0.164 + 1.96 \times 0.0104 \times 1.414 = 0.19$$

and the detection limit to $y = 0.22$. According to the calibration obtained in § 6.5.1 and formula 33, these values correspond to K contents of 0.4 and 0.6 mg L^{-1}, respectively

Bibliography

Statistics
AFNOR. 1996. *Méthodes statistiques*, vol. 1: *vocabulaire et symboles*. AFNOR, Paris, 408 pp.
AFNOR. 1996. *Méthodes statistiques*, vol. 2: *estimations et tests statistiques*. AFNOR, Paris, 404 pp.
AFNOR. 1996. *Méthodes statistiques*, vol. 3: *contrôle statistique d'acceptation*. AFNOR, Paris, 590 pp.

AFNOR. 1996. *Méthodes statistiques*, vol. 4: *maîtrise statistique des processus*. AFNOR, Paris, 250 pp.

AFNOR. 1996. *Méthodes statistiques*, vol. 5: *traitement des résultats*. AFNOR, Paris, 574 pp.

Caulcutt R. and Boddy R. 1983. *Statistics for Analytical Chemists*. Chapman and Hall, 253 pp.

CEA, CETAMA. 1986. *Statistique appliquée à l'exploitation des mesures*. Masson, 2nd ed., 443 pp.

Dagnelie P. 1970, 1975, 1978, 1980, 1984, 1986. *Théorie et méthodes statistiques*.2 vols. Presses agronomiques de Gembloux.

Dixon W.J. 1986. Extraneous values. In: A. Klute (ed.), *Methods of Soil Analysis*. American Society of Agronomy, Madison WI, USA, pp. 83-90.

Kempthorne O. and Allmaras R.R. 1986. Errors and variability of observations. In: A. Klute (ed.), *Methods of Soil Analysis*. American Society of Agronomy, Madison WI, USA, pp. 1-31.

Miller J.C. and Miller J.N. 1993. *Statistics for Analytical Chemistry*. Ellis Horwood Ltd, 3rd ed., 256 pp.

Neuilly M. 1989. *Application des méthodes statistiques à l'analyse. Techniques de l'ingénieur*, Paris, **P 260**, 20 pp.

Neuilly M., CETAMA. 1996. *Précision des dosages de traces*. Lavoisier, Tec & Doc.

Neuilly M., CETAMA. 1998. *Modélisation et estimation des erreurs de mesure*. Lavoisier, Tec & Doc.

Rocchiccioli-Deltcheff. C. 1968. *Application des méthodes statistiques à l'analyse. Techniques de l'ingénieur*, Paris, **P 5000**, 14 pp.

Tomassone R., Dervin C. and Masson J.P. 1993. *Biométrie—Modélisation des phénomènes biologiques*. Masson, 551 pp.

Wernimont G.T. and Spendley W. (eds.), 1985. *Use of Statistics to Develop and Evaluate Analytical Methods*. AOAC International.

Youden W.J. 1984. *Statistical Manual of the Association of Official Agricultural Chemists*. AOAC International.

Youden W.J. and Steiner E.H. 1975. *Statistical Manual of the AOAC: Statistical Techniques for Collaborative Tests*. Association of Official Agricultural Chemists, Arlington VA, USA.

Chemometrics

Beebe K.R., Pell R.J. and Seasholtz M.B. 1998. *Chemometrics: A Practical Guide* (Wiley-Interscience Series on Laboratory Automation). John Wiley and Sons.

Brereton R.G. 1990. *Chemometrics: Applications of Mathematics and Statistics to Laboratory Systems* (Ellis Horwood Series in Chemical Computation, Statistics and Information). Ellis Horwood, 307 pp.

Einax J.W. (ed.) 1995. *Chemometrics in Environmental Chemistry: Statistical Methods* (The Handbook of Environmental Chemistry Series). Springer Verlag.

Haswell S.J. 1992. *Practical Guide to Chemometrics*. Marcel Dekker, 324 pp.

Kowalski B.R. 1993. *Chemometrics: Mathematics and Statistics in Chemistry*. NATO ASI Series C, 485 pp.

Massart D.L. 1987. *Chemometrics: A Textbook* (*Data Handling in Science and Technology*, Vol. 2). Elsevier Science Ltd.

Milan M., Militky M. and Forina M. 1993. *Chemometrics for Analytical Chemistry: PC-Aided Statistical Data Analysis* (Ellis Horwood Series in Analytical Chemistry). Ellis Horwood Ltd.

Morgan E. 1995. *Chemometrics: Experimental Design* (Analytical Chemistry by Open Learning). John Wiley and Sons, 294 pp.

Sharaf M.A., Kowalski B.R. and Illman D.L. 1986. *Chemometrics* (*Chemical Analysis*, Vol. 82). John Wiley and Sons, 332 pp.

Zwanziger H.W., Geib S. and Einax J.W. 1997. *Chemometrics in Environmental Analysis*. John Wiley and Sons, 404 pp.

Laboratory quality, good laboratory practice

AFNOR. 1996. *Métrologie—gérer et maîtriser les équipements de mesure*. AFNOR, Paris.

AFNOR. 1998. *Essais et analyses—maîtriser la conception et la réalisation.* AFNOR, Paris.

Crosby N.T. and Prichard E. 1995. *Quality in the Analytical Chemistry Laboratory* (Analytical Chemistry by Open Learning). John Wiley and Sons, 334 pp.

Dux J.P. 1990. *Handbook of Quality Assurance for the Analytical Chemistry Laboratory.* Chapman and Hall, 2nd ed.

NF ISO 9000. 1996. Normes fondamentales d'assurance qualité. In: *Gérer et assurer la qualité.* AFNOR, Paris, 720 pp.

Nilsen C. 1996. *Managing the Analytical Laboratory: Plain and Simple.* Interpharm Printers.

Ratliff T.A., Jr. 1997. *The Laboratory Quality Assurance System: A Manual of Quality Procedures with Related Forms.* John Wiley and Sons, 2nd ed., 334 pp.

Revoil G., 1996. *Assurance qualité dans les laboratoires d'analyses et d'essais.* AFNOR, Paris.

Weinberg S. (ed.). 1995. *Good Laboratory Practice Regulations.* Marcel Dekker, 2nd ed.

Field sampling

Abbaspour K.C. and Moon D.E. 1992. Relationships between conventional field information and some soil properties measured in the laboratory. *Geoderma,* **55:** 119-140.

Boivin P. 1990. *Géostat-PC. Logiciel interactif pour calcul géostatistique.* ORSTOM, Bondy, France.

Brown J.R. (ed..) 1987. *Soil Testing: Sampling, Correlation, Calibration and Interpretation (SSSA Special Publication 21).* Soil Science Society of America, Madison WI, USA.

Byrnes M.E. 1994. *Field Sampling Methods for Remedial Investigations.* Lewis Publishers.

Gascuel-Odoux C., Boivin P. and Walter C. 1994. Éléments de géostatistique. In: H. Laudelout, C. Cheverry and R. Calvet (eds.), *Modélisation mathématique des processus pédologiques.* Actes éditions, Rabat, pp. 217-248.

Gomez A., Leschbar R. and L'Hermite P. (eds.). 1986. *Sampling Problems for the Chemical Analysis of Sludge, Soils and Plants.* Elsevier Applied Science.

Hodgson J.M. 1978. *Soil Sampling and Soil Description.* Clarendon Press, Oxford.

Iris J.M. 1986. Analyse et interprétation de la variabilité spatiale de la densité apparente dans 3 matériaux ferrallitiques. *Science du Sol,* **24:** 245-256.

Matheron G. 1965. *Les variables régionalisées et leur estimation.* Masson, Paris, 305 pp.

Petersen R.G. and Calvin L.D. 1986. Sampling. In: A. Klute (ed.), *Methods of Soil Analysis.* American Society of Agronomy, Madison WI, USA, pp. 33-51.

Russell-Boulding J. 1994. *Description and Sampling of Contaminated Soils: a Field Guide.* Lewis Publishers, Inc.

Sabbe W.E. and Marx D.B. 1987. Soil sampling: spatial and temporal variability. In: *Soil Testing: Sampling, Correlation, Calibration and Interpretation (SSSA Special Publication 21).* Soil Science Society of America, Madison WI, USA, pp. 1-14.

Tan K.H. 1996. *Soil Sampling, Preparation and Analysis.* Marcel Dekker.

Warrick A.W., Myers D.E. and Nielsen D.R. 1986. Geostatistical methods applied to soil science. In: A. Klute (ed.), *Methods of Soil Analysis.* American Society of Agronomy, Madison WI, USA, pp. 53-82.

Laboratory sampling

Gy P. 1988. *Hétérogénéité, échantillonage, homogénéisation, ensemble cohérent de théories.* Masson, 607 pp.

Gy P. 1992. *Sampling of Particulate Materials: Theory and Practice.* Elsevier, 431 pp.

Gy P. 1992. *Sampling of Heterogeneous and Dynamic Material Systems: Theories of Heterogeneity, Sampling and Homogenizing* (Data Handling in Science and Technology). Elsevier Science Ltd, 654 pp.

Gy P. 1998. *Sampling for Analytical Purposes.* John Wiley and Sons, 153 pp.

Ingamells C.O. and Switzer P. 1973. A proposed sampling constant for use in geochemical analysis. *Talanta,* **20:** 547-568.

Jaffrezic H. 1976. L'estimation de l'erreur introduite dans le dosage des éléments à l'état de traces dans les roches liée aux caractéristiques statistiques de leur répartition. *Talanta,* **23:** 497-501.

Wilson A.D. 1964. The sampling of silicate rock powders for chemical analysis. *Analyst,* **89:** 18-30.

Appendix

APPENDIX 1

Classification of Analytical Techniques used in Study of Soils

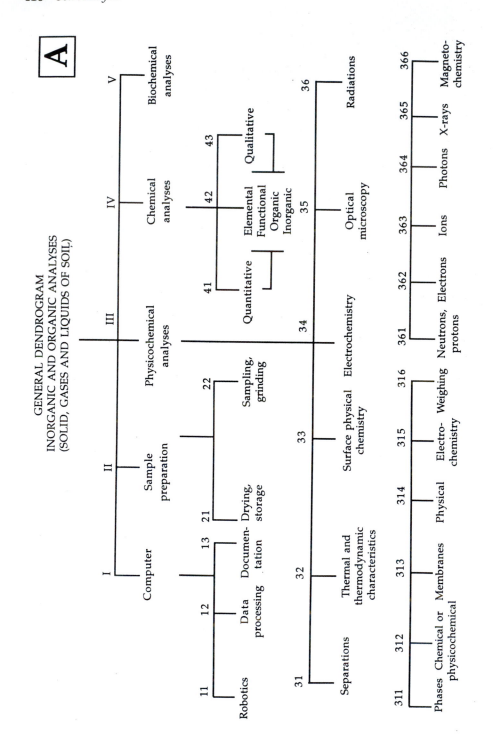

GENERAL DENDROGRAM
INORGANIC AND ORGANIC ANALYSES
(SOLID, GASES AND LIQUIDS OF SOIL)

DENDROGARM OF DETAILED SUBUNITS

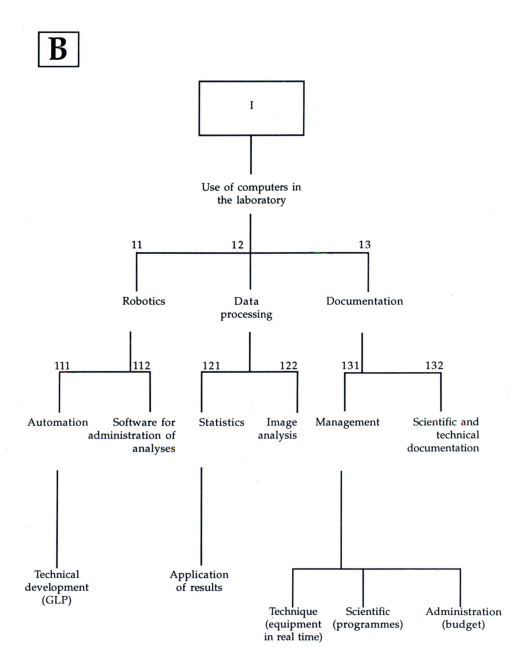

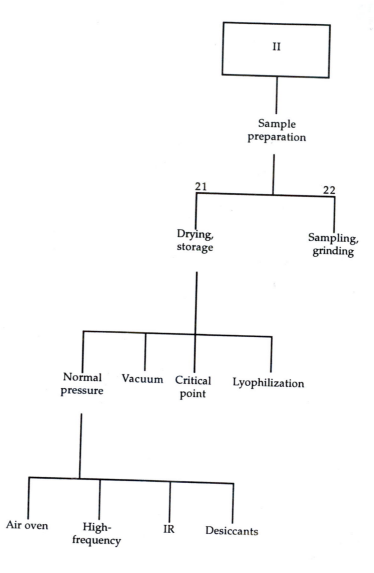

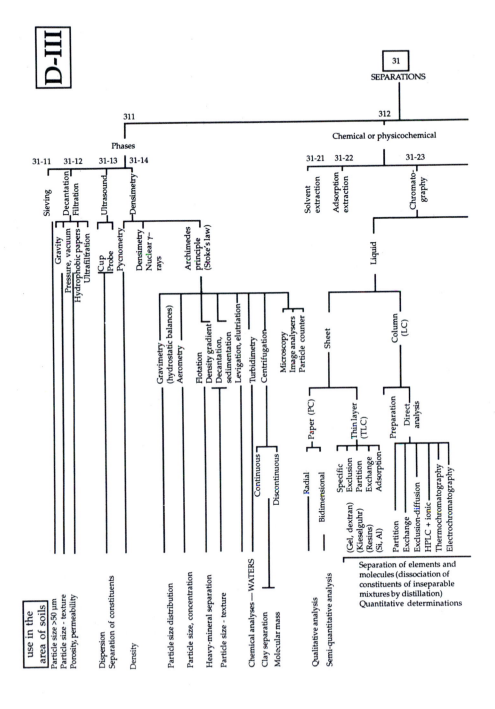

D-III contd

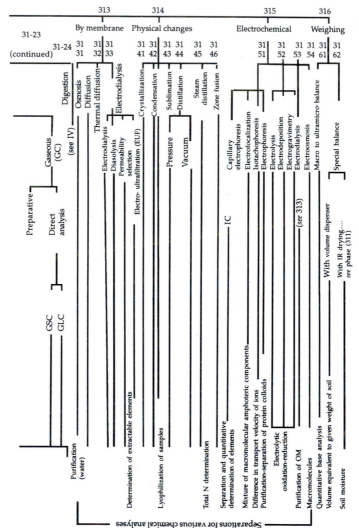

313   314   315   316

By membrane   Physical changes   Electrochemical   Weighing

31-23
(continued)

31-24

31 31 31   31 31 31 31   31 31   31   31 31 31   31 31
31 32 33   41 42 43 44   45 46   51   52 53 54   61 62

Preparative

Gaseous (GC)

Direct analysis

Digestion (see IV)

Osmosis
Diffusion
Thermal diffusion
Electrodialysis
Diasolysis
Permeability selection
Electrodialysis
Electro-ultrafiltration (EUF)

Crystallization
Condensation
Sublimation
Distillation
Pressure
Vacuum
Steam distillation
Zone fusion

Capillary electrophoresis
Electrofocalization
Isotachophoresis
Electrophoresis
Electrolysis
Electrodeposition
Electrogravimetry
Electrodialysis
Electroosmosis

Macro to ultramicro balance
Special balance

GSC
GLC

Purification (water)

Determination of extractable elements

Lyophilization of samples

Total N determination

IC

(see 313)

With volume dispenser
With IR drying......
see phase (311)

Separation and quantitative determination of elements
Mixture of macromolecular amphoteric components
Difference in transport velocity of ions
Purification-separation of protein colloids
Electrolytic oxidation-reduction
Purification of OM
Macromolecules
Quantitative base analysis
Volume equivalent to given weight of soil
Soil moisture

Separations for various chemical analyses

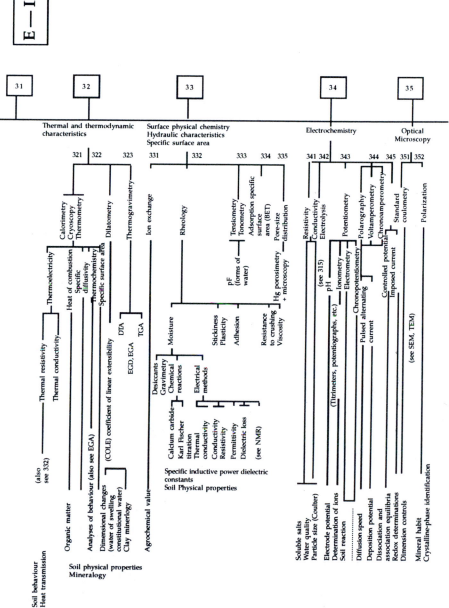

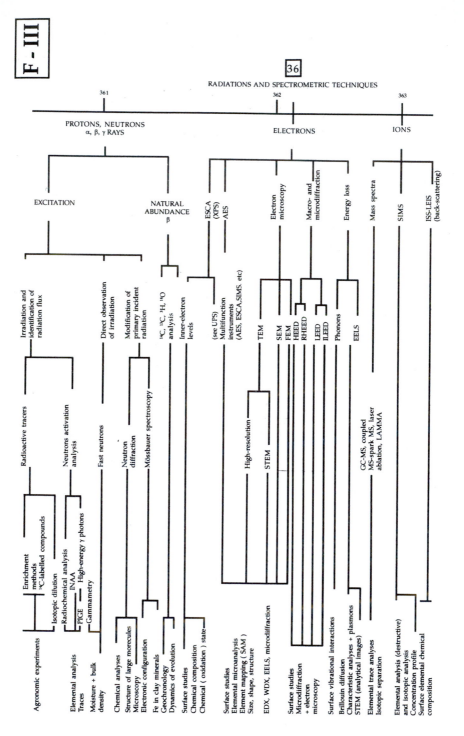

F - III

36

RADIATIONS AND SPECTROMETRIC TECHNIQUES

361 · PROTONS, NEUTRONS α, β, γ RAYS

362 · ELECTRONS

363 · IONS

EXCITATION

NATURAL ABUNDANCE β

ESCA (XPS)
AES

Electron microscopy

Macro- and microdiffraction

Energy loss

Mass spectra

SIMS

ISS-LEIS (back-scattering)

Irradiation and identification of radiation flux

Radioactive tracers

Neutrons activation analysis

Direct observation of irradiation

Modification of primary incident radiation

¹⁴C, ¹³C, ²H, ¹⁸O analysis

Inner-electron levels

(see UPS)

Multifunction instruments (AES, ESCA, SIMS, etc)

TEM

SEM
FEM
HEED
RHEED

LEED
ILEED

Phonons

EELS

Fast neutrons

Neutron diffraction

Mössbauer spectroscopy

High-resolution

STEM

GC-MS, coupled
MS-spark MS, laser ablation, LAMMA

Enrichment methods
¹⁴C-labelled compounds

Isotopic dilution

Radiochemical analysis
INAA
PIGE — High-energy γ photons
Gammametry

Agronomic experiments

Elemental analysis
Traces

Moisture + bulk density

Chemical analyses
Structure of large morecules
Microscopy
Electronic configuration

Fe in clay minerals
Geochronology
Dynamics of evolution

Surface studies
Chemical composition
Chemical (oxidation) state

Surface studies
Elemental microanalysis
Element mapping (SAM)
Size, shape, structure

EDX, WDX, EELS, microdiffraction

Surface studies
Microdiffraction
+ electron microscopy

Surface vibrational interactions

Brillouin diffusion
Characteristic analyses + plasmons
STEM (analytical images)

Elemental trace analyses
Isotopic separation

Elemental analysis (destructive)
and isotopic analysis
Concentration profile
Surface elemental chemical composition

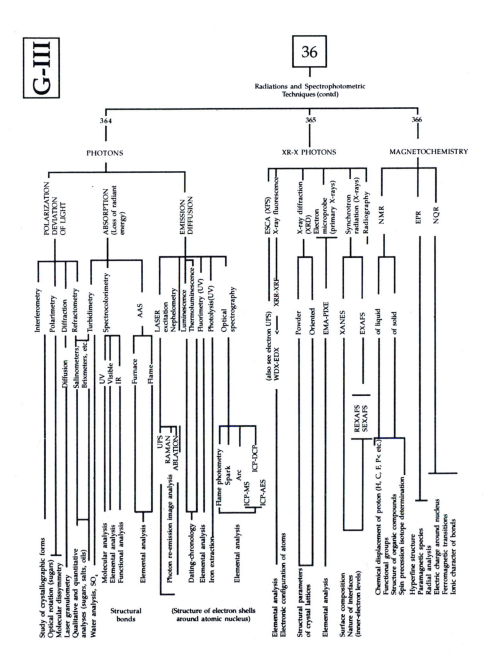

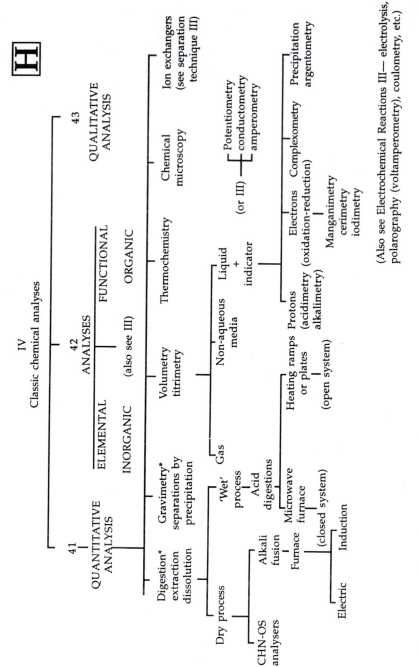

*—digestion, extraction applying the process of transfer of protons, electrons, ions, etc. (solubility product). —similarly, gravimetry, etc.

Analytical Equipment and Techniques Bilingual Glossary of Abbreviations, Symbols and Acronyms

Equipment and techniques are often denoted by abbreviations that allow, particularly in publications, avoidance of long repetitions.

This glossary brings together most of the abbreviations and symbols, of which some have become acronyms in colloquial language to the extent that the significance has often been forgotten (for example, LASER = Light Amplification by Stimulated Emission of Radiation).

As techniques are in a perpetual state of flux, some of these abbreviations are ephemeral and disappear; others, on the contrary, are perennial after progressive elimination of synonyms and possible redundancies in other fields. Some names of commercial instruments are sometimes adopted in practice to the detriment of generic terms, which complicates the glossaries.

This list cannot be exhaustive, but for ease of reference, the acronyms adopted in English are given along with the French equivalent. For example,

NMR   Nuclear Magnetic Resonance *(see RMN)*
RMN   *Résonance Magnétique Nucléaire* (see NMR).

In many cases scientific reviews in English are more prolix, so that many English abbreviations still do not have French equivalents.

A

AA   Atomic Absorption
AA   *Absorption Atomique*
AAS   Atomic Absorption Spectrometry *(see SAA)*
ACD   *Analyse Calorimétrique Différentielle* (see DCA)
ACSP   Alternating Current Superimposed Polarography
ADC   Analog Digital Converter *(see CAN)*

AE	Atomic Emission
AEAPS	Auger Electron Appearance Potential Spectroscopy
AEM	Analytical Electron Microscopy
	Auger Electron Microscopy
AES	Atomic Emission Spectroscopy/Spectrometry
	Auger Electron Spectrography/Spectroscopy
AF	Atomic Fluorescence
AFM	Atomic Force Microscopy
AFS	Atomic Fluorescence Spectroscopy (*see SFA, SEAF*)
AGE	*Analyse des Gaz Émis* (see EGA)
AIM	Adsorption Isotherm Measurements
AINR	Accelerator Induced Nuclear Reaction
ALCHEMI	Atom Location by Channelling Enhanced Microanalysis
ALICISS	Alkali Ion Scattering Spectroscopy (in impact collision mode)
AMS	Accelerator Mass Spectrometry
AOS	Acousto-Optic Spectrograph
API	Atmospheric Pressure Ionization
APIMS	Atmospheric Pressure Ionization Mass Spectrometry
APS	Appearance Potential Spectroscopy
ARAES	Angular Resolved Auger Electron Spectroscopy
ARASS	Angle Resolved Auger Surface Spectroscopy
ARC	Accelerating Rate Calorimeter
ARMS	Angle-Resolved Mass Spectrometry
ARP	Angular Resolved Photo-Emission
ARUPS	Angular-Resolved Ultraviolet Photo-Electron Spectroscopy
	Angular-Resolved Ultraviolet Photo-Emission Spectroscopy
ARXES	Angle Resolved X-ray Emission Spectroscopy
ASEM	Analytical Scanning Electron Microscopy
ASPID	Adsorption Spectroscopy by Polarization-Induced Desorption
ASV	Anodic Stripping Voltametry
ASW	Acoustic Surface-Wave Measurements
ATD	*Analyse Thermique Différentielle* (see DTA)
ATG	*Analyse Thermo Gravimétrique* (see TGA)
ATR	Attenuated Total Reflectance

B

BE	Back-Scattered Electrons
BEA	Binary Encounter Approximation
BESI	Back-Scattered Electron Scanning Image
BET	Brunauer-Emmett-Teller (method for surface area)
BIA	Batch Injection Analysis
BIS	Bremsstrahlung Isochromato Ion Spectrography
BJH	Barrett-Joyner-Halenda (method for pore texture, pore size distribution)

BLE	Bombardment-Induced Light Emission
BSE	Back-Scattered Electrons

C

CAD	Collision Activated Dissociation
	Computer Aided Design
CAM	Computer Aided Manufacture
CAMS	Constant Angle Mie Scattering
CAN	*Convertisseur Analogique Numérique* (see ADC)
CAO	*Conception Assistée par Ordinateur*
CARS	Coherent Antistokes Raman Spectroscopy *(see DRASC)*
CAT	Computer Aided Titrimeter
CBED	Convergent Beam Electron Diffraction
CC	Column Chromatography
CC	*Chromatographie sur Colonne*
CCC	Counter Current Chromatography
CCC	*Chromatographie à Contre Courant*
CCD	Coupled Charge Device (camera)
CCM	*Chromatographie Couche Mince* (see TLC)
CCMHP	*Chromatographie Couche Mince Pressurisée Haute Performance* (see HPTLC)
CCMP	*Chromatographie Couche Mince Pressurisée* (see PTLC)
CD	Coincidence Detection
CELS	Characteristic Electron Loss Spectroscopy
CEM	Constant Emission Method
	Conventional (transmission) Electron Microscopy
CEMS	Conversion Electron Mössbauer Spectroscopy
CF	*Chromatographie sur Feuille*
CFA	Continuous Flow Analysis (non-segmented)
CGL	*Chromatographie Gaz Liquide* (see GLC)
CGO	*Création Graphique par Ordinateur*
CGS	*Chromatographie Gaz Solide* (see GSC)
CG-SM	*Chromatographie en phase Gazeuse-Spectrométrie de Masse* (see GC-MS)
CI	Chemical Ionization
CIA	Capillary Ion Analysis
CID	Collision Induced Decomposition
CI-EI	Chemical Ionization—Electron Impact
CIMS	Chemical Ionization Mass Spectrometry
CIP	Constant Impulse Polarography
CIR	Cylindrical Internal Reflectance (IR)
CIS	Characteristic Isochromat Spectroscopy
CL	Cathodo Luminescence
CLBP	*Cathodo Luminescence Basse Pression*

CLG	*Chromatographie Liquide Gel* (see LGC)
CLHP	*Chromatographie Liquide Haute Performance* (see HPLC)
CLIBS	Collinear Laser Ion Beam Spectroscopy
CLL	*Chromatographie Liquide Liquide* (see LLC)
CLOA	*Combinaisons Linéaires des Orbitales Atomiques* (see LCAO)
CLOM	*Combinaisons Linéaires des Orbitales Moléculaires* (see LCMO)
CLS	Core Level Characteristic Loss Spectroscopy
CLS	*Chromatographie Liquide Solide* (see LSC)
CL-SM	*Chromatographie Liquide (couplée) à Spectrométrie de Masse* (see LC-MS)
CMA	Cylindrical Mirror Analyser
CMDE	Crossing Mercury Drop Electrode
CMP	Capacitively Coupled Microwave Plasma
CMS	Capillary Mössbauer Spectroscopy
CNA	*Convertisseur Numérique Analogique* (see DAC)
CP	*Chromatographie sur papier* (see PC)
CPAA	Charged Particle Activation Analysis
CPAS	Correlation Photo Acoustic Spectroscopy
CPC	Centrifugal Partitioning Chromatography
CPC	*Chromatographie de Partage Centrifuge*
CPD	Contact Potential Difference
CPG	*Chromatographie eu Phase Gazeuse* (see GC)
CPL	*Chromatographie eu Phase Liquide* (see LC)
CPM	Counts Per Minute
CPMAR-NMR	Cross Polarization Magic Angle Rotation—Nuclear Magnetic Resonance Spectroscopy
CPV	*Chromatographie Phase Vapeur* (see VPC)
CPXE	Charged Particle X-ray Emission
CS	*Chromatographie de Surface* (see SC)
CSFA	Continuous Segmented Flow Analysis
CSL	*Compteur à Scintillation Liquide* (see LSC)
CSRS	Coherent Stokes Raman Scattering
CTEM	Conventional Transmission Electron Microscopy
CVAAS	Cold Vapour Atomic Absorption Spectroscopy
CVD	Chemical Vapour Deposition
CZE	Capillary Zone Electrophoresis

D

DAC	Digital Analog Converter (*see CNA*)
DAO	*Dessin Assisté par Ordinateur* (see CAD)
DAPS	Disappearance Potential Spectroscopy
DARSS	Diode Array Rapid Scan Spectrometer
DBC	Digital Beam Control
DCA	Differential Calorimetric Analysis

DCEMS	Differential Conversion Electron Mössbauer Spectroscopy
DCI	Direct Chemical Ionization
	Desorption Chemical Ionization
DCM	*Dichroïsme Circulaire Magnétique*
DCP	Direct Current Plastic
	Direct Current Argon Plasma Emission Spectrometry—ICP
DEG	Detection of Emitted Gases
DEL	*Diffraction des Electrons Lents* (see SED)
DEMS	Differential Electrochemical Mass Spectroscopy
DES	De-Excitation Electron Spectroscopy
DGE	*Détection des Gaz Émis*
DIAL	Differential Adsorption Light Detection and Ranging (Raman)
DID	Diffusion Induced Disordering
DIL	Dehydration Induced Luminescence
DIS	*Détecteur d'Ion Secondaire*
	Détecteur d'Ion Spécifique
	Dissociation Induite de Surface
DLEPC	Discriminative Low Energy Photon Counting
DLI	Doppler Laser Interferometry
DLTS	Deep Level Transient Spectroscopy
DMA	Dynamic Mechanical Analysis
DME	Dropping Mercury Electrode
DPM	Disintegrations Per Minute
DPM	*Désintegrations Par Minute*
DPP	Differential Pulse Polarography
DRASC	*Diffusion Raman Anti-Stokes Cohérente* (see CARS)
DRIFT	Diffuse Reflectance Infrared Fourier Transform (spectroscopy)
DRM	*Dispersion Rotatoire Magnétique* (see MRD)
DRX	*Diffraction des rayons X* (see XRD)
DSC	Differential Scanning Calorimetry (enthalpy)
DSRS	Detected Stimulated Raman Spectroscopy
DTA	Differential Thermal Analysis *(see ATD)*
DTG	Differential Thermo Gravimetry *(see ATG)*
DXRD	Differential X-ray Diffraction

E

EA	Energy Analysis
EAAS	Electrothermal Atomic Absorption Spectrometry
EB	Electron Beam
EBG	Eyraud-Bricoud-Grillet Method (permeability of porous membranes)
EBIC	Electron Beam Induced Current
EBIS	Electron Beam Ion Source

ECON	Energy Dispersive X-ray Analysis of Carbon Oxygen Nitrogen
ECS	*Electrode Calomel Saturée* (see SCE)
ED	Energy Dispersion
EDAX	Energy Dispersive Analysis of X-ray (=EDXRA)
EDL	Electrodeless Discharge Lamp
EDPA	Electron Diffraction Pattern Analysis
EDS	Energy Dispersive Spectroscopy
EDXR	Energy Dispersive X-ray
EDXRA	Energy Dispersive X-ray Analysis
EDXRF	Energy Dispersive X-ray Fluorescence Spectroscopy
EEAES	Electron Excited Auger Electron Spectroscopy
EELFS	Electron Energy Loss Fine Structure
EELS	Electron Energy Loss Spectrometry/Spectroscopy
EFM	*Electrode à Film de Mercure*
EGA	Evolved Gas Analysis *(see AGE)*
EGD	Evolved Gas Detection
EGMC	*Electrode à Goutte de Mercure Croissante*
EGMP	*Electrode à Goutte de Mercure Pendante*
EI	Electron Impact
EI	*Electrode Indicatrice* (see IE)
EID	Electron Induced Desorption
	Electron Impact Desorption
EIID	Electron Induced Ion Desorption
EL	Electro Luminescence
ELCD	Electrolytic Conductivity Detector
ELDOR	Electron Double Resonance
ELNES	Energy Loss Near Edge Spectroscopy
ELS	Electron Energy Loss Spectrometry (*see* EELS)
	Energy Loss Spectroscopy
EM	Electron Microprobe (Emission Spectroscopy)
EMA	Electron Microprobe Analysis (= EMP = EMPA)
EMIRS	Electrochemical Modulated IR Spectroscopy
EMMA	Electron Microscope Microprobe Analyser/Analysis (combined transmission microscope and probe)
EMP	Electron Micro Probe (analysis) (= EMA = EPMA)
	Electron Micro Probe
EMPA	Electron Micro Probe Analyser (= EMA = EPMA)
ENAA	Epithermal Neutron Activation Analysis
ENDOR	Electron Nuclear Double Resonance
ENH	*Electrode Normale à Hydrogéne*
EPES	Elastic Peak Electron Spectroscopy
EPMA	Electron Probe Micro Analysis (= EMA = EMPA)
EPR	Electron Paramagnetic Resonance (= ESR)

EPXMA	Electron Probe X-ray Micro Analysis
ER	*Electrode de Référence (de potentiel)*
ERD	Electron Recoil Detection
ERDA	Electron Recoil Detection Analysis
ES	*Electrode Solide*
ESCA	Electron Spectroscopy for Chemical Analysis (= XPS)
ESD	Electron Stimulated (Stimulation of) Desorption
ESDI	Electron Stimulated Desorption of Ions
ESDIAD	Electron Stimulated Desorption Angular Distribution
ESDN	Electron Stimulated Desorption of Neutrals
ESI-MS	Electron Spray Ionization—Mass Spectrometry
ESR	Electron Spin Resonance (= EPR; *see RPE*)
ETA	Electro Thermal Atomization
ETAAS	Electro Thermal Atomic Absorption Spectrometry
ETDL	Electron Transfer Dye Laser
EUF	Electro Ultra Filtration
EXAFS	Extended X-ray Absorption Fine Structure
EXELFS	Extended Energy Loss Fine Structure
EXF	Energy Dispersive X-ray Fluorescence

F

FAAS	Flame Atomic Absorption Spectrometry
FAB	Fast Atom Bombardment
FABMS	Fast Atom Bombardment Mass Spectrometry
FAES	Flame Atomic Emission Spectrometry
FAFS	Flameless Atomic Fluorescence Spectroscopy
FAO	*Fabrication Assistée par Ordinateur* (see CAM)
FD	Field Desorption
FDM	Field Desorption Microscopy
FDS	Field Desorption Spectroscopy
FEED	Field Emission Energy Distribution
FEEM	Field Electron Emission Microscopy
FEES	Field Electron Emission Spectroscopy
FEG	Field Emission Gun
FEM	Field Emission Microscopy (emission or field effect microscope)
FEMO	Free Electron Molecular Orbital
FET	Field Effect Transistor
FFF	Field Flow Fractionation
FI	Field Injection
FIA	Flow Injection Analysis
FID	Flame Ionization Detector
FIM	Field Ion Microscope/Microprobe (field ion emission microscope)

FIMAP	Field Ion Microscopy Atom Probe
FIMAPS	Field Ion Microscope Atom Probe Spectroscopy
FIMS	Fast Ion Mass Spectrometer
FINS	*Fractionnement Isotopique Naturel Spécifique*
FIS	Field Ion Spectroscopy
FLNS	Fluorescence Line Narrowing Spectrometry
FMS	Frequency Modulation Spectroscopy
FNAA	Fast Neutron Activation Analysis
FOCS	Fibre Optic Chemical Sensor = Optode = Optrode
FODA	Fibre Optic Doppler Anemometry (laser particle measurement)
FT	Fourier Transform
FTICR	Fourier Transform Ion Cyclotron Resonance
FTIR	Fourier Transform Infra Red Spectroscopy
FTMS	Fourier Transform Mass Spectrometry
	Flight Time Mass Spectrometry
FTNMR	Flight Time NMR
FTPLS	Flight Time Photo Luminescence Spectroscopy (NIR)

G

GAA	Gamma Activation Analysis
GC	Gas Chromatography
GC-MS	Gas Chromatography—Mass Spectrometry
GDL	Glow Discharge Lamp
GDMS	Gas Discharge Mass Spectrometry
	Glow Discharge Mass Spectrometry
GES	Gas Emission Sputtering
GFAAS	Graphite Furnace Atomic Absorption Spectrometry (*see SAAE*)
GFC	Gel Filtration Chromatography
GIXD	Grazing Incidence X-ray Diffraction
GLC	Gas-Liquid Chromatography
GNR	Gamma Nuclear Resonance (Mössbauer Effect)
GPC	Gel Permeation Chromatography
GSC	Gas-Solid Chromatography

H

HA	Heat of Adsorption
HCL	Hollow Cathode Lamp
HCP	Hollow Cathode Plume
HDC	Hydro Dynamic Chromatography
HDS	Helium Desorption Spectroscopy
HEED	High Energy Electron Diffraction
HEIB	High Energy Ion Beam

HEIS	High Energy Ion Scattering Spectrometry
HF	High Frequency
HGMS	High Gradient Magnetic Separation
HIIMS	Heavy Ion Induced Mass Spectrometry
HIRS	Heavy Ion Rutherford Scattering
HMDE	Hanging Mercury Drop Electrode
HMRS	Hard Mode Raman Spectroscopy
HOLZ	High Order Lave Zone
HPLC	High Performance Liquid Chromatography (*see CLHP*)
	High Pressure Liquid Chromatography
HPRPLC	High Pressure Reversed Phase Liquid Chromatography
HPTLC	High Performance Thin Layer Chromatography
HREELS	High Resolution Electron Energy Loss Spectroscopy
HREM	High Resolution Electron Microscopy
HRLC	High Resolution Liquid Chromatography
HRMS	High Resolution Mass Spectrometry
HRS	*Haute Résolution du Solide*
HRTEM	High Resolution Transmission Electron Microscopy
HVEM	High Voltage Electron Microscope ($\geq$1,000,000 eV) (*see MEHT*)
HVTEM	High Voltage Transmission Electron Microscopy

I

IAO	*Ingénierie Assistée par Ordinateur*
IBA	Ion Beam Analysis
IBOS	Ionic Bombardment Optical Spectroscopy
IC	Ion Chromatography
	Integrated Circuit
ICEMS	Integral Conversion Electron Mössbauer Spectroscopy
ICES	Ion Chromatography Element Suppression
ICLAS	Intra Cavity Laser Absorption Spectroscopy
ICP	Inductively Coupled Plasma (*see PIHF*)
ICP-AES	Inductively Coupled Plasma Atomic Emission Spectroscopy
ICP-ES	Inductively Coupled Plasma Emission Spectroscopy
ICP-MS	Inductively Coupled Plasma Mass Spectroscopy
ICP-OES	Inductively Coupled Plasma Optical Emission Spectroscopy
ICRMS	Ion Cyclotron Resonance Mass Spectrometry
IDES	Image Dissector Echelle Spectrometer
IE	Indicator Electrode
IETS	Inelastic Electron Tunnelling Spectroscopy
IIRS	Ion Impact Radiation Spectroscopy
IIX	Ion Induced X-rays
IIXE	Ion Induced X-ray Emission
IIXS	Ion Induced X-ray Spectroscopy

ILEED	Inelastic Low Energy Electron Diffraction
IMMA	Ion Microprobe Mass Analysis (ARL®)
IMXA	Ion Microprobe X-ray Analysis
INAA	Instrumental Neutron Activation Analysis
INEPT	Insensitive Nuclei Enhanced by Polarization Transfer
INS	Ion Neutralization Spectroscopy (low-energy ions)
IP	Ionization Potential
IR	Infra Red
IR	*Infra Rouge*
IR-ERS	Infra Red External Reflection Spectroscopy
IRM	Infra Red Microscopy
IRM	*Imagerie par RMN*
IRRAS	Infra Red Reflection Absorption Spectroscopy
IRS	Internal Reflection Spectroscopy
IS	Ionization Spectroscopy
ISD	Ion Stimulated Desorption
ISE	Ion Selective Electrode
ISS	Ion Scattering Spectrometry (low energy-LEIS)
	Ion Slow Scattering
ITD	Ion Trap Detector
ITS	Inelastic Tunnelling Spectroscopy

L

LA	Laser Ablation
LAMA	Laser Ablation Mass Analysis
LAMMA	Laser Microprobe Mass Analysis
LAMMS	Laser Micro Mass Spectrometry (LAMMA)
LAS	Laser Absorption Spectrometer
LASER	Light Amplification by Stimulated Emission of Radiation
LASV	Linear Anodic Stripping Voltametry
LC	Liquid Chromatography
LCAO	Linear Combination of Atomic Orbitals (*see CLOA*)
LCMO	Linear Combination of Molecular Orbitals
LC-MS	Liquid Chromatography—Mass Spectrometry
LD	Laser Desorption
LDA	Laser Doppler Anemometry
LDMS	Laser Desorption Mass Spectrometry
LEED	Low Energy Electron Diffraction
LEEIXS	Low Energy Electron Induced X-ray Spectroscopy
LEI	Laser Enhanced Ionization
LEIS	Low Energy Ion Scattering (= ISS)
LEEM	Low Energy Electron Microscopy
LEERM	Low Energy Electron Reflection Microscopy
LESS	Laser Excited Shpol'skii Spectrometry

LGC	Liquid-Gel Chromatography
LIDAR	Light Detection And Ranging (Raman)
LIFS	Laser Induced Fluorescence Spectroscopy
LIMS	Laboratory Information Management System
LIOS	Laser Induced Optoacoustic Spectroscopy
LLC	Liquid Liquid Chromatography
LMIS	Liquid Metal Ion Source
LMMS	Laser Microprobe Mass Spectrometry
LMP	Laser Microprobe (*see* LAMMA)
LOES	Laser Optical Emission Spectroscopy
LOI	Loss On Ignition
LPCL	Low Pressure Cathodo Luminescence
LPCVD	Low Pressure Chemical Vapour Deposition
LPMS	Laser Probe Mass Spectrometer
LRM	Laser Raman Microprobe
LRMA	Laser Raman Micro Analysis
LS	Light Scattering
LSC	Liquid Scintillation Counter
	Liquid Solid Chromatography
LSV	Linear Scanning Voltamperometry
LTA	Low Temperature Ashing

M

MAO	*Maintenance Assistée par Ordinateur*
MAS	Molecular Absorption Spectrometry
	Magic Angle Spinning (NMR)
MASE	*Micro Analyseur à Sonde Electronique* (see EPMA)
MASNMR	Magic Angle Spinning NMR
MBSRS	Molecular Beam Surface Reactive Scattering
MBSS	Molecular Beam Surface Scattering
MCD	Magnetic Circular Dichroism
MDS	Metastable De-excitation Spectroscopy
MEB	*Microscope Electronique à Balayage en réflexion* (see SEM)
MEBT	*Microscope Electronique à Balayage en Transmission* (see STEM)
MECA	Molecular Emission Cavity Analysis
MEES	Medium Energy Electron Spectroscopy
MEHR	*Microscope Electronique à Haute Résolution* (see HREM)
MEHT	*Microscope Electronique Haute Tension* (see HVEM)
MEIS	Medium Energy Ion Scattering
MER	*Microscope Electronique à Réflexion* (see REM)
MET	*Microscope Electronique à Transmission* (see TEM)
MFD	Multi Function Detector
MFE	Mercury Film Electrode
MIKES	Mass Analysed Ion Kinetic Energy Spectroscopy

MIP	Microwave Induced Plasma
MIPOES	Microwave Induced Optical Emission Spectrometry
MIRFTIRS	Multiple Internal Reflection Fourier Transform Infra Red Spectroscopy
MOS	Metal Oxide Semiconductor (transmission)
MOSS	Mössbauer Spectroscopy
MPI	Multi Photon Ionization
MPIRS	Multi Photon Ionization Resonance Spectroscopy
MRD	Magnetic Rotary Dispersion
MS	Mass Spectrometry
MSID	Mass Spectrometric Isotope Dilution
MSTA	Mass Spectral Thermal Analysis
MVEM	Medium Voltage Electron Microscope

N

NAA	Neutron Activation Analysis
NBS	Neutron Bombardment Spectrometry
	Nuclear Back Scattering
NIP	Normal Impulse Polarography
NIR	Near Infrared Reflectance
NIRA	Near Infrared Reflectance Analysis
NIRS	Neutron Impact Radiation Spectroscopy
NMR	Nuclear Magnetic Resonance (*see* RMN)
NPP	Normal Pulse Polarography
NPSV	Normalized Potential Sweep Voltametry
NQR	Nuclear Quadrupole Resonance
NR	Nuclear Reaction
NRA	Neutron Reflection Analysis
	Nuclear Reaction Analysis
NRS	Nuclear Reaction Spectroscopy
NTTS	Near Threshold Translational Spectrometry

O

OED	Oxidation Enhanced Diffusion
OES	Optical Emission Spectrometry
OFDR	Optical Frequency Domain Reflectometry
OM	Optical Microscopy
OM	*Orbitale Moléculaire*
OPLC	Over Pressure Layer Chromatography
ORD	Oxidation Retarded Diffusion
ORTL	Optical Resonance Transfer Laser

P

| PA | Photo Acoustic |
| PAA | Photo Activation Analysis |

PAO	*Publication Assistée par Ordinateur*
PAS	Photo Acoustic Spectrometry/Spectroscopy
PASS	Parametric Amplification Sampling Spectroscopy
PBL	*Polarographie à Balayage Linéaire*
PC	Personal Computer
	Paper Chromatography
PC	*Polarographie Classique*
PCI	Positive Chemical Ionization
PCS	Photon Correlation Spectrometry
PD	Photo Desorption
PDA	Photo Diode Array
PDMS	Plasma Desorption Mass Spectrometry
PDS	Photothermal Deflection Spectroscopy
PECB	Piezo Electric Crystal Balance
PEEM	Photo Electron Emission Microscopy
PEM	Photo Electron Microscopy
PEPSIOS	Poly Etalon Pressure Scanned Interferometry Optical Spectrometry
PES	Photo Electron Spectroscopy
	Photo Emission Spectroscopy
PET	Positron Emission Tomography
PGNAR	Prompt Gamma Neutron Activation Analysis
PI	*Polarographie Impulsionelle*
PIC	*Polarographie à Impulsions Constantes*
PID	*Polarographie Impulsionelle Différentielle*
PIES	Penning Ionization Electron Spectroscopy
PIGE	Particle Induced Gamma Emission
	Proton Induced Gamma Emission
PIHF	*Plasma Induit par Haute Fréquence* (see ICP)
PIN	*Polarographie Impulsionelle Normale*
PIPPS	Particle Induced Prompt Photo Spectrometry
PIS	*Polarographie à Impulsions Sous-imposées*
	Polarographie à Impulsions Surimposées
PIXE	Particle Induced X-Ray Emission
PL	Photo Luminescence
PLAP	Pulsed Laser Atom Probe
PLRR	Polarized Light Raman Resonance
PM	Photo Multiplier
PMP	Proton Micro Probe
PMT	Photo Multiplier Tube
PPDI	Primary/Pristine Potential Determining Ion (surface)
PPZC	Primary/Positive Point of Zero Charge
PRFS	Phase Resolved Fluorescence Spectroscopy
PSA	Potentiometric Stripping Analysis

PSC	Power Scanning Colorimetry
PSMS	Plasma Source Mass Spectrometry
PTAS	*Polarographie à Tension Alternative Surimposée*
PTLC	Pressurized Thin Layer Chromatography
PTS	Photo Thermal Spectroscopy
PVD	Physical Vapour Deposition
PWBA	Plain Wave Born Approximation
PYS	Photoemission Yield Spectroscopy

Q

QCM	Quartz Crystal Microbalance
QMIKES	Quadrupole Mass Resolved Ion Kinetic Energy Spectra
QMS	Quadrupole Mass Spectrometer

R

RAM	Random Access Memory
RASER	*RMN Laser*
RBS	Rutherford Back-Scattering Spectrometry (= HEIS)
RDE	Rotating Disc Electrode
REM	Reflection Electron Microscopy *(see MER)*
REMPI	Resonance Enhanced Multi Photon Ionization
REXAFS	Reflection Enhanced X-Ray Absorption Fine Structure
RF	Radio Frequency
RFA	Rapid Flow Analysis
RFS	Radio Frequency Scanning
	Radio Frequency Spectroscopy
RGN	*Résonance Gamma Nucléaire* (Mössbauer effect)
RHEED	Reflection High Energy Electron Diffraction (diffraction of fast electrons)
RIA	Radio Immuno Assay *(see RID)*
RID	*Radio Immuno Dosage* (see RIA)
RIMS	Resonant Ionization Mass Spectrometry
RIS	Resonance Ionization Spectrometry
RMN	*Résonance Magnétique Nucléaire* (see NMR)
RNAA	Radiochemical Neutron Activation Analysis
ROM	Read Only Memory
RPE	*Résonance Paramagnétique Electronique* (see ESR, EPR)
RQN	*Résonance Quadripolaire Nucléaire*
RRDE	Rotating Ring Disc Electrode
RRLP	*Résonance Raman en Lumiére Polarisée*
RRS	Raman Resonance Scattering
RRX	*Reflectométrie RX*
RS	Raman Spectroscopy
RX	*Rayon X (see XR)*

S

SAA	*Spectrométrie d'Absorption Atomique* (see AAS)
SAAE	*Spectrométrie d'Absorption Atomique Electrothermique* (see GFAAS)
SAED	Selected Area Electron Diffraction
SAEM	Scanning Auger Electron Microprobe
SAES	Scanning Auger Electron Spectrometry
SAM	Scanning Acoustic Microscope
	Scanning Auger Microscope
SANS	Small Angle Neutron Scattering
SAW	Surface Acoustic Wave Sensor
SAXS	Small Angle X-Ray Scattering
SBN	*Spectrométrie par Bombardement de Neutrons*
SC	Surface Capacitance
	Sheet Chromatography
SCAM	Scanning Acoustic Microscopy
SCANIIR	Surface Composition by Analysis of Neutron and Ion Impact Radiation
SCE	Saturated (KCl) Calomel Electrode
SCFA	Segmented Continuous Flow Analysis
SCIC	Single Column Ion Chromatography
SCOM	Scanning Optical Microscopy
SDAM	Scanning Desorption Molecular Microscopy
SE	Secondary Electrons
SEAF	*Spectrométrie d'Émission Atomique de Flamme*
SEC	Size Exclusion Chromatography
SED	Slow Electron Diffraction
SEE	Secondary Electron Emission
SEELFS	Surface Extended Loss Fine Structure
SEELS	Slow Electron Energy Loss Spectrometry
SEI	Secondary Electron Image
SEM	Scanning Electron Microscopy *(see MEB)*
SEM-EDXRA	Scannning Electron Microscopy + Energy Dispersive X-Ray Analysis
SERRS	Surface Enhanced Resonance Raman Scattering
SERS	Surface Enhanced Raman Spectroscopy
	Surface Exalted Raman Spectroscopy
SEXAFS	Surface Extended X-Ray Absorption Fine Structure
SFA	Segmented Flow Analysis
SFA	*Spectroscopie de Fluorescence Atomique*
SFC	Supercritical Fluid Chromatography
SFFF	Sedimentation Field Flow Fractionation
SFX	*Spectroscopie de Fluorescence X* (see XRF)

SHRIMP	Sensitive High Mass Resolution Ion Microprobe
SI	Surface Ionization
SID	Secondary Ion Detector
	Specific Ion Detector
	Surface Induced Dissociation (mass spectra)
SIIMS	Secondary Ion Imaging Mass Spectrometry
SIM	Secondary Ion Monitoring
	Selected Ion Monitoring
SIMAAC	Simultaneous Multielement Atomic Absorption Continuum (source)
SIMS	Secondary Ion Mass Spectrometry
SIP	Superimposed Pulse Polarography
SIPS	Secondary Induced Particle Spectroscopy
SLAM	Scanning Laser Acoustic Microscopy
SLEEP	Scanning Low-Energy Electron Probe
SM	*Spectrométrie de Masse*
SMA	*Spectrométrie de Masse par Accélérateur*
SME	*Spectrométrie de Masse à Etincelles* (see SMS)
SMS	Spark Mass Spectrometry *(see SME)*
SNIFTIRS	Subtractively Normalized Interfacial Fourier Transform Infra Red Spectroscopy
SNMS	Secondary Neutron Mass Spectrometry
	Sputter Neutron Mass Spectrometry
SNR	Signal Noise Ratio
SOBI	*Spectroscopie Optique par Bombardement Ionique*
SOM	Scanning Optical Microscopy
SPA	*Spectroscopie Photo Acoustique* (see PAS)
SPAIRS	Single Potential Alteration Surface Infra Red Spectroscopy
SPDI	Secondary Potential Determining Ions
SPT	*Spectroscopie Photo Thermique*
SQUID	Superconducting Quantum Interference Device
SRF	Scanning Radio Frequency
	Spectroscopy Radio Frequency
SRIXE	Synchrotron Radiation Induced X-Ray Emission
SRM	Selected Reaction Monitoring
SRS	Stimulation Raman Scattering
	Stimulation Raman Spectroscopy
	Surface Reflectance Spectroscopy
SRXFA	Synchrotron Radiation X-Ray Fluorescence Analysis
SRXFS	Synchrotron Radiation X-Ray Fluorescence Spectroscopy
SSMS	Spark Source Mass Spectrometry
STEM	Scanning Transmission Electron Microscopy *(see MEBT)*
STIM	Scanning Transmission Ion Microscopy
STIPE	Scanning Tunnelling Inverse Photo Emission

STM	Scanning Tunnelling Microscopy
STOM	Scanning Tunnelling Optical Microscopy
STPF	Stabilized Temperature Platform Furnace (with Zeeman system)
SXAPS	Soft X-Ray Appearance Potential Spectroscopy
SXES	Soft X-Ray Emission Spectroscopy

T

TA	Thermal Analysis
TAMS	Tandem Accelerator Mass Spectrometry
TD	Thermal Desorption
	Thermal Dilatometry
TDS	Thermal Desorption Spectroscopy
	Transmission Dispersion Spectroscopy
TE	Thermoionic Emission
TED	Transmission Electron Diffraction
	Trapped Electron Detector
TEELS	Transmission Electron Energy Loss Spectrometry
TEM	Transmission Electron Microscopy *(see MET)*
TF	*Transformée de Fourier* (see FT)
TFFF	Thermal Field Flow Fractionation
TG	Thermo Gravimetry
TGA	Thermo Gravimetric Analysis
TGD	*Thermo Gravimétrie Dérivée* (see DTG)
TID	Time Interval Digitizer
TIRED	Transient Infra Red Digitizer
TL	Thermo Luminescence
TLC	Thin Layer Chromatography
TM	Thermo Magnetometry
TMA	Thermo Mechanical Analysis
TMS	Thermoionization Mass Spectrometer
TOF	Time of Flight
TOFMS	Time of Flight Mass Spectrometer
TPD	Temperature Programmed Desorption
TPSCA	Triple Potential Step Chrono Amperometry
TRD	Transmission Elution Diffraction
TRRS	Time Resolved Raman Spectroscopy
TSMS	Thermal Source Mass Spectroscopy
TXRF	Total Reflection X-Ray Fluorescence

U

UHF	Ultra High Frequency
UHV	Ultra High Vacuum
UPS	Ultraviolet Photoelectron Spectroscopy

	Ultraviolet Photoemission Spectroscopy
UV	Ultra Violet

V

VASE	Variable Angle Spectroscopic Ellipsometry
VBL	*Voltampérométrie à Balayage Linéaire*
VGAAS	Vapour Generation Atomic Absorption Spectrometry
VLSI	Very Large Scale Integration
VPC	Vapour Phase Chromatography
VPO	Vapour Pressure Osmometry
VSA	Velocity Selection Analyser

W

WD	Wavelength Dispersion
WDS	Wavelength Dispersion Spectroscopy (= WDXRA = WDX)
WDX	Wavelength Dispersive X-Ray
WDXRA	Wavelength Dispersive X-Ray Analysis

X

XAFS	X-Ray Analysis Fine Structure
XANES	X-Ray Analysis Near Edge Structure
XEAES	X-Ray Excited Auger Electron Spectroscopy
XEM	Exo Electron Microscopy
XES	Exo Electron Spectroscopy
XFS	X-Ray Fluorescence Spectrometry
XPED	X-Ray Photo Electron Diffraction
XPS	X-Ray Photoelectron Spectra/Spectroscopy
	X-Ray Photoemission Spectroscopy
XR	X Ray (*see RX*)
XRD	X Ray Diffraction (*see DRX*)
XRF	X Ray Fluorescence (*see SFX*)
XRFA	X Ray Fluorescence Analysis
XRR	X Ray Reflection/Reflectometry

APPENDIX 3

Soil Chemistry and the International System of Units (SI)[1]

---

[1] Text by Michel Misset, IRD, 32 rue Henri Varagnat, 93143 Bondy Cedex, France.

1. Measures of Length

The familiar term micron (μ) is no longer acceptable and should be replaced by micrometre ($\mu m = 10^{-6}$ m).

For measuring very small lengths the following are used:
—nanometre (nm = 10^{-9} m)
—picometre (pm = 10^{-12} m).

The familiar term Ångstrom (Å = 10^{-10} m) is also not acceptable because its index is not a multiple of three.

2. Measures of Volume

The basic units are the m^3 and its derivatives (cm^3, dm^3) and also
litre (L = dm^3 = 10^{-3} m^3)
millilitre (mL = cm^3 = 10^{-6} m^3)
microlitre (μL = mm^3 = 10^{-9} m^3).

Therefore, the units of salt concentration (*see* § 3 below) of waters are, for example,

1 mol/m^3 becomes 1 mol m^{-3} = 1 mmol L^{-1} (instead of 1 mmol/L).

It can be concluded that it is preferable to give every unit occurring in the denominator in the form of the product of the same unit with a negative index in the numerator, and removing the oblique slash.

The relative atomic mass, Ar, formerly designated atomic weight, is the mean mass of an atom of the element compared to 1/12 of the mass of the atom of carbon 12. The value of the relative atomic mass is a dimensionless ratio, and so is the molecular mass, Mr.

3. Concentration

The symbol c is the quantity of substance divided by the volume of the solution containing it. It is measured in mol m^{-3} if the molar mass is known and, if not, in kg m^{-3} or in g L^{-1}. It is also acceptable to use mol L^{-1} (= mol dm^{-3}).

Example: a solution of concentration 0.1 mol L^{-1} is said to be a 0.1 *M* or 0.1 M solution.

mmol L^{-1} is identical with mol m^{-3}.

Note: Molality (expressed in mol kg^{-1}) is preferred for non-isothermal conditions because, unlike molarity (in mol L^{-1}), it is independent of temperature.

The recommended definition of the *equivalent* is currently the following. The equivalent is the entity that will be equal:
—in an acid-base reaction to a titratable entity of H^+ ions or protons (p^+),
—in an oxidation-reduction reaction to an entity of electrons, e^-.

Thus,

1 equivalent of Cl^- will be established as 1 mole of 1 Cl^-,

1 equivalent of H_2SO_4 as 1 mole of $1/2$ H_2SO_4,

1 equivalent of Al^{3+} as 1 mole of $1/3$ Al^{3+}, etc.

The ratios 1, $1/2$, $1/3$. etc., are the equivalent factors. The equivalent can be realized from knowledge of the equivalent factor and of the chemical formula of the species.

The cation exchange capacity (CEC) and anion exchange capacity (AEC), traditionally computed in equivalents per 100 g, become in the international system of units:

mole (p^+) kg^{-1} and mole (e^-) kg^{-1} referring to the cationic or anionic charges neutralizing the exchange sites.

Let us consider a soil containing 4 meq exchangeable K^+ and 6 meq exchangeable Ca^{2+} per 100 g soil with CEC of 11 meq per 100 g; in the new system, we should write:

40 mmol (K^+) kg^{-1} or 4 cmol (K^+) kg^{-1} and 60 mmol ($1/2$ Ca^{2+}) kg^{-1} or 6 cmol ($1/2$ Ca^{2+}) kg^{-1}; the sum of exchangeable bases is:

100 mmol (+) kg^{-1} or 10 cmol (+) kg^{-1}, and the cation exchange capacity: 110 mmol (+) kg^{-1} or 11 cmol (+) kg^{-1}.

4. Pressure

The standard atmosphere is that which balances a column of 0.76 m mercury.

1 atm = $0.76 \times 13.595 \times 10^3$ kg $m^{-3} \times 9.806$ m s^{-2}

= 101,300 N m^{-2}

1 pascal, Pa = 1 N m^{-2}, and therefore,

1 atm should be written 101,300 Pa = 101.3 kPa = 1013 hPa = 1013 mb.

1 bar = 10^5 Pa. But the bar is totally unacceptable for the same reason as the Ångstrom.

5. Conductivity

Until now we have used the mmho cm^{-1}. The electrical conductivity of solutions is currently computed in siemens per metre: S m^{-1}.

1 decisiemens per m = 1 millisiemens per cm;

thus, 1 mmho cm^{-1} becomes 1 dS m^{-1} = 1 mS cm^{-1}.

As the micromho/cm will be equal to 10^{-3} dS m^{-1} (10^{-1} mS m^{-1}), its use is not acceptable because the factor is outside the international norms.

Table 1   Factors for converting non-conventional units to SI units.

Non-international units	Multiplied by	To give international units
Acre	0.405	hectare, ha (10^4 m^2)
Atmosphere	0.101	Megapascal, MPa (10^6 Pa)
Bar	10^{-1}	Megapascal, MPa (10^6 Pa)
Calorie	419	joule, J
Degree (angle)	1.75×10^{-2}	radian, rad
Dyne	10^{-5}	newton, N
Erg	10^{-5}	joule, J
Foot	0.305	metre, m
Gallon (Imperial)	4.546^{-1}	litre, L (10^{-3} m^3)
Millimho per centimetre	1	decisiemens per metre, dS m^{-1}
Pound	0.454	gram, g
Quart (liquid, Imperial))	0.946	litre, L
Temperature (°F – 32)	0.555	Temperature (°C)

Table 2   Example of preferred units for expressing physical parameters in journals of the ASA (American Society of Agronomy)

Parameter	Application	Unit	Symbol
Area	Specific surface area of soil	square metre per kilogram	m^2 kg^{-1}
	Other areas	square centimetre	cm^2
		square metre	m^2
		hectare	ha
Exchange capacity	Soil	mole (+) per kilogram	mol (+) kg^{-1}
Concentration	Liquid with known molecular mass	mole per litre	mole L^{-1}
	Liquid with unknown molecular mass	gram per litre	g L^{-1}
	Fertilizer	kilogram per hectare	kg ha^{-1}
	Gas	gram per cubic metre	g m^{-3} (mg dm^{-3})
Osmotic pressure		pascal	Pa
Mass per unit volume	Bulk density of soil	Megagram per cubic metre	Mg m^{-3}
Electrical conductivity	Salt tolerance	decisiemen per metre	dS m^{-1}
Length	Soil depth	metre	m
		centimetre	cm

Bibliography

AFNOR, 1994. Noms et symboles des unités de mesure du systéme international d'unités (SI), X02-004, 7 pp.
AFNOR, 1994. Le système international d'unités (SI), X02-006, 25 pp.
Lafaye P. 1984. Unités légales et facteurs de conversion. *Techniques de l'ingénieur*, **24**: 1-8.
Thien S.J. and Oster J.D. 1981. The international system of units and its particular application to soil chemistry. *J. Agron. Educ.* **10**: 62-71.

Appendix 4

Statistical Tables

Normal Distribution Table (Bilateral Test)

The boxed values in bold characters are the characteristic values indicated in Chapter 18, §3.2. For a normal population of mean μ and standard deviation σ, 68% of the values are located in the range $\mu \pm \sigma$ ($u = 1$), 95% of the values in the range $\mu \pm 1.96\sigma$ ($u = 1.96$) and 99.7% in the range $\mu \pm 3\sigma$ ($u = 3$).

u	0.00	0.01	0.02	0.03	0.04	0.05	0.06	0.07	0.08	0.09
0.0	0.0000	0.0080	0.0160	0.0239	0.0319	0.0399	0.0478	0.0558	0.0638	0.0717
0.1	0.0797	0.0876	0.0955	0.1034	0.1113	0.1192	0.1271	0.1350	0.1428	0.1507
0.2	0.1585	0.1663	0.1741	0.1819	0.1897	0.1974	0.2051	0.2128	0.2205	0.2282
0.3	0.2358	0.2434	0.2510	0.2586	0.2661	0.2737	0.2812	0.2886	0.2961	0.3035
0.4	0.3108	0.3182	0.3255	0.3328	0.3401	0.3473	0.3545	0.3616	0.3688	0.3759
0.5	0.3829	0.3899	0.3969	0.4039	0.4108	0.4177	0.4245	0.4313	0.4381	0.4448
0.6	0.4515	0.4581	0.4647	0.4716	0.4778	0.4843	0.4907	0.4971	0.5035	0.5098
0.7	0.5161	0.5223	0.5285	0.5346	0.5407	0.5467	0.5527	0.5587	0.5646	0.5705
0.8	0.5763	0.5821	0.5878	0.5935	0.5991	0.6047	0.6102	0.6157	0.6211	0.6265
0.9	0.6319	0.6372	0.6424	0.6476	0.6528	0.6579	0.6629	0.6680	0.6729	0.6778
1.0	**0.6827**	0.6875	0.6923	0.6970	0.7017	0.7063	0.7109	0.7154	0.7199	0.7243
1.1	0.7287	0.7330	0.7373	0.7415	0.7457	0.7499	0.7540	0.7580	0.7620	0.7660
1.2	0.7699	0.7737	0.7775	0.7813	0.7850	0.7887	0.7923	0.7959	0.7995	0.8029
1.3	0.8064	0.8098	0.8132	0.8165	0.8198	0.8230	0.8262	0.8293	0.8314	0.8355
1.4	0.8385	0.8415	0.8444	0.8473	0.8501	0.8529	0.8557	0.8584	0.8611	0.8638
1.5	0.8664	0.8690	0.8715	0.8740	0.8764	0.8789	0.8812	0.8836	0.8859	0.8882
1.6	0.8904	0.8926	0.8948	0.8969	0.8990	0.9011	0.9031	0.9051	0.9070	0.9090
1.7	0.9109	0.9127	0.9146	0.9164	0.9181	0.9199	0.9216	0.9233	0.9249	0.9265
1.8	0.9281	0.9297	0.9312	0.9328	0.9342	0.9357	0.9371	0.9385	0.9399	0.9412
1.9	0.9426	0.9439	0.9451	0.9464	0.9476	0.9488	**0.9500**	0.9512	0.9523	0.9534
2.0	0.9545	0.9556	0.9566	0.9576	0.9586	0.9596	0.9606	0.9615	0.9625	0.9634
2.1	0.9643	0.9651	0.9660	0.9668	0.9676	0.9684	0.9692	0.9700	0.9707	0.9715
2.2	0.9722	0.9729	0.9736	0.9743	0.9749	0.9756	0.9762	0.9768	0.9774	0.9780
2.3	0.9786	0.9791	0.9797	0.9802	0.9807	0.9812	0.9817	0.9822	0.9827	0.9832
2.4	0.9836	0.9840	0.9845	0.9849	0.9853	0.9857	0.9861	0.9865	0.9869	0.9872
2.5	0.9876	0.9879	0.9883	0.9886	0.9889	0.9892	0.9895	0.9898	0.9901	0.9904
2.6	0.9907	0.9909	0.9912	0.9915	0.9917	0.9920	0.9922	0.9924	0.9926	0.9929
2.7	0.9931	0.9933	0.9935	0.9937	0.9939	0.9940	0.9942	0.9944	0.9946	0.9947
2.8	0.9949	0.9950	0.9952	0.9953	0.9955	0.9956	0.9958	0.9959	0.9960	0.9961
2.9	0.9963	0.9964	0.9965	0.9966	0.9967	0.9968	0.9969	0.9970	0.9971	**0.9972**

Student Table

The values of t in bold characters are those of Table 18.4 in Chapter 18.

| Prob. α | 0.0010 | 0.010 | 0.020 | 0.040 | 0.050 | 0.100 | 0.200 | 0.400 |
υ $\alpha/2$	0.0005	0.005	0.010	0.020	0.025	0.050	0.100	0.200
1	636.618	**63.657**	31.821	15.895	**12.706**	6.314	3.078	1.376
2	31.599	**9.925**	6.965	4.849	**4.303**	2.920	1.886	1.061
3	12.924	**5.841**	4.541	3.482	**3.182**	2.353	1.638	0.978
4	8.610	**4.604**	3.747	2.999	**2.776**	2.132	1.533	0.941
5	6.869	**4.032**	3.365	2.757	**2.571**	2.015	1.476	0.920
6	5.959	3.707	3.143	2.612	2.447	1.943	1.440	0.906
7	5.408	3.499	2.998	2.517	2.365	1.895	1.415	0.896
8	5.041	3.355	2.896	2.449	2.306	1.860	1.397	0.889
9	4.781	3.250	2.821	2.398	2.262	1.833	1.383	0.883
10	4.587	**3.169**	2.764	2.359	**2.228**	1.812	1.372	0.879
11	4.437	3.106	2.718	2.328	2.201	1.796	1.363	0.876
12	4.318	3.055	2.681	2.303	2.179	1.782	1.356	0.873
13	4.221	3.012	2.650	2.282	2.160	1.771	1.350	0.870
14	4.140	2.977	2.624	2.264	2.145	1.761	1.345	0.868
15	4.073	2.947	2.602	2.249	2.131	1.753	1.341	0.866
16	4.015	2.921	2.583	2.235	2.120	1.746	1.337	0.865
17	3.965	2.898	2.567	2.224	2.110	1.740	1.333	0.863
18	3.922	2.878	2.552	2.214	2.101	1.734	1.330	0.862
19	3.883	2.861	2.539	2.205	2.093	1.729	1.328	0.861
20	3.850	**2.845**	2.528	2.197	**2.086**	1.725	1.325	0.860
21	3.819	2.831	2.518	2.189	2.080	1.721	1.323	0.859
22	3.792	2.819	2.508	2.183	2.074	1.717	1.321	0.858
23	3.768	2.807	2.500	2.177	2.069	1.714	1.319	0.858
24	3.745	2.797	2.492	2.172	2.064	1.711	1.318	0.857
25	3.725	2.787	2.485	2.167	2.060	1.708	1.316	0.856
26	3.707	2.779	2.479	2.162	2.056	1.706	1.315	0.856
27	3.690	2.771	2.473	2.158	2.052	1.703	1.314	0.855
28	3.674	2.763	2.467	2.154	2.048	1.701	1.313	0.855
29	3.659	2.756	2.462	2.150	2.045	1.699	1.311	0.854
30	3.646	**2.750**	2.457	2.147	**2.042**	1.697	1.310	0.854
40	3.551	2.704	2.423	2.123	2.021	1.684	1.303	0.851
60	3.460	2.660	2.390	2.099	2.000	1.671	1.296	0.848
120	3.373	2.617	2.358	—	1.980	1.658	1.289	0.845
∞	3.291	**2.576**	2.326	—	**1.960**	1.645	1.282	0.842

F Test Critical Values:
Unilateral Test for 5% Confidence Interval.

v2							Degree(s) of freedom v1								
	1	2	3	4	5	6	7	8	9	10	12	15	20	60	∞
1	161.4	199.5	215.7	224.6	230.2	234.0	236.8	238.9	240.5	241.9	243.9	246.0	248.0	252.2	254.3
2	18.51	19.00	19.16	19.25	19.30	19.33	19.35	19.37	19.38	19.40	19.41	19.43	19.45	19.48	19.50
3	10.13	9.55	9.28	9.12	9.01	8.94	8.89	8.85	8.81	8.79	8.74	8.70	8.66	8.57	8.53
4	7.71	6.94	6.59	6.39	6.26	6.16	6.09	6.04	6.00	5.96	5.91	5.86	5.80	5.69	5.63
5	6.61	5.79	5.41	5.19	5.05	4.95	4.88	4.82	4.77	4.74	4.68	4.62	4.56	4.43	4.36
6	5.99	5.14	4.76	4.53	4.39	4.28	4.21	4.15	4.10	4.06	4.00	3.94	3.87	3.74	3.67
7	5.59	4.74	4.35	4.12	3.97	3.87	3.79	3.73	3.68	3.64	3.57	3.51	3.44	3.30	3.23
8	5.32	4.46	4.07	3.84	3.69	3.58	3.50	3.44	3.39	3.35	3.28	3.22	3.15	3.01	2.93
9	5.12	4.26	3.86	3.63	3.48	3.37	3.29	3.23	3.18	3.14	3.07	3.01	2.94	2.79	2.71
10	4.96	4.10	3.71	3.48	3.33	3.22	3.14	3.07	3.02	2.98	2.91	2.85	2.77	2.62	2.54
12	4.75	3.89	3.49	3.26	3.11	3.00	2.91	2.85	2.80	2.75	2.69	2.62	2.54	2.38	2.30
15	4.54	3.68	3.20	3.06	2.90	2.79	2.71	2.64	2.59	2.54	2.48	2.40	2.33	2.16	2.07
20	4.35	3.49	3.10	2.87	2.71	2.60	2.49	2.45	2.39	2.35	2.28	2.20	2.12	1.95	1.84
60	4.00	3.15	2.76	2.53	2.37	2.25	2.17	2.10	2.04	1.99	1.92	1.84	1.75	1.53	1.39
∞	3.84	3.00	2.60	2.37	2.21	2.10	2.01	1.94	1.88	1.83	1.75	1.67	1.57	1.32	1.00

Unilateral
Test for 1% Confidence Interval

$v2$	\multicolumn{15}{c}{Degree(s) of freedom $v1$}
	1	2	3	4	5	6	7	8	9	10	12	15	20	60	∞
1	4052	5000	5403	5625	5764	5859	5928	5982	6022	6056	6106	6157	6209	6313	6366
2	98.50	99.00	99.17	99.25	99.30	99.33	99.36	99.37	99.39	99.40	99.42	99.43	99.45	99.48	99.50
3	34.12	30.82	29.46	28.71	28.24	27.91	27.67	27.49	27.35	27.23	27.05	26.87	26.69	26.32	26.13
4	21.20	18.00	16.69	15.98	15.52	15.21	14.98	14.80	14.66	14.55	14.37	14.20	14.02	13.65	13.46
5	16.26	13.27	12.06	11.39	10.97	10.67	10.46	10.29	10.16	10.05	9.89	9.72	9.55	9.20	9.02
6	13.75	10.92	9.78	9.15	8.75	8.47	8.26	8.10	7.98	7.87	7.72	7.56	7.40	7.06	6.88
7	12.25	9.55	8.45	7.85	7.46	7.19	6.99	6.84	6.72	6.62	6.47	6.31	6.16	5.82	5.65
8	11.26	8.65	7.59	7.01	6.63	6.37	6.18	6.03	5.91	5.81	5.67	5.52	5.36	5.03	4.86
9	10.56	8.02	6.99	6.42	6.06	5.80	5.61	5.47	5.35	5.26	5.11	4.96	4.81	4.48	4.31
10	10.04	7.56	6.55	5.99	5.64	5.39	5.20	5.06	4.94	4.85	4.71	4.56	4.41	4.08	3.91
12	9.33	6.93	5.95	5.41	5.06	4.82	4.64	4.50	4.39	4.30	4.16	4.01	3.86	3.54	3.36
15	8.68	6.36	5.42	4.89	4.56	4.32	4.14	4.00	3.89	3.80	3.67	3.52	3.37	3.05	2.87
20	8.10	5.85	4.94	4.43	4.10	3.87	3.70	3.56	3.46	3.37	3.23	3.09	2.94	2.61	2.42
60	7.08	4.98	4.13	3.65	3.34	3.12	2.95	2.82	2.72	2.63	2.50	2.35	2.20	1.84	1.60
∞	6.63	4.61	3.78	3.32	3.02	2.80	2.64	2.51	2.41	2.32	2.18	2.04	1.88	1.47	1.00

Bilateral
F Test for 5% Confidence Interval.

v_2	Degree(s) of freedom v_1														
	1	2	3	4	5	6	7	8	9	10	12	15	20	60	∞
1	647.8	799.5	864.2	899.6	921.8	937.1	948.2	956.7	963.3	968.6	976.7	984.9	993.1	1010.0	1018.0
2	38.51	39.00	39.17	39.25	39.30	39.33	39.36	39.37	39.39	39.40	39.41	39.43	39.45	39.48	39.50
3	17.44	16.04	15.44	15.10	14.88	14.73	16.62	14.54	14.47	14.42	14.34	14.25	14.17	13.99	13.90
4	12.22	10.65	9.98	9.60	9.36	9.20	9.07	8.98	8.90	8.84	8.75	8.66	8.56	8.36	8.26
5	10.01	8.43	7.76	7.39	7.15	6.98	6.85	6.76	6.68	6.62	6.52	6.43	6.33	6.12	6.02
6	8.81	7.26	6.60	6.23	5.99	5.82	5.70	5.60	5.52	5.46	5.37	5.27	5.17	4.96	4.85
7	8.07	6.54	5.89	5.52	5.29	5.12	4.99	4.90	4.82	4.76	4.67	4.57	4.47	4.25	4.14
8	7.57	6.06	5.42	5.05	4.82	4.65	4.53	4.43	4.36	4.30	4.20	4.10	4.00	3.78	3.67
9	7.21	5.71	5.08	4.72	4.48	4.32	4.20	4.10	4.03	3.96	3.87	3.77	3.67	3.45	3.33
10	6.94	5.46	4.83	4.47	4.24	4.07	3.95	3.85	3.78	3.72	3.62	3.52	3.42	3.20	3.08
12	6.55	5.10	4.47	4.12	3.89	3.73	3.61	3.51	3.44	3.37	3.28	3.18	3.07	2.85	2.72
15	6.20	4.77	4.15	3.80	3.58	3.41	3.29	3.20	3.12	3.06	2.96	2.86	2.76	2.52	2.40
20	5.87	4.46	3.86	3.51	3.29	3.13	3.01	2.91	2.84	2.77	2.68	2.57	2.46	2.22	2.09
60	5.29	3.93	3.34	3.01	2.79	2.63	2.51	2.41	2.33	2.27	2.17	2.06	1.94	1.67	1.48
∞	5.02	3.69	3.12	2.79	2.57	2.41	2.29	2.19	2.11	2.05	1.94	1.83	1.71	1.39	1.00

Bilateral
Test for 1% Confidence Interval.

$v2$	$\cdot 1$	2	3	4	5	6	7	8	9	10	12	15	20	60	∞
						Degree(s) of freedom $v1$									
1	16211	20000	21615	22500	23056	23437	23715	23925	24091	24224	24426	24630	24836	25253	25465
2	198.5	199.0	199.2	199.2	199.3	199.3	199.4	199.4	199.4	199.4	199.4	199.4	199.4	199.5	199.5
3	55.55	49.80	47.47	46.19	45.39	44.84	44.43	44.13	43.88	43.69	43.29	43.08	42.78	42.15	41.83
4	31.33	26.28	24.26	23.15	22.46	21.97	21.62	21.35	21.14	20.97	20.70	20.04	20.17	19.61	19.32
5	22.78	18.31	16.53	15.56	14.94	14.51	14.20	13.96	13.77	13.62	13.38	13.15	12.90	12.40	12.14
6	18.63	14.54	12.92	12.03	11.46	11.07	10.79	10.57	10.39	10.25	10.03	9.81	9.59	9.12	8.88
7	16.24	12.40	10.88	10.05	9.52	9.16	8.89	8.68	8.51	8.38	8.18	7.97	7.75	7.31	7.08
8	14.69	11.04	9.60	8.81	8.30	7.95	7.69	7.50	7.34	7.21	7.01	6.81	6.61	6.18	5.95
9	13.61	10.11	8.72	7.96	7.47	7.13	6.88	6.69	6.54	6.42	6.23	6.03	5.83	5.41	5.19
10	12.83	9.43	8.08	7.34	6.87	6.54	6.30	6.12	5.97	5.85	5.66	5.47	5.27	4.86	4.64
12	11.75	8.51	7.23	6.52	6.07	5.76	5.52	5.35	5.20	5.09	4.91	4.72	4.53	4.12	3.90
15	10.80	7.70	6.48	5.80	5.37	5.07	4.85	4.67	4.54	4.42	4.25	4.07	3.88	3.48	3.26
20	9.94	6.99	5.82	5.17	4.76	4.47	4.26	4.09	3.96	3.85	3.68	3.50	3.32	2.92	2.69
60	8.49	5.79	4.73	4.14	3.76	3.49	3.29	3.13	3.01	2.90	2.74	2.57	2.39	1.96	1.69
∞	7.88	5.30	4.28	3.72	3.35	3.09	2.90	2.74	2.62	2.52	2.36	2.19	2.00	1.53	1.00

Dixon Tables
Values of $(r_1)p$ (for $n \leq 10$)

P n	0.95	0.99
3	0.941	0.988
4	0.765	0.889
5	0.642	0.780
6	0.560	0.698
7	0.507	0.637
8	0.468	0.590
9	0.437	0.555
10	0.412	0.527

Values of $(r_2)p$ (for $n > 10$)

P n	0.95	0.99
11	0.637	0.745
12	0.600	0.704
13	0.570	0.670
14	0.546	0.641
15	0.525	0.616
16	0.507	0.595
17	0.490	0.577
18	0.475	0.561
19	0.462	0.547
20	0.450	0.535
21	0.440	0.524
22	0.430	0.514
23	0.421	0.505
24	0.413	0.497
25	0.406	0.489
26	0.399	0.486
27	0.393	0.475
28	0.387	0.469
29	0.381	0.463
30	0.376	0.457

Soil Classification and Reference Base

Ecological Classification: Table of Classes and Subclasses[1]

I. Immature Soils

1. Climatic immature soils
2. Erosional immature soils
3. Depositional (alluvial, colluvial) immature soils

II. Desaturated humiferous weakly developed soils (AC profile)

Rapid insolubilization of abundant organo-metallic complexes (insolubilized humic compounds).
1. Without allophanes or allophane-poor: Ranker
2. Allophane-rich: Andosols

III. Calcimagnesic soils

Humification blocked at an early stage by calcium carbonate; strong incorporation of immature humus in the profile.
1. Humiferous: A_1C — Rendzinas and Pararendzinas
2. Weakly humiferous: well developed (B) of weathering. Brunified Calcimagnesic soils
3. Highly humiferous, $A_0 A_1C$ or $A_1(B)C$ profile.

IV. Isohumic soils

Deep biological incorporation of organic matter stabilized by prolonged climatic maturation.
1. With saturated exchange complex, A_1Ca: Chernozems, Chestnut soils
2. With desaturated exchange complex, brunified intergrade A(B)C or AB_tC profile: Brunizems

---

[1] C.P.C.S. 1967. *Classification des sols.* Ecole Nationale Supérieure Agronomique, Grignon, 87 pp. Duchaufour P. 1977. *Pédologie. I. Pédogénèse et classification.* Masson, Paris, pp. 190-192.

3. Intergrade isohumic-fersiallitic: Subarid Chocolate Brown soils
4. With arid regime: Sierozems.

V. Vertisols

Soils with swelling clay: deep incorporation of very stable dark-coloured mineral-organic complexes by vertic movements.
1. Dark Vertisols
   —immature (inherited clays);
   —mature (neoformed clays)
2. Coloured Vertic soils (intergrade or degraded)
   —weakly marked vertic character
   —strongly marked vertic character.

VI. Brunified soils with A(B)C or AB$_t$C profile

Humus with rapid turnover resulting from insolubilization by iron.
1. Brown soils with (B) horizon of weathering
2. Clay-leached soils with argillic type B$_t$
3. Continental or Boreal Clay-leached soils.

VII. Podzolized soils

Immature organic matter, forming mobile organo-mineral complexes: weathering by complexolysis predominant.
1. Non-hydromorphic or weakly hydromorphic Podzolized soils
2. Hydromorphic Podzolized soils (with groundwater).

VIII. Hydromorphic soils

Soils with local segregation of iron by oxidation-reduction processes.
1. Soils with marked oxidation-reduction processes (groundwater soils): Pseudogley — Stagnogley — Gley — Peats.
2. Oxidation-reduction processes often attenuated; hydromorphy by capillary imbibition of clayey material and surficial impoverishment of clay: Pélosols — Planosols.

IX. Fersiallitic soils

Particular evolution of iron oxides (rubefaction); 2/1 clay minerals predominant (transformation and neoformation).
1. Incomplete rubefaction: Fersiallitic Brown soils
2. Complete rubefaction, saturated or almost saturated complex: Fersiallitic Red soils
3. Partial desaturation and degradation of exchange complex: Acid Fersiallitic soils.

X. Ferruginous soils

Abundance of crystalline oxides of iron (goethite or haematite): incomplete weathering of primary minerals, 1/1 clay minerals (neoformed) predominant.
  1. Persistence of primary minerals and 2/1 clay minerals in all horizons: Ferruginous soils *sensu stricto*
  2. Complete weathering of primary minerals at least at the top of the profile: Ferrisols.

XI. Ferrallitic soils

Complete weathering of primary minerals (except quartz); exclusively 1/1 clay minerals; high content of sesquioxides; crystalline oxides of iron and aluminium.
  1. Ferrallitic soils *sensu stricto*: kaolinite predominant.
  2. Ferrallite: sesquioxides (gibbsite and iron oxides predominant).
  3. Ferrallitic soils with hydromorphic segregation of iron.

XII. Salsodic soils

Evolution conditioned by the Na^+ ion in its two forms.
  1. As salt: Saline soils
  2. As exchangeable sodium: Alkali soils.

List of Units of the Reference Base[1]

Typic Alocrisols
Humic Alocrisols

Andosols and Vitrosols
Perhydric Silandosols
Humic Silandosols
Eutric Silandosols
Dystric Silandosols
Perhydric Aluandosols
Humic Aluandosols
Haplic Aluandosols
Vitrosols

Anthroposols
Transformed Anthroposols
Artificial Anthroposols
Reconstituted Anthroposols

Arenosols

Saturated Brunisols
Mesosaturated Brunisols
Oligosaturated Brunisols
Resaturated Brunisols

Carbonated and saturated sola

Rendosols
Rendisols
Calcosols
Dolomitosols
Calcisols
Magnesisols
Calcarisols

Haplic Chernosols
Typic Chernosols
Melanoluvic Chernosols
Colluviosols

Histic Cryosols
Mineral Cryosols

Carbonated Fersialsols
Calcic Fersialsols

Unsaturated Fersialsols
Eluvic Fersialsols

Immature Fluviosols
Typic Fluviosols
Brunified Fluviosols

Thalassosols

Eluvic Grisols
Degraded Grisols
Haplic Grisols
Haplic Gypsosols
Petrogypsic Gypsosols

Leptic Histosols
Fibric Histosols
Mesic Histosols
Sapric Histosols
Composite Histosols
Buried Histosols
Flowing Histosols

Typic Reductisols
Stagnic Reductisols
Duplic Reductisols
Redoxisols

Leptismectisols

Lithosols

Luvisols
Neo Luvisols
Typic Luvisols
Degraded Luvisols
Dernic Luvisols
Truncated Luvisols
Quasi-Luvisols
Pseudo-Luvisols

Calcareous Organosols
Calcic Organosols
Unsaturated Organosols
Tangelic Organosols

1. INRA. 1995. *Référentiel pédologique*. INRA. Techniques *et pratiques*, p. 323.

Typic Pelosols, etc.
Brunified Pelosols
Differentiated Pelosols

Stony Peyrosols
Gravelly Peyrosols

Haplic Phaeosols
Melanoluvic Phaeosols

Typic Planosols
Distal Planosols
Structural Planosols

Duric Podzosols
Eluvic Podzosols
Humo-duric Podzosols
Unconsolidated Podzosols
Placic Podzosols
Ochric Podzosols
Humic Podzosols
Post-Podzosols

Rankosols

Regosols

Salsodic sola
Chloruro-sulphated Salisols
Carbonated Salisols

Undifferentiated Sodisols
Solonetzic Sodisols
Solodized Sodisols
Sodisalisols
Salisodisols

Thiosols and Sulphatosols
Thiosols
Sulphatosols

Veracrisols
Vertisols

Lithovertisols

Topovertisols

Haplic Paravertisols
Planosolic Paravertisols

Appendix 6

Suppliers of Analytical Equipment and Instruments

Field Equipment (Chapter 1)

Bonne Espérance, matériels de sondage et de forage, 11 rue Gries, 67 240
  Bischwiller, France
  Tél. 03 88 63 24 25
Eijkelkamp, P.O Box 4, 6987 ZG Giesbeek, The Netherlands
  Tél. 31 (08 336) 1941   Télex 35 416 EYKEL NL
Objectif K (ex Nardeux), B.P. 121, 11 rue de Granges Galand, F-37 552 Saint-
  Avertin Cedex, France
  Tél. 02 47 28 74 00   Fax 02 47 28 90 22
SDEC (ex Nardeux), BP 4233, 19 rue Edouart Vaillant, 37 000 Tours, France
  Tél. 02 47 92 22 00   Fax 02 47 92 86 16
A And L Agricultural Laboratory Inc., 1311 Woodland Avenue, Modesto,
  California 95 350, USA
  Tél. (209)529 4080
Ele International Ltd., Eastman Way, Hemel Hempstead, Hert Fordshire
  HP27HB United Kingdom
Lindqvist International S.A., Z.I. La Marinière Bondoufle, 5 rue Gutemberg,
  BP 1207, 91 912 Evry Cedex 9, France
  Tél. 01 60 86 44 72   Fax 01 60 86 40 23
Daiki Soil And Moisture, 60-3 Nishiogu 7-Chome, Arakawa-Ku, Tokyo 116,
  Japan
Kiya Seisakushu Ltd., 20-8 Mukogaoka 1-Chome, Bunkyo-Ku, Tokyo 113,
  Japan
Soil Moisture, 801 S. Kellogg Avenue, Goleta CA 93 117, USA
  PO Box 30 025 Santa Barbara, CA 93 105, USA
Seditech, 8 rue des Bruyères, 78 770 Thoiry, France
  Tél. 01 34 87 59 59

Field Tests (Chapter 2)

Soil-Water Field Analysis Kits

Palintest Ltd., Palintest House, Kingsway, Team Valley Estate, Gateshead,
  Tyne and Wear, NE 110 NS, United Kingdom
  Tél. 44-91-4910808   Fax 44-91-4825372
HNU Systems Inc., 160, Charlemont Street, Newton Ma 02161, USA
  Fax 1-617-965 5812
Inforlab Chimie, 2, Allée des Hirondelles, 77500 Chelles, France
Horiba Ltd. (Cardy), Miyanoshigash, Kisshoin Minami-KU, Kyoto (Japan)
The Tintometer Ltd., Waterloo rd. Salisbury, WILTS SP1 2 JY, United
  Kingdom
  Tél. (0722) 327 242   Fax (0722) 412 322

Merck-Clevenot, 5-9 rue Anquetil, 94736 Nogent-sur-Marne Cedex, France
    Tél. 01 43 94 54 00
Hach Europe S.A., B.P. 229 B-5000 Namur, Belgium
    Tél. (32) (81) 44-53-81   Fax (32) (81) 44 13 00
Blender and Laukart, Mundelheimer str. 45, 4100 Duisburg 25, Germany
Whatman Inc., 9 Bridewell Pl., Clifton NH 07014, USA
    Fax 201-472-6949
Lamotte Thomas scientific, 99 High Hill Road zt 1-295 PO box 99, Swedesborg
    NH 08085-009, USA
    Tél. 1-800-345-2100
Ward's, 5100 West Henrietta Road, PO box 92912
    Rochester New-York 14692-9012, USA
    Tél. 716-359-2502   Fax 716-334-6174

Light Autonomous Instruments for Field Measurements

Miscellaneous

Testoterm, 27 A. rue Nationale, B.P. 100, 57602 Forbach Cedex, France
    Tél. 03 87 29 29 00   Fax 03 87 87 40 79
Horiba-Cardy France, rue L. et A. Lumière, Technoparc, 01630 St Genis-
    Pouilly, France
    Tél. 04 50 42 27 63   Fax 04 50 42 07 74
Astro-Med, Parc d'activités de Pissaloup, 1 rue Édouard Branly, 78190
    Trappes, France
    Tél. 01 34 82 09 00   Fax 01 34 82 05 71
Ponselle, 16; av. de la Pépinière, ZAE 78220 Viroflay, France
    Tél. 01 30 24 62 62   Fax 01 30 24 31 85
Solomat, 16, rue Jacques Tati, B.P. 187, 91006 Evry Cedex, France
    Tél. 01 60 77 89 90   Fax 01 60 77 93 73
IBA-Corning Analytical, Colchester Road, Halstead, Essex CO 92 DX, United
    Kingdom
    Tél. (0787) 475 088
Orion Research Inc. 840 Memorial Drive, Cambridge MA 0 2139, USA
    Tél. 800 225 1480
ELE International Lim. Eastman way, Hemel Hempstead, Hertforshire HP 2
    7HB, United Kingdom
Shott Gerate GMBH, D-6238 Hofheim, Germany
Cole-Parmer Instr. 7425 north Oak Park, Chicago, Illinois 60 648, USA
Interlab Instrument, av. du Général de Gaulle, 93118 Rosny-sous-Bois Cedex,
    France
    Tél. 01 45 28 35 91
Omnidata Dept. ss, Pro Box 3489, Logan Utah 84 321, USA
    Tél. (801) 753 7760

Gas chromatography

Photovac Inc., 134 Dow Caster, av. Unit 2, Thornhill (Ontario) L3t 123, Canada

X-ray fluorescence

Shlumberger, rue de la Roseraie, Parc des Tanneries, 67380 Lingolsheim,
France
Tél. 03 88 78 57 10

Texas Nuclear-Ramsey, PO Box 9267, Austin Texas 78 766, USA
Tél. (512) 836 0801

Gamma-ray Spectroscopy

E.G Road G. Ortec., 100 Midland Rd. Oak Ridge, TN 37831, USA
Tél. (615) 482 4411   Fax (615) 483 0396

Optical microscopy

Mc Arthur-K.W.Kirck and Sons Ltd., Winship Industrial Estate Hilton
Cambridge CB4 4BC, United Kingdom
Tél. (0223) 420 102   Fax (0223) 420 175

Mass Spectrometry (mobile)

Brucker, 34, rue de l'Industrie, 67160 Wissembourg, France
Tél. 03 88 73 68 00   Fax 03 88 73 68 79

Grinding And Sieving (Chapter 4)

Grinders

Fritsch, 12 Chaussée Jules César, B P 519, 95520 Osny (Cergy-Pontoise),
France
Tél. 01 30 73 84 00   Télex 609 919
Industriestrasse 8, D-6580, Idar Oberstein 1, Germany

Retsch F. Kurt Et Cie (pompes Verder) Parc activité de la Danne, BP 40,
95610 Eragny-sur-Oise, France
Tél. 01 34 53 31 88   Télex 605-429   Fax 01 34 64 44 50
Reinische strasse 36, PO Box 1554, D-5657 Haan 1, Germany
Tél. 02 129 5561-0   Télex 859 445   Fax 02-129-8702

Stokes-Pennwalt, BP 204, 90-94 rue d'Estienne d'Orves, 95502 Rueil-
Malmaison Cédex, France

Rocklabs, PO Box 18-142, 187 Morin road Knox industrial Estate, Auckland
6, New Zealand

IKA, Post Fach 1165, D-7813, Janke et Kunnel strasse, Staufen, Germany
Labo-Moderne, 37 rue Dombasle, 75015 Paris, France
Tél. 01 45 32 62 54   Télex 203 813   Fax 01 45 32 01 09

Perstorp Analytical (Tecator), Box 70, 12 chaussée Jules César, S-26301, Hoganas, Sweden

BP 13, 1 rue Jean Carrasso, 97872 Bezons Cédex, France
Tél. 01 34 23 38 38   Télex 605 947   Fax 01 34 23 39 03

Spex Inds. Inc., 3880 Park avenue, 08820, Edison, NJ USA Jobin-Yvon, BP 118, 16-18 rue du Canal, 91163 Longjumeau Cédex, France

Custom Lab., PO Box 667, 32763 Orange City, FL, USA

Mc Crone Research Associates Lt, 2 Mc Crone Mews, Belsize Lane NW3 5BG London, United Kingdom
Tél. 071-435-2282   Fax 071435 5270

Daiki Rika Kogyo, 60-3 Nishiogu 7-Chome, Arakawa-Ku Tokyo 116, Japan
Tél. 03-810-2181   Fax 03-810-2185

Soil Test, 2205 Lee Str., 60202 Evanston, Illinois, USA
Tél. (312) 869-5500   Telex 72 4496

Thomas Scientific (Wileymill), PO Box 99, 99 High Hill Road at I-295, 08085-099, Swedesboro, USA
Télex 685 1166   Fax 609-467-3087

Sieving machines (all types)

Fritsch, Industrie Strasse 8, D 6581, Idar Oberstein 1, Germany

Gilson, BP 45, 72 rue Gambetta, Villiers-le-Bel, 95400, France
Tél. 01 34 29 50 50   Télex 606 682   Fax 01 34 29 50 60

Gilson, 3000 Beltline Hwm., 53 562, Middleton WI
Tél. (608) 836 1551

Retsch F, PO Box 1554, Rhinische Strasse 36, D-5637 Haan 1, Germany
Tél. 02 129 55610   Télex 859 445   Fax 02 1298702

Thomas, PO Box 99, 99 High Hill Road at I-295, 08085-0099 Swedles Bord NJ USA
Tél. 710-991-8749   Télex 685 1156   Fax 609-467-3087

Sieves

Saulas Et Cie, 16 rue du Buisson St Louis, 75010 Paris, France
Tél. 01 42 05 58 50

Ele International, Eastman Way, Hemel Hempstead Hertforshire, HPZ 7 HB, United Kingdom
Tél. (0442) 218 355   Télex 825239   Fax (0442) 52474

Seal Tamis Tram, 20 - 22 rue des Groseillers, 93100 Montreuil, France
Tél. 01 42 87 26 23   Télex 231-291   Fax 01 42 87 89 89

Endecotts, Lombard Road Morden Factory Estate, 19 3 BR, London SW, United Kingdom

Retsch F. Kurt Et Cie, PO Box 1554, Rheinische Strasse 36, D - 5657 Haan 1, Germany.
Tél. 02 129 55610   Télex 859-445   Fax 02-1298702

Analytical Balances (Chapter 6)

Cahn Instr. (Bibby), ZI du rocher vert, BP 79, 77793 Nemours Cédex, France
Tél. 01 64 28 88 89 et 01 64 28 17 74
Mettler-Toledo, 18-20 av. de la Pépinière, 78220 Viroflay, France
Tél. 01 30 97 17 17   Fax 01 30 97 16 16
Precisa Oerlikon (S.N.L), 6 rue Gutenberg, Z.I de la Marinière, 91111 Evry
Cédex 9, France
Tél. 01 60 86 85 80   Fax 01 60 86 74 74
Prolabo, 12 rue Pelée-BP 369-75526 Paris Cédex 11, France
Tél. 01 49 23 15 00   Fax 01 49 23 15 15
Sartorius,11 av. du 1er Mai-91127 Palaiseau Cédex, France
Tél. 01 69 20 93 11   Fax 01 69 20 09 22
Setaram, 7 rue de 1'Oratoire-BP 34-69641 Caluire Cédex, France
Tél. 04 78 29 38 38   Fax 04 78 29 63 65
Ainsworth, A and D, Ohaus, Sauter Shimadzu, Stanton, Voland, etc.

Separations on Filter Papers and Membranes (Chapter 7)

Amicon, Grace S.A. -53, rue Saint-Denis - 28231 Epernon Cédex, France
Chrompack - P.O Box 3-NL 4330 AA-Middelburg, The Netherlands
Tél. 31-1180-11251
Filtron Technology, 42, route Nationale 10-78310 Coignières, France.
Gelman Sciences, Parc club de la Haute Maison-Cité Descartes-16, rue Galilée-
Champs-sur-Marne
77438 Marne La Vallée Cédex 2, France
Tél. 01 64 68 30 81   Fax 01 64 68 29 56
Hamilton, PO Box 10030-Reno NV 89510, USA
Hettich-Ceralabo, 153, av. Jean-Jaurès-93307 Aubervilliers Cédex, France.
Tél. 01 48.34 00 74   Fax 01 48 33 15 22
Inceltech, Rue Haute - 21310 Tanay, France.
Lida-Interchim, BP 15-213, av. J. -F. Kennedy - 03103 Montlucon, France.
Tél. 04 70 03 88 55   Fax 04 70 03 82 60
Merck E., Frankfurter strasse 250-D-6100 Darmstadt, Germany
Microgon Inc. (Dynagard) Laguna Hills - CA 926553, USA
Millipore-Waters, Rue Jacques Monod/Rond-point des Sangliers-78280
Guyancourt, France
Tél. 01 30 12 70 00   Fax 01 30 12 71 82
Norton-Pampus, 4 rue de Salonique-BP 132-95103 Argenteuil Cédex, France.
Tél. 01 39 82 02 62
Poretics (Serlabo) BP 82-94382 Bonneuil sur Marne Cédex, France.
Sartorius AC: PO Box 3243-D-3400 Goettingen, Germany
Tél. (551) 308-0   Fax (551) 308-289

Schleicher Et Schuell, PO Box 4-D-3354 Dassel, Germany
    Tél. 05561-791-0   Fax 05564-2309
Schott, Geschäftsbereich Chemie-Postfach 2480-W-6500 Mainz 1, Germany
SGF, 15 allées de Bellefontaine - 31100 Toulouse, France
    Tél. 05 61 40 85 85   Fax 05 61 41 51 78
Whatman Scientific Ltd., Whatman House - St Leonard Road 20/20-Kent
    ME 16 OLS, United Kingdom
    Tél. 44 622 676670   Fax 44 622 677611

Molecular Spectrometry (Chapter 9)

Major Suppliers in France

Visual comparator-colorimeters and simple modular colorimeters
Cifec, 12, bis rue du Commandant Pilot - 92000 Neuilly-sur-Seine, France.
    Tél. 01 46 37 54 02
Pierron, 4, rue Gutenberg, BP 609-57206 Sarreguemines, France

Spectrophotometry
I.L.O., avenue du Général de Gaulle. Tour de bureaux-Rosny II - 93118
    Rosny-sous-Bois, France.
    Tél. 01 49 35 19 44
Bioblock, rue Sébastien Brant - 67400 Illkirch, France.
    Tél. 01 88 67 14 14
OSI, 141 rue de Javel - 75739 Paris Cédex 15, France.
    Tél. 01 45 54 97 31
Roucaire (distributeur Methrom): BP 65-78143 Vélizy-Villacoublay, France
    Tél. 01 30 67 75 00
Secomam (ex Jean Et Constant), 4 rue des Charpentiers, BP 106, 95 334
    Domont Cédex, France
    Tél. 01 39 35 42 00   Fax 01 39 91 30 18

Other Suppliers

Near IR-visible-near UV Spectrometers
Alltech Associates France, BP 11, 59242 Templeuve, France
    Tél. 03 20 79 25 25   Fax 03 20 59 33 69
Beckman Instruments, 92-94 chemin des Bourdons, 93220 Gagny, France
    Tél. 01 43 01 70 00   Fax 01 43 81 16 80
Fischer Scientific CO, 711, Forbes ave. Pittsburg P.A 15219, USA
    Tél. 1 412 562 8300
GBC Scientific Equipment, 22, Brooklyn av., Dandenong, Victoria 3175,
    Australia

Hewlett-Packard, 1, av. du Canada, 91947 Les Ulis Cédex, France
Tél. 01 69 82 60 60   Fax 01 69 82 60 61
Hitachi Instruments, 24-14 Mishi Shimbashi, 1-Chome Minatoku, Tokyo, Japan
Kontron Instruments, 2 av. du Manet, B.P. 81, 78185 St-Quentin-en-Yvelines Cédex, France
Tél. 01 30 57 66 00   Fax 01 30 44 23 57
Nicolet, 16, av. Jean d'Alembert, Z.I. de Pissaloup, BP 118, 78192 Trappes, France
Tél. 01 30 66 33 30   Fax 01 30 66 70 36
Perkin-Elmer, BP 304, Montigny le Bretonneux, 78054 St-Quentin-en-Yvelines Cédex, France
Tél. 01 30 85 63 63   Fax 01 30 85 63 00
Pharmacia-LKB France, BP 210, 78051 St Quentin-en-Yvelines Cédex, France
Bioritech, 7, voie del'Ormaille, Janville sur Juine, 91510 Lardy, France
Tél. 01 60 82 60 44   Fax 01 60 82 29 40
Shimadzu Corp. Europe, Albert Hahn Strass, D-4100 Duisburg 2/9 6-10, Postfach 29 02 60, Germany
Spectronics Corp., 956, Brush Hollow Road, P.O Box 483, Westbury, New York, 11590, USA
Varian, 7, av. des Tropiques, B.P. 12, 91941 Les Ulis Cédex, France
Tél. 01 69 86 38 38   Fax 01 69 28 23 08
Bio-Rad, 94, 96, rue Victor-Hugo, 94200 Ivry Sur Seine, Frnace
Tél. 01 49 60 68 34   Fax 01 46 71 24 67
Gilford products - CIBA-CORNING, 132. Artino St., Oberlin OH 44074, USA
                              Fax 216-774-3939
Malvern, 30, rue J. Rostan, 91893 Orsay Cédex, France
Tél. 01 60 19 02 00   Fax 01 60 19 13 26
Milton Roy, E. Vlietimackstraat 20, B-8400, Oostende, Belgium
Omega optical, P.O. Box 573, 3, Grove St, Brattle boro VT 05301, USA
Fax 802 254 3937
The Tintometer-Lovibond-OSI, Schleefstrasse 84, D-4600 Dortmund 41, Germany
Bacharach-Coleman, 625 Alpha drive, Pittsburgh PA 15328, USA
Fax (412) 963 2091
Instruments with flexible immersible optical guide (visible)
Hewlett-Packard, 1, av. du Canada, 91947 Les Ulis Cédex, France
Tél. 01 69 82 60 60   Fax 01 69 82 60 61
Guided Wave Inc., 5190 Golden Foothill Parkway, EI Dorado Hills CA 95630, USA
Methrom, Oberdorstrasse 68, CH-9 100 Erisau. Appenzell, Switzerland

Instruments with flexible immersible optical guide (IR)
Axiom Analytical Inc, 18103 C-Sky Park south, Irvine CA 92714, USA
Fax 1-714-7579306

Perstorp (NIRS System), 1-3, rue Jean Carrasso, B.P. 13, 95871 Bezons Cédex, France

    Tél. 01 34 23 38 38   Fax 01 34 23 39 03

Bruker, 34, rue de l'Industrie, 67160 Wissembourg, France

    Tél. 03 88 73 68 00   Fax 03 88 73 68 79

Oriel-Lot, 9, av. de Laponie. Z.A. de Courtaboeuf, 91951 Les Ulis Cédex, France

    Tél. 01 69 07 20 20   Fax 01 69 07 23 57

Euro-Labo, 35, rue de Meaux, 75019 Paris, France

    Tél. 01 42 08 01 28   Fax 01 42 08 13 65

Atomic Absorption Spectrometers (Chapter 10)

Fisons-ARL, 85, av. Aristide Briand, 94110 Arcueil, France

    Tél. 01 47 40 48 40   Fax 01 45 46 22 50

Hitachi (voir SKALAR), 24-14, Mishi Shimbashi-Chome Minato Ku, Tokyo, Japan

Perkin-Elmer, BP 304, Montigny-le-Bretonneux, 78054 St-Quentin-en-Yvelines Cédex, France

    Tél. 01 30 85 63 63   Fax 01 30 85 63 00

Philips-Unicam, 105, rue de Paris, BP 122, 93003 Bobigny Cédex, France

    Tél. 01 49 42 81 55   Fax 01 49 42 80 30

Shimadzu Europe (Touzart et Matignon) Albert Hahnstrasse 6-10, D-4100 Duisburg 29, Germany

    Postfac 29 02 60

Thermo-Jarrell Ash, 8E Forge Parkway, Franklin MA 02038, USA

Thermo Jarrel Ash, 13, rue de la Perdrix, BP 50321, 95940 Roissy CDG, France

    Tél. 48 63 78 00   Fax 48 63 27 60

Varian, 7, av. des Tropiques, BP 12, 91941 Les Ulis, France

    Tél. 01 69 86 38 38   Fax 01 69 28 23 08

Spectra Instrumentation, 21, rue Marc Seguin, Z.I Mitry-Compans, 77292 Mitry-Mory, France

    Tél. 01 64 27 22 11   Fax 01 64 27 37 61

GBC Scientific equipment (SKALAR), 22, Brooklyn Av., Dandenong. Victoria 3175, Australia

Emission Spectrometers (Chapter 11)

Flame Photometers

Bioblock Scientific, Parc d'innovation 67 403 Illkirch Cédex, France

    Tél. 03 88 67 14 14   Fax 03 88 67 11 68

Lange, 19 bd Georges Bidault, 77 183 Croissy Beaubourg, France

    Tél. 01 64 62 07 17   Fax 01 64 62 24 14

(*See* also Chapter 10-Atomic Absorption)

Spark and Spark Spectrometry

Baird Analytical Instr., 125 Middlesex Thpk, Bedford, MA 01780, USA
   Fax 617-276-6510
Baird Analytical Instr, Les Rives de Seine, 10 quai de la borde, bâtiment C,
   91 130 Ris-Orangis, France
   Tél. 01 69 43 44 55   Fax 01 69 43 23 49
Shimadzu Europe (Touzart et Matignon)Albert Hahnstrasse 6-10, D-4100
   Duisburg 29, Germany
   Postfac 29 02 60
Thermo-Jarrell Ash, 8E Forge Parkway, Franklin MA 02038, USA
Thermo Jarrel Ash, 13, rue de la Perdrix, BP 50321, 95940 Roissy CDG,
   France
   Tél. 01 48 63 78 00   Fax 01 48 63 27 60

ICP and ICP-MS Spectrometers

Baird Analytical Instr., 125 Middlesex Thpk, Bedford, MA 01780 USA
   Fax 617-276-6510
Baird Analytical Instr, Les Rives de Seine, 10 quai de la borde, bâtiment C,
   91 130 Ris-Orangis, France
   Tél. 01 69 43 44 55   Fax 01 69 43 23 49
Finnigan Mat, 355 River Oaks Pkwy, San Jose, CA 95 134, USA
   Fax 408 433 4823
Finnigan Mat, Parc club Orsay Université, 2 place J. Monod, 91893 Orsay
   Cédex, France
   Tél. 01 69 41 98 00   Fax 01 69 41 98 16
Fisons-Arl, 85, av. Aristide Briand, 94110 Arcueil, France
   Tél. 01 47 40 48 40   Fax 01 45 46 22 50
GBC Scientific equipment, 22, Brooklyn Av., Dandenong Victoria 3175,
   Australia
Hewlett-Packard, 3495 Deer Creek Rd., Palo Alto, CA 94 304, USA
Hewlett-Packard, 1 Avenue du Canada, 91 947 Les Ulis Cédex, France.
   Tél. 01 69 82 60 60   Fax 01 69 82 60 61
Jobin Yvon, 16-18 Rue du Canal, 91165 Longjumeau Cédex, France
   Tél. 01 64 54 13 00   Fax 01 69 09 90 88
Perkin-Elmer, BP 304, Montigny-le-Bretonneux, 78054 St-Quentin-en-Yvelines
   Cédex, France
   Tél. 01 30 85 63 63   Fax 01 30 85 63 00
Questron Corp., PO Box 2387, Princeton, NJ 08543, USA
   Fax 609 587 0513

Thermo Jarrel Ash, 13, rue de la Perdrix, BP 50321, 95940 Roissy CDG, France
    Tél. 01 48 63 78 00   Fax 01 48 63 27 60
Varian, 7, av. des Tropiques, B.P 12, 91941 Les Ulis, France
    Tél. 01 69 86 38 38   Fax 01 69 28 23 08

Ionometry, Ion Electrodes and Optodes (Chapter 12)

Altech Associates France, B.P. 11, 59242 Templeuve, France
    Tél. 03 20 79 25 25   Fax 03 20 59 33 69
Beckman Instruments, 92-94, chemin des Bourdons, 93220 Gagny, France
    Tél. 01 43 01 70 00   Fax 01 43 81 16 80
Bran and Luebbe, Analyzing Technologies Inc. (Technicon)
    Z.I. Le Chêne sorcier, BP 61, 78340 Les Clayes-sous-Bois, France
    Tél. 01 30 81 81 81   Fax 01 30 55 96 94
Brinkman Instruments (Metrohm), One Cantiague Rd. P.O Box 1019 Westbury NY 11590, USA
    Fax 516-334-7506
Dionex Corp., 103, av. Pierre Grenier, 92100 Boulogne-Billancourt, France
    Tél. 01 46 21 66 66   Fax 01 46 21 13 69
Fluka-L'isle d' Abeau Chesnes-BP 701, 38297, St-Quentin-Fallavier Cédex, France
Ingold, Prolabo Labo-moderne, Siemensstrasse 9, D-6374 Steubbac/TS, Germany
Lachat Instrument, 6645 W-Mill Road, Milwaukee WI 53 209, USA
    Fax 414-358-4206
Perstorp Analytical-Tecator, 1-3, rue Jean Carrasso, BP 13, 95871 Bezons Cédex, France
    Tél. 01 34 23 38 38   Fax 01 34 23 39 03
Radiometer-Tacussel, 8, rue Edmond Michelet, BP 80, ZA La Fontaine du Vaisseau, 93360 Neuilly Plaisance, France
    Tél. 01 43 09 81 60   Fax 01 43 09 81 41
Skalar Analytique, 40, quai d'Issy-les-Moulineaux, 75015 Paris, France
    Tél. 01 45 54 75 25   Fax 01 45 57 69 04
Waters-Millipore, BP 307, 78054 Saint-Quentin-Yvelines Cédex, France
    Tél. 01 30 12 70 00   Fax 01 30 12 71 80
Horiba Ltd Kyoto, Japan
    Tél. (81) 75-313-8123
Labo Moderne, 37, rue Dombasle, 75015 Paris, France
    Tél. 01 45 32 62 54   Fax 01 45 32 01 09
Orion Research, 529, Main Street, Boston MA 02129, USA
Mettler-Toledo, 18-20, av de la Pépinière, 78220 Viroflay, France
    Tél. 01 30 97 17 17   Fax 01 30 97 16 16

Corning-Bibby, Z.I du Rocher Vert, B.P 79, 77793 Nemours Cédex, France
Tél. 01 64 28 88 89   Fax 01 64 88 17 74
Nova Biomedical France, 15, av. du Québec, BP 632, 91965 Les Ulis, France
Tél. 01 69 07 01 05   Fax 01 69 07 01 60
OSI, 141, rue de Javel, 75739 Paris Cédex 15, France
Tél. 01 45 54 97 31   Fax 01 45 54 26 28
Hach Europe S.A. BP 229, 5000 Namur, Belgium
Tél. 32 81 44 53 81   Fax 32 81 441 300
Prolabo, 54 rue Roger Salengro, 94 126 Fontenay-sous-Bois Cédex, France
Tél. 01 45 14 85 00   Fax 01 45 14 85 15
Metrohm, Oberdorfstrasse 68, Herisau Appenzell CH-9100, *Switzerland*
Roucaire, 20, av. de l' Europe, BP 65, 78 143 Velizy Cédex, France
Tél. 01 30 67 75 00   Fax 01 30 70 87 20

Consumables for Chromatography (Chapter 13)

Merck-Clevenot, 5-9, rue Anquetil, BP 8, 94736 Nogent-sur-Marne Cédex, France
Tél. 01 43 94 54 00   Fax 01 48 76 58 15
Alltech Associates Inc., 2051 Waukegan Road, Deerfield IL 60015, USA
Tél. 312-948-8600   Fax 313-948-1078
Bio-Rad S.A., 94-96 rue Victor Hugo, BP 220, 94203 Ivry-sur-Seine, France
Tél. 01 49 60 68 34   Fax 01 46 71 24 67
Whatman Inc., St Léonard Road, Maidstone Kent GB-ME 16025, United Kingdom
Tél. 0622 67 66 70   Fax 0622 677 011
distributed by Prolabo; Touzart Et Matignon; OSI; Poly-Labo...
LKB-Produk Tex A.B-Pharmacia, Box 305, S-16126 Bromma, Sweden
Tél. 46(8) 98 00 40   Telex 10492
distributed by Pharmacia-France, BP 210, 78051 St Quentin Yvelines Cédex, France
Interchim, B.P 15, 213, av. J.F. Kennedy, 03103 Montlucon, France
Tél. 04 70 03 88 55   Fax 04 70 03 82 60
Hewlett-Packard, Parc d'Activité du Bois Briard, 2, av. du Lac, 91040 Evry Cédex, France
Spectra-Physic, Autolab Div. 3333 N. First Street, San José CA 95134, USA
Fax 408-432-0203
Spectra-Physics, 2 avenue de Scandinavie, Z.A. de Courtaboeuf, BP 28, 91946 Les Ulis Cédex, France
Tél. 01 69 07 99 56   Fax 01 69 07 99 56
Bruker, 34, rue de l'Industrie, 67160 Wissembourg, France
Tél. 03 88 73 68 00   Fax 03 88 73 68 79
Gilson Electronics, 3000 W. Beltine Hwy. Middleton WI 53562, USA
Fax 608-831-4451

Gilson Medical Electronics, 72, rue Gambetta, BP 45, 95400 Villiers-le-Bel,
  France
  Tél. 01 34 29 50 50   Fax 01 34 29 50 60
Millipore-Waters Co., Bedford MA 01730, USA
  Tél. (617) 275-9200
Millipore-Waters, BP 307, 78054 Saint-Quentin-Yvelines Cédex, France
  Tél. 01 30 12 70 00
Analytichem International (France: Prolabo), 24201 Frampton Ave. Harbor
  City CA. 90710, USA
Pharmacia-LKB, Box 175, S-75104 Uppsala 1, Sweden
Supelco Inc., Supelco park, Bellafonte PA 16823, USA
  Fax 814-359-3044
Supelco, 20 quater, rue Schnapper, 78101 Saint-Germain-en-Laye Cédex,
  France
  Tél. 01 34 51 12 13   Fax 01 39 73 62 00
Matheson Gas Products Inc., 30, Seaview Dr. Secausus NJ 07096, USA
  Fax 201-867-4572
Fluka Chemical Corp., 27, rue des trois frontières, 68110 Illzach-Ile Napoléon,
  France
  Tél. 03 89 61 75 65   Fax 03 89 61 77 76
Hamilton Co (France: Touzart et Matignon), 4970 Energy way, Reno NV
  89502, USA
                              Fax 702-323-7259
Isco Inc. Box 5347, 4700 Superior Ave, Lincoln NE 68505, USA
                              Fax 402-464-4543
J.T. Bader Inc., 222, Red School Ln., Phillipsburg NJ 08865, USA
                              Fax 908-859-9318
Schleicher and Schuell Inc. postfach 4, D-3354 Dassel, Germany
Schleicher and Schuell, 5 rue des Fontanelles, ZAI du Petit Parc, 78920
  Ecquevilly, France
Cera-Labo, 153, av. Jean-Jaurès, 93307 Aubervilliers Cédex, France
  Tél. 01 48 34 00 74   Fax 01 48 33 15 22

Gas and Supercritical Phase Chromatography (Chapter 14)

Gas Chromatography

Carlo-Erba-Fisons Instruments, Bishop meadox Rd., Lough Borough, Leics,
  United Kingdom Le11 ORG
                              Fax 0509-23 1893
Perkin-Elmer Corp. 761. Main avd. Norwalk CT 06859, USA
                              Fax 203-762-6000
Perkin-Elmer, BP 304 Montigny Le Bretonneux 78054 St-Quentin -en-Yvelines,
  France
  Tél. 01 30 85 63 63   Fax 01 30 85 63 00

Shimadzu Scientific Instruments Inc. 3-Kanda-Nishikicho 1-Chome (Roucaire) Chiyoda-Ku Tokyo, Japan
Fax 03 32 19 5710
Varian Instruments, 220 Humboldt Cr., Sunnywale CA 94089, USA
Fax 408-744-0261
Varian, 7 av. des Tropiques, BP 12, 91941 Les Ulis, France
Tél. 01 69 86 38 38   Fax 01 69 28 23 08
Hewlett-Packard Co., 3495 Deer Creek Rd. Palo Alto CA. 94 304, USA
Hewlett-Packard, 1 av. du Canada, 91947 Les Ulis Cédex, France
Tél. 01 69 82 60 60   Fax 01 69 82 60 61
Alltech Associates, 2051, Waukegan Rd. Deerfield IL. 60015, USA
Tél. 1-708-948
Nicolet Instruments Corp. 5225-1 Verona Road Madison WI 53711, USA
Fax 608-273-5046
Nicolet France, 16, av. Jean d'Alembert, Z.I de Pissaloup 78192 Trappes, France
Tél. 01 30 66 33 30   Fax 01 30 66 70 36
Chrompack International, Kuipersweg 6, PO Box 8033 Middelburg, The Netherlands
Chrompac France, 5, rue de la Terre de Feu. Z.A de Courtaboeuf BP 20 91941 Les Ulis, France
Tél. 01 69 07 36 52   Fax 01 69 07 75 40

Supercritical Phase Chromatography

Dionex S.A, 103, av. Pierre Grenier 92100 Boulogne-Billancourt, France
Tél. 01 46 21 66 66
Dionex Co. 1228 Titan Way, PO Box 3603, Sunny wale CA 94088, USA
Tél. (408) 737-0700   Fax (408) 730-9403
Isco Inc. Box 5347, 4700 Superior Ave. Lincoln NE 68505, USA
Fax 402-464-4543
Suprex Corporation, 125, William Pitt Way, Pittsburg P.A 15238, USA
Touzart Et Matignon, 8, rue Eugène Henaff, BP 52, 94403 Vitry-sur-Seine Cédex, France
Tél. (1) 46 80 85 21   Fax (1) 46 80 45 18
Nicolet-France (*See* GC above)

Liquid Chromatography (Chapter 15)

Miscellaneous (HPLC)

Pharmacia-LKB, S-75182, Uppsala, Sweden
Pharmacia, B.P 210, 78051 St-Quenti-en-Yvelines, France
Tél. 01 30 64 34 00   Fax 01 30 43 44 45

Alltech Associates, 2051, Waukegan Rd. Deerfield IL. 60015, USA
    Tél. 1-708-948
Alltech France, BP 11, 59242 Templeuve, France
    Tél. 03 20 79 25 25   Fax 03 20 59 33 69
Dionex Corp., 1228 Titan Way Sunnyvale CA 94086, USA
Diónex, 103, av. Pierre Grenier, 92100 Boulogne-Billancourt, France
    Tél. 01 46 21 66 66   Fax 01 46 21 13 69
Fisons Instruments (Carlo-Erba), Bishop meadox Rd., Lough borough, Leics,
    United Kingdom LE11 ORG
    Fax 0509-23 1893
Fisons Instruments, 85, av. Aristide Briand, 94110 Arcueil, France
    Tél. 01 47 40 48 40   Fax 01 45 46 22 50
Gilson Electronics, 3000 W. Beltine Hwy. Middleton WI 53562, USA
    Fax 608-831-4451
Gilson Medical Electronics, 72, rue Gambetta, BP 45, 95400 Villiers-le-Bel,
    France
    Tél. 01 34 29 50 50   Fax 01 34 29 50 60
Hewlett-Packard Co., 3495 Deer Creek Rd, Palo Alto CA. 94304, USA
Hewlett-Packard, 1, av. du Canada, 91947 Les Ulis Cédex, France
    Tél. 01 69 82 60 60   Fax 01 69 82 60 61
Isco Inc. Box 5347, 4700 Superior Ave. Lincoln NE 68505, USA
    Fax 402-464-4543
Roucaire, 20, av. de l'Europe, BP 65, 78143 Velizy-Villacoublay Cédex, France
    Tél. 01 30 67 75 00   Fax 01 30 70 87 20
Nicolet Instruments Corp. 5225-1 Verona Road Madison WI 53 711, USA
    Fax 608-273-5046
Nicolet, 16, av. Jean d'Alembert, Z.I de Pissaloup 78192 Trappes, France
    Tél. 01 30 66 33 30   Fax 01 30 66 70 36
Perkin-Elmer Corp. 761 Main ave. Norwalk CT 06859, USA
    Fax 203-762-6000
Perkin-Elmer, BP 304 Montigny-Le-Bretonneux 78054 St-Quentin-en-Yvelines,
    France
    Tél. 01 30 85 63 63   Fax 01 30 85 63 00
Shimadzu Scientific Instruments Inc. 3-Kanda-Nishikicho 1-Chome, Chiyoda-
    Ku Tokyo, Japan
    Fax 03 32 19 5710
Spectra-Physics, Autolab Div. 3333 N. First Street, San Jose CA 95134, USA
    Fax 408-432-0203
Spectra-Physics, 2 av. de Scandinavie Z.A de Courtaboeuf B.P 28, 91941 Les
    ULIS Cédex, France
    Tél. 01 69 07 99 56   Fax 01 69 07 60 93
Tremetrids Inc. 2215 Grand Ave. PKWY Austin TX 78728, USA
    Fax 512-251-1597
Varian Instruments Grup, 220 Humboldt Cr., Sunnywale CA 94089, USA
    Fax 408-744-0261

Varian, 7, av. des Tropiques, BP 12, 91941 Les Ulis, France
Tél. 01 69 86 38 38   Fax 01 69 28 23 08.

Ion Chromatography

Dionex Corp., 1228 Titan Way, P.O Box 3603, Sunnyvale CA 94088, USA
Tél. (408) 737-0700   Fax (408) 730-9403
Dionex, 103, av. Pierre Grenier, 92100 Boulogne-Billancourt, France
Tél. 01 46 21 66 66   Fax 01 46 21 13 69
Metrohm, Oberdorfstrasse 68 CH-9101 Herisau Appenzell, Switzerland
Tél. 071/53 11 33   Fax 071/52 11 14
Waters-Millipore, 34, Maple street, Midford Massachussetts 01757, USA
Tél. 617-478-2000
Millipore, B.P 307, 78054 St-Quentin-en-Yvelines Cédex, France
Tél. 01 30 12 70 00   Fax 01 30 12 71 80
Wescan Instruments Inc. (Alltech France) 3018 Scott Bld. Santa Clara CA
95050, USA
Tél. (408) 727-3519   Telex 171627 ATTN: Wesca
Alltech Associates, 2051, Waukegan Rd. Deerfield IL. 60015, USA
Tél. 1-708-948
Perkin-Elmer Corp. 761 Main avd. Norwalk CT 06859, USA
Fax 203-762-6000
Perkin-Elmer, BP 304 Montigny-Le-Bretonneux, 78054 St-Quentin-en-
Yvelines, France
Tél. 01 30 85 63 63   Fax 01 30 85 63 00

CHN-OS Analysers (Chapter 16)

Fusion Analysers

Antek (Sopares): 51 bis rue Raymond Lefévre, 94257 Gentilly Cédex, France
Tél. 01 49 86 44 01   Fax 01 45 47 07 77
Dany Spa (Stang Instruments) - 42, viale Elvezia, 1-20.0520 Monza (Milan),
Italy
45, allées des Platanes-93320 Les-Pavillons-sous-Bois, France
Tél. 01 48 50 07 18   Telex STANG 232-579
Dohrmann (Rosemount Analytical) 3240 Scott Boulevard, Santa Clara-CA
95 054, USA
Tél. (800) 538-7708   Fax (408) 727-1601
Fisons Instruments S.A (ex. Erba-Sciences, Carlo-Erba)-VG Instruments, ARL,
85, av. Aristide Briand- 94117 Arcueil Cédex, France
Tél. 01 47 40 48 40   Fax 01 45 46 22 50
LECO France-ZAC « Les Doucettes », 22, av. des Morillons, BP 74—95144
Garges-les-Gonesses, France
Tél. 01 39 93 98 00   Fax 01 39 86 41 05

Perkin-Elmer (Coleman)1, av. Franklin-Montigny-le-Bretonneux, BP 304-
78054 Saint-Quentin-Yvelines Cédex, France
Tél. 01 30 85 63 63   Fax 01 30 85 63 00
Techmation, 20 quai de la Marne, Paris, France
Tél. 01 42 00 11 05   Fax 01 42 40 37 80
Wösthoff H. (Bran-Lubbe) Z.I. « Le Chêne sorcier », BP 61-78340 Les-
Clayessous-Bois, France
Tél. 01 30 81 81 81   Fax 01 30 55 96 94

Wet-digestion Analysers

Büchi (Roucaire): Postfach CH-9230 Flawil. Switzerland 20, av. de l'Europe-
BP 65-78143 Vélizy-Villacoublay Cédex, France
Tél. 01 30 67 75 00   Fax 01 30 70 87 20
Gerhardt (OSI) 141, rue de Javel-75739 Paris Cédex 15, France
Tél. 01 45 54 97 31   Fax 01 45 54 26 28
Hach Europe (Prolabo) Chaussée de Namur, 1-B-5150 Floriffoux, Belgium
54, rue Roger Salengro-94 126 Fontenay-sous-Bois Cédex, France
Tél. 01 45 14 85 00   Fax 01 45 14 85 15
Heraeus-Foss-Leybold (+ UIC USA) Z.A. Courtaboeuf, 4, av. de l'Atlantique-
91941 Les Ulis Cédex, France
Tél. 01 69 07 65 00   Fax 01 69 28 12 45
Hewlett-Packard, Route 41-Starr road-Avondale P.A-19311, USA 1, av. du
Canada-91947 Les Ulis Cédex, France
Tél. 01 69 82 60 60   Fax 01 69 82 60 61
Labconco Corp.-8811, Prospect ave.-Kansas City-MO 64132, USA
Tél. 1-816-33-8811   Fax 816-363 0130
Mitsubishi Chemical Industries Limited (Prolabo, European Prosumer, Touzart-
Matignon) S-2, Marunouchi, 2, Chome, Chiyuda-Ku—Tokyo 100, Japan
Tél. (0,) 455-4220   Fax (03) 455-4233
TECATOR (Perstorp Analytical) 1-3, rue Jean Carrasso, BP 13-95871 Bezons
Cédex, France
Tél. 01 34 23 38 38   Fax 01 34 23 39 03
VELP (Labo-moderne) 37, rue Dombasle-75015 Paris, France
Tél. 01 45 32 62 34   Fax 01 45 32 01 09
Verre Et Technique (Bicasa) 90, av. de la Convention, 94117 Arcueil Cédex,
France
Tél. 01 54 73 30 30   Télex 20 22 79 F
Wescan Instr. In. (All Tech France) 2051 Waukegan road, Deerfield-IL 60015,
USA
BP II-59242 Templeuve, France
Tél. 03 20 79 25 25   Fax 03 20 59 33 69

Automation And Robotics (Chapter 17)

Segmented Continuous-Flow Analysis

Alliance instruments, ZA les bosquets 4, BP 31, 95540 Mery-sur Oise, France
Tél. 01 30 36 24 24   Fax 01 30 36 24 20
Alpkem-Perstorp Analytical Co., PO Box 1260 Clackamas, OK 97015, USA
Tél. 503 657 3010   Fax 503 657 5288
Alpkem-Perstorp An. Co., 1-3 rue J. Carrasso, BP 13, 95871 Bezons Cédex,
France
Carlo Erba-Fisons, 85 avenue Aristide Briand, 94110 Arcueil, France
Tél. 01 47 40 48 40   Fax 01 45 46 22 50
Fontenille S.A, 71 grande rue St Michel, 31400 Toulouse, France,
Neotechnic, 15 rue Guyton de Morveau, 75013 Paris, France
DF Technologie Diffusion France, Zone d'activités de Sautes, rue de
l'Industrie. 11800 Trébes, France
Tél. 01 68 78 77 43
Roche, 52 Bd du Parc, 92521 Neuilly-sur-Seine Cédex, France
Tél. 01 46 40 50 00   Fax 01 46 40 52 92
Skalar Analytical, Spinveld 62, NL-4815 Breda HT, The Netherlands
Skalar Analytique, 40 quai d'Issy-les-Moulineaux, 75015 Paris, France
Tél. 01 45 54 75 25   Fax 01 45 57 69 04
Radiometer-Tacussel, 8 rue Edmond Michelet, BP 80, ZA La fontaine du
Vaisseau, 93360 Neuilly-Plaisance, France
Tél. 01 43 09 81 60   Fax 01 43 09 81 41
Mettler-Toledo, 18-20 avenue de la Pépinière, 78220 Viroflay, France
Tél. 01 30 97 17 17   Fax 01 30 97 16 16
Alfa Laval-Bran-Luebbe, Zi « Le Chêne sorcier », BP 61, 78340 Les Clayesous-
Bois, France
Tél. 01 30 81 81 81   Fax 01 30 55 96 94
Norton Plastics, P.O Box 350 Akron Ohio 44309, USA

FIA Equipment

Chemlab Instr, Ltd Hornminster House, 129 Upminster road, Hornchurch,
Essex RM 11 3X, United Kingdom
Dionex Corp., 103, av. Pierre Grenier, 92100, Boulogne-Billancourt, France
Tél. 01 46 21 66 66   Fax 01 46 21 13 69
Lachat, 10500 N Port Washington Rd., Mequon, W.I. 53092, USA
Tél. 414-241-3872   Fax 414-241-5128.
Perstorp Analytical; 1-3, rue Jean Carrasso, B.P. 13, 95871 Bezons Cédex,
France
Tél. 01 34 23 38 38   Fax 01 34 23 39 03
Roucaire-Eppendorf, 20 av. de l'Europe, BP 65, 78143 Vélizy Villacoublay
Cédex, France
Tél. 01 30 67 75 00   Fax: 01 30 70 87 20.

Skalar USA Inc., 40 quai d'Issy-les-Moulineaux, 75015 Paris, France
Tél. 01 45 54 75 25   Fax 01 45 57 69 04

Robotic Arms

Anatech-Itec: 42, av. Chanoine Cartellier-69230- St-Genis-Laval, France
Tél. 04 72 39 97 88   Télex: 306 612
Arl-Fisons Instr., 85, av. Aristide Briand-94110 Arcueil, France
Tél. 01 47 40 48 40   Fax 01 45 46 22 50
Hewlett-Packard S.A., Meyrin, Switzerland
Tél. (022) 78 08 111   Fax (022) 78 08 542
Hewlett-Packard, 1 av. du Canada-91947 Les Ulis Cédex, France
Tél. 01 69 82 60 60
Microbot (Minimover-S), Menlo Park CA, USA
Perkin-Elmer, BP 304-Montigny-le-Bretonneux-78054 St-Quentin-en-Yvelines
Cédex, France
Tél. 01 30 85 63 63   Fax 01 30 85 63 00
Prolabo-Aid, 54 rue Roger Salengro, 94 126 Fontenay-sous-Bois Cédex, France
Tél. 01 43 55 44 88
Zymark, ZAC Paris-NordII-13, rue de la Perdrix-BP 40016-95911 Roissy
Charlesde-Gaulle Cédex, France
Tél. 01 48 63 71 35   Fax 01 48 63 71 53
Tecan Inc., PO Box 13953, Research Triangle PK, NC 27 709, USA
Tél. 919 361 5200   Fax 919 362 5201
Scilog Inc., 7 778 Noll Valley Rd., Verona WI 53 593, USA
Tél. 608 798 1280   Fax 608 798 1281

Various Automated Equipment for Soil Analysis

Beckman Bioanalytique, 92-94, chemin des Bourdons-93220 Gagny, France
Tél. 01 43 01 70 00
Berthold Analyse Instrumentation (Tecan), 6, rue du Maréchal Ferrant-78990
Elancourt, France
Tél. 01 30 62 31 12
Centurion International Inc., PO Box 82846-Lincoln Nebraska 68501, USA
Custom Equipment Inc., 205, E. Michigan av.-Orange city-Floride 32763,
USA
                                        Fax: 904 775 9890
Euro/DPC Ltd, 31, Station Lane-Witney-Oxon OX8 6AN- United Kingdom
Tél. (0993) 702 977   Fax (0993) 778 155
GBC Scientific Equipement, 22, Brooklyn avenue-Dandenong Victoria,
Australia 3175
Tél. 03 793 1448

Gilson Medical Electronics, 72, rue Gambetta-BP 45-95400 Villiers-le-Bel. France
Tél. 01 34 29 50 50

Hamilton-Touzart Et Matignon, 8, rue Eugène Hénaff-BP 52-94403 Vitry-sur-Seine Cédex, France
Tél. 01 46 80 85 21   Fax 01 46 80 45 18

Isabert, 49, rue Ernest Renan-95320 Saint-Leu-la-Forêt, France
Tél. 01 39 60 74 31

Jobin-Yvon, 16-18, rue du canal- BP 118-91163 Longjumeau Cédex, France
Tél. 01 64 54 13 00   Fax 01 69 09 03 21

Lisabio, 12, av. Charles de Gaulle-91420 Morangis, France
Tél. 01 65 54 85 28

Malvern, 30, rue J. Rostand-91893 Orsay Cédex, France
Tél. 01 60 19 02 00   Fax 01 60 19 13 26

Metrohm-Roucaire: 20, av. de l'Europe-BP 65 - 78143 Vélizy Villacoublay Cédex, France
Tél. 01 30 67 75 00   Fax 01 30 70 87 20

Mettler-Toledo, 18-20, av. de la Pépinière-78220 Viroflay, France
Tél. 01 30 97 17 17   Fax 01 30 97 16 16

Micromeritics, 181, rue Henri Bessemer, France-60100 Creil, France
Tél. 03 44 24 23 02

Radiometer-Tacussel, 8, rue Edmond Michelet-BP 80-ZA La Fontaine du Vaisseau-93360 Neuilly Plaisance, France
Tél. 01 43 09 81 60   Fax 01 43 09 81 41

Sartorius, 11, av. du 1er mai-91127 Palaiseau Cédex, France
Tél. 01 69 20 93   Fax 69 20 09 22

Schott-Gerate France, 8, rue Fournier-92110 Clichy, France
Tél. 01 40 87 39 00   Fax 01 42 70 73 22

Seishin Enterprise Co, Nippon Brunswick Bldg 5-27-7-Sendagaya-Shibuya-Ku, Tokyo, Japan
Tél. 03 (350) 5771-8   Télex: 232-4242 SEISINJ-   Telfax 03 (350) 5779

GBX SARL, Quartier les Horizons-26300 Pizancon, France
Tél. 04 75 02 37 32   Fax 04 72 02 39 17

Skalar Analytique, 40, quai d'Issy-les-Moulineaux-75015 Paris, France
Tél. 01 45 54 75 25   Fax 01 45 57 69 04

Spectra-Physics France, av. de Scandinavie-Z.A de Courtaboeuf-BP 28-91941 Les Ulis Cédex, France
Tél. 01 69 07 99 56   Fax 01 69 07 60 93

Tacussel-Solea, 72, rue d'Alsace-69627 Villeurbanne Cédex, France
Tél. 04 78 68 01 22   Fax 04 78 68 88 12

Technologie Diffusion France-Lisabio, 12, av. Charles de Gaulle-91420 Morangis, France
Tél. 01 64 54 85 28

Quality-Control Organizations (Chapter 18)

Association francaise de normalisation (AFNOR), Tour Europe, La Défense, 92 080 Paris Cédex 7, France
Tél. 01 42 91 55 55

Association of Official Analytical Chemists 1111., North nineteenth street, Suite 210, Arlington, VA 22209, USA

BIPEA-Bureau interprofessionnel d'études analytiques, 9 à 14 Avenue Louis Roche, 92 230 Gennevilliers, France.

Bureau of analysed samples Ltd, Newham Hall, Newby, Middlesbrough, Cleveland T58 GEA, United Kingdom.

Canada Center of mineral and energy technology (CANMET), 555 Booth Street, Ottawa, Ontario K1A 061, Canada.

Clay Mineral Society, Univ. Missouri-Columbia, College of Arts and Science, Dept. of Geology, Geology Building, Columbia, Missouri 65211, USA.

COFRAC-Comité Français d'accréditation, 37 rue de Lyon, 75012 Paris, France.

Commission of the European Communities, Programme mesures et essais, Rué de la loi 200, B-1049, Bruxelles, Belgium.
Tél. (02) 235 50 14

Geological Survey of Japan, 1-3 Higashi 1-chome, Yatebemachi, T Sukuba-Gun, Ibaragi, Japan.

Gosstand of Russia, 9 Leninski prospekt, 11704 Moscow, Russia

I.S.E. (International Soil Analytical Exchange), CO Dr V. Houba, Dep. of Soil Sci. and Plant Nutrition Wageningen Agricultural University, PO Box 8005, 6700 EC Wageningen, The Netherlands.

Instituto de Pesquisas Technologicas do estado de Sao Paulo, Disisao de Quimica de engenharia quimica, 01000 Sao Paulo SP, Brazil.

National Bureau of Standard (NBS), Standard reference materials, Room B111, Chemistry Building, Washington, DC 20 234 USA.

NIST-SRM, National Institute of Standards and Technology, Standard Reference Material, Building 202, Room 204, Gaithersburg, MD 20 899, USA.
Fax (301) 948-3730

South African Bureau of Standards, Private BAG x191, Pretoria, Transvaal, South Africa.

US Geological Survey, National Center 972, Reston, VA 22 092, USA.

Ward's 5100 West Henrietta Road, PO Box 92 912, Rochester NY, USA.
Fax 716-334-6174

AMT für standardissierung und Warenprüfung (ASMW), 102 Berlin, Wallstrasse 16, Germany.

Geostandards, CRPG-CNRS, 15 rue ND des pauvres, BP 20, 54 501 Vandoeuvre-les-Nancy Cédex, France.

Periodic Table of the Elements

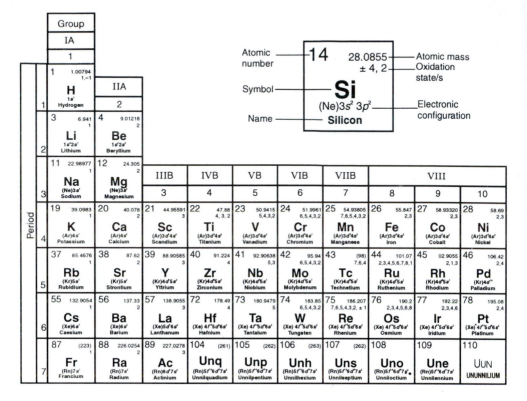

					Group
					0
					18

IIIA	IVA	VA	VIA	VIIA	2 4.002602 0 **He** 1s¹ Helium
13	14	15	16	17	

$$
\begin{array}{|c|c|c|c|c|c|}
\hline
\text{IIIA} & \text{IVA} & \text{VA} & \text{VIA} & \text{VIIA} &
\begin{matrix} 2 & 4.002602 \\ & 0 \\ \mathbf{He} \\ 1s^2 \\ \text{Helium} \end{matrix} \\
\hline
13 & 14 & 15 & 16 & 17 & \\
\hline
\begin{matrix} 5 & 10.811 \\ & 3 \\ \mathbf{B} \\ 1s^2 2s^2 2p^1 \\ \text{Boron} \end{matrix} &
\begin{matrix} 6 & 12.011 \\ & \pm,4,2 \\ \mathbf{C} \\ 1s^2 2s^2 2p^2 \\ \text{Carbon} \end{matrix} &
\begin{matrix} 7 & 14.0067 \\ & \pm,3,5,4,\pm,2,1 \\ \mathbf{N} \\ 1s^2 2s^2 2p^3 \\ \text{Nitrogen} \end{matrix} &
\begin{matrix} 8 & 15.9994 \\ & -2 \\ \mathbf{O} \\ 1s^2 2s^2 2p^4 \\ \text{Oxygen} \end{matrix} &
\begin{matrix} 9 & 18.998403 \\ & -1 \\ \mathbf{F} \\ 1s^2 2s^2 2p^5 \\ \text{Fluorine} \end{matrix} &
\begin{matrix} 10 & 20.179 \\ & 0 \\ \mathbf{Ne} \\ 1s^2 2s^2 2p^6 \\ \text{Neon} \end{matrix} \\
\hline
\end{array}
$$

			IIIA	IVA	VA	VIA	VIIA	**He**

Periodic table (lower portion):

IB	IIB	13 (IIIA)	14 (IVA)	15 (VA)	16 (VIA)	17 (VIIA)	18 (0)
11	12						
		13 26.98154 3 **Al** (Ne)3s²3p¹ Aluminium	14 28.0855 ±,4,2 **Si** (Ne)3s²3p² Silicon	15 30.97376 ±,3,5,4 **P** (Ne)3s²3p³ Phosphorus	16 32.066 ±,2,3,4,6 **S** (Ne)3s²3p⁴ Sulphur	17 35.453 ±,1,4,5,6,7 **Cl** (Ne)3s²3p⁵ Chlorine	18 39.948 0 **Ar** (Ne)3s²3p⁶ Argon
29 63.546 2,1 **Cu** (Ar)3d¹⁰4s¹ Copper	30 65.39 2 **Zn** (Ar)3d¹⁰4s² Zinc	31 69.723 3 **Ga** (Ar)3d¹⁰4s²4p¹ Gallium	32 72.61 4,2 **Ge** (Ar)3d¹⁰4s²4p² Germanium	33 74.92159 ±,3,5 **As** (Ar)3d¹⁰4s²4p³ Arsenic	34 78.96 -2,4,6 **Se** (Ar)3d¹⁰4s²4p⁴ Selenium	35 79.904 ±,1,4,5 **Br** (Ar)3d¹⁰4s²4p⁵ Bromine	36 83.80 0 **Kr** (Ar)3d¹⁰4s²4p⁶ Krypton
47 107.8682 1,2,3 **Ag** (Kr)4d¹⁰5s¹ Silver	48 112.41 2 **Cd** (Kr)4d¹⁰5s² Cadmium	49 114.82 3 **In** (Kr)4d¹⁰5s²5p¹ Indium	50 118.710 4,2 **Sn** (Kr)4d¹⁰5s²5p² Tin	51 121.75 ±,3,5 **Sb** (Kr)4d¹⁰5s²5p³ Antimony	52 127.60 -2,4,6 **Te** (Kr)4d¹⁰5s²5p⁴ Tellurium	53 126.9045 ±,1,4,5,7 **I** (Kr)4d¹⁰5s²5p⁵ Iodine	54 131.29 0 **Xe** (Kr)4d¹⁰5s²5p⁶ Xenon
79 196.96654 3,1 **Au** (Xe)4f¹⁴5d¹⁰6s¹ Gold	80 200.59 2,1 **Hg** (Xe)4f¹⁴5d¹⁰6s² Mercury	81 204.3833 3,1 **Tl** (Xe)4f¹⁴5d¹⁰6s²6p¹ Thallium	82 207.2 4,2 **Pb** (Xe)4f¹⁴5d¹⁰6s²6p² Lead	83 208.98037 3,5 **Bi** (Xe)4f¹⁴5d¹⁰6s²6p³ Bismuth	84 (209) 2,4 **Po** (Xe)4f¹⁴5d¹⁰6s²6p⁴ Polonium	85 (210) ±,1,3,5,7 **At** (Xe)4f¹⁴5d¹⁰6s²6p⁵ Astatine	86 (222) 0 **Rn** (Xe)4f¹⁴5d¹⁰6s²6p⁶ Radon
111 **Uuu** UNUNUNIUM	112 **Uub** UNUNBIUM	113 **Uut** UNUNTRIUM	114 **Uuq** UNUNQUADIUM	115 **Uup** UNUNPENTIUM	116 **Uuh** UNUNHEXIUM	117 **Uus** UNUNSEPTIUM	118 **Uuo** UNUNOCTIUM

Lanthanides (partial):

66 162.50 3 **Dy** (Xe)4f¹⁰5d⁰6s² Dysprosium	67 164.93032 3 **Ho** (Xe)4f¹¹5d⁰6s² Holmium	68 167.26 3 **Er** (Xe)4f¹²5d⁰6s² Erbium	69 168.93421 3 **Tm** (Xe)4f¹³5d⁰6s² Thulium	70 173.04 3,2 **Yb** (Xe)4f¹⁴5d⁰6s² Ytterbium	71 174.967 3 **Lu** (Xe)4f¹⁴5d¹6s² Lutetium

Actinides (partial):

98 (251) 3 **Cf** (Rn)5f¹⁰6d⁰7s² Californium	99 (252) **Es** (Rn)5f¹¹6d⁰7s² Einsteinium	100 (257) **Fm** (Rn)5f¹²6d⁰7s² Fermium	101 (258) **Md** (Rn)5f¹³6d⁰7s² Mendelevium	102 (259) **No** (Rn)5f¹⁴6d⁰7s² Nobelium	103 (260) **Lr** (Rn)5f¹⁴6d¹7s² Lawrencium

Index